OXFORD RESEARCH STUDIES IN GEOGRAPHY

The Geomorphology of Rock Coasts

ALAN S. TRENHAILE

CLARENDON PRESS · OXFORD
1987

Oxford University Press, Walton Street, Oxford OX2 6DP
Oxford New York Toronto
Delhi Bombay Calcutta Madras Karachi
Petaling Jaya Singapore Hong Kong Tokyo
Nairobi Dar es Salaam Cape Town
Melbourne Auckland
and associated companies in
Beirut Berlin Ibadan Nicosia

Oxford is a trade mark of Oxford University Press

Published in the United States
by Oxford University Press, New York

British Library Cataloguing in Publication Data

Trenhaile, Alan S.
The geomorphology of rock coasts.
—(Oxford research studies in geography)
I. Coasts 1. Title
551.4'57 GB451.2
ISBN 0-19-823279-9

Library of Congress Cataloging in Publication Data

Trenhaile, Alan S.
The geomorphology of rock coasts.
(Oxford research studies in geography)
Bibliography: p. Includes index.
1. Coasts. 2. Erosion. 3. Petrology.
I. Title. II. Series.
GV451.2.T74 1987 551.4'57 8-21695
ISBN 0-19-823279-9

Set by Spire Print Services, Ltd
Printed in Great Britain
by Butler and Tanner Ltd
Frome, Somerset

Preface

A large proportion of the coasts of the world is rocky, and even many sandy and stony beaches are backed by rock cliffs and underlain by shore platforms. Nevertheless, there is a growing tendency for books, book chapters, and review articles, purporting to represent all types of coasts, to be primarily or exclusively concerned with depositional landforms. There has been a great deal of research on, or relevant to, various aspects of rock coasts, but the literature is widely scattered among the journals of numerous disciplines. Valuable contributions have been made, for example, by marine biologists, engineers, geologists, phyical geographers and oceanographers. A great deal of this work is not specifically concerned with rock coasts, but it can be interpreted in a geomorphological context.

Despite the use of sophisticated methods of chemical and physical analysis, geochronometric dating, computers, physical and theoretical models, and the careful study and measurement of processes and erosion rates, we can still only speculate on the mode of development of rock coasts. This is partly because of the importance of events of high intensity and low frequency, which makes it particularly difficult to determine the relative significance of the various erosional processes. In any case, even if the most important processes could be identified and measured, they are not necessarily the same, or of the same intensity, as those which sculptured the coast in the past. This is because the nature and relative importance of erosive processes change with variations in relative sea level, climate, geology, and coastal gradient. Most rock coasts change very slowly, retaining vestiges of former sea levels and environmental conditions. Nevertheless, it needs to be emphasized that they are dynamic elements of a landscape, in the process of adjusting to the contemporary morphogenic environment.

There has been no comprehensive survey of the current state of knowledge on rock coasts. Furthermore, some important topics have not been discussed in any detail in the English language. In writing this book, I became aware not only of the enormous amount of work which has been done on rock coasts, but also of the vast amount which remains to be done. This book was written to provide a convenient source of reference for researchers, senior students, and instructors, to identify deficiencies in our knowledge, and to encourage and facilitate new investigation.

I thank my parents and friends for their encouragement in writing this book, and particularly my wife and children, Sue, Rhys and Lynwen, who not only provided unstinting encouragement but also over the years, invaluable assistance in the field, often in less than ideal conditions. I also thank my friends and colleagues in the University of Windsor, Professor Peter Halford of the French Department, who helped translate numerous articles as well as researching the origin of the term 'tafoni', and Dr V. C. Lakhan of the Department of Geography who made useful comments on chapter one. Thanks are also due to the staff of the Oxford University Press, the Interlibrary Loan Department of the University of Windsor, and the Natural Sciences and Engineering Research Council of Canada who have funded my research.

The author thanks the following for permission to reproduce figures: Gebrüder Borntraeger for Figs 4.2 and 6.3; The Geologists' Association for Fig. 11.1; Edward Arnold for Figs 9.4 and 9.6; The Waikato Geological and Lapidary Society for Fig. 4.4; and the US Army Corps of Engineers for Figs 1.4, 1.7, and 1.8. A fee has been paid to the Institute of British Geographers for permission to use Figs 1.12 and 1.13.

A.S.T.

Contents

I. PROCESSES

1

Mechanical Wave Erosion

Mechanical wave action is the dominant erosional agent in many parts of the world. In the storm and in the more vigorous of the swell wave environments, chemical and physical weathering and biological activity are usually only significant in sheltered areas, above the high tidal level, and where the rock is particularly resistant to wave action. Even in many polar, tropical, and other low energy environments, wave action plays an important role in quarrying and washing away weathered material. This chapter considers the magnitude and distribution of wave induced pressures, and its relevance to geomorphological problems. It is particularly concerned with the following questions:

1. What type of waves exert the greatest pressures?
2. At what elevation are the maximum pressures generated?

The concern of coastal engineering with the pressures exerted by waves on jetties, breakwalls, and other vertical structures is represented by a considerable body of literature. Because very little is known about the processes which erode rock coasts, we cannot be sure that there is any relationship between wave pressure and erosional efficacy. Nevertheless, it will be argued later in this chapter that the most important erosive processes probably operate most effectively where wave pressures are high.

Any discussion of the role of wave action on rock coasts must distinguish between standing, breaking, and broken waves. Tidal oscillations, submarine topography, deep water wave characteristics and other factors determine the type of wave that reaches a coast, which in turn determines the magnitude and distribution of the forces exerted on coastal structures.

Standing Waves or Clapotis (Non-breaking Waves)

When waves impinge on a vertical structure standing in deep water, reflection of the incident wave causes standing waves to develop in front of it. If the reflected and incident waves are of nearly equal magnitude, a vertical jet of water will alternately rise and then collapse upon itself, as kinetic energy is converted into potential energy and back again. The wave peaks are stationary in the horizontal plane, with the nodes appearing every half wavelength (Fig. 1.1). The kinetic and potential energy of

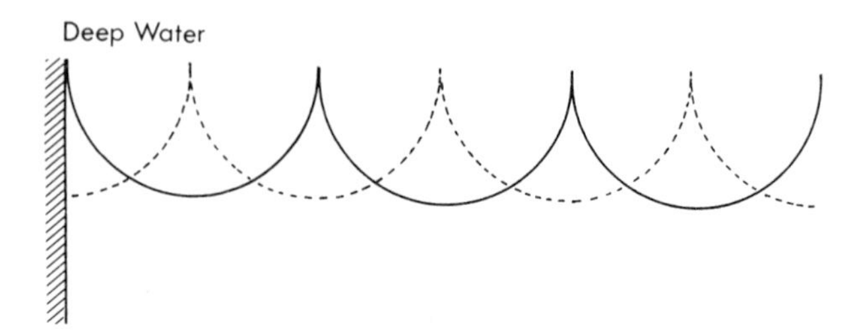

Figure 1.1. True standing waves or clapotis (Bagnold 1939).

clapotis are twice those of progressive waves, and unlike breaking waves their energy is not dissipated as turbulence.

The water at the base of cliffs and other natural features is usually not deep enough to permit the development of a true clapotis. The reflected wave is generally of smaller amplitude than the advancing wave because of the loss of energy incurred in travelling over a shallow bottom, or because the structure is not perfectly vertical. Collapse of the clapotis produces travelling waves which move away from it in a landward and a seaward direction. Other clapotis are subsequently formed by the collison of these travelling waves. Where the reflected wave meets an incident wave crest, an antinode develops; and where the reflected wave crest meets an incident wave trough, a node is formed. Nodes and antinodes appear at intervals of about one-quarter of the wavelength of the incoming waves. The effects on the structure is therefore the same as that owing to a series of single travelling waves advancing towards it (Fig. 1.2).

The reflection coefficient is the ratio of the height of the reflected wave to the height of the incident wave. It has been determined by measuring the profiles of the wave crest envelopes (Fig. 1.3). Examples of this envelope, for different degrees of reflection, have been provided by Wallet and

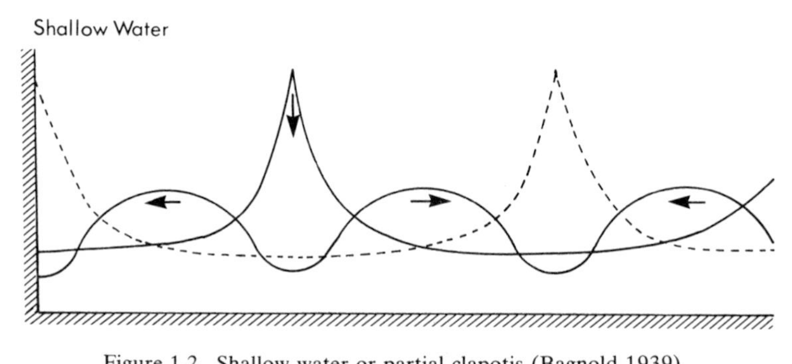

Figure 1.2. Shallow water or partial clapotis (Bagnold 1939).

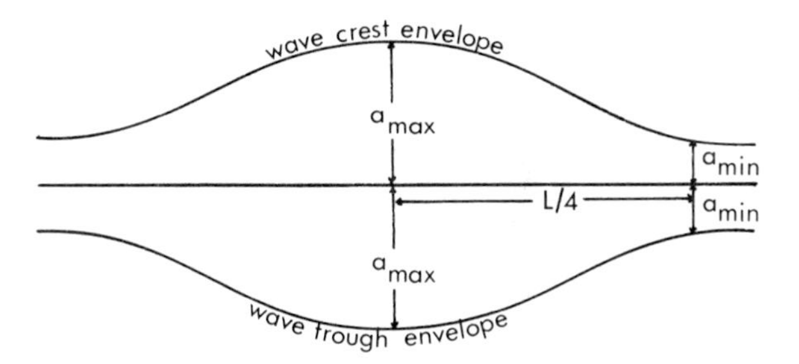

Figure 1.3. The envelopes of partial standing waves (Horikawa 1978).

Ruellan (1950). According to Healy (1953), the reflection coefficient (R_f) is given by:

$$R_f = (a_{max} - a_{min})/(a_{max} + a_{min}),$$

where a_{max} and a_{min} are the amplitudes of the antinode and the node, respectively. If reflectance is perfect, a_{min} is zero, and the reflection coefficient is equal to unity.

More complex techniques have been developed to consider the spectral nature of waves in the field (Goda and Abe 1968, Thornton and Calhoun 1972). Miche (1951) found that the reflection coefficient is given by:

$$R_f = x_1 x_2,$$

where x_1 is a coefficient determined by the roughness and permeability of the reflecting surface, and x_2 is a function of the inclination of the surface and the steepness of the waves in deep water. Miche's conclusions have been verified by Healy (1953) and Greslou and Mahe (1955). Healy (1953) proposed that the reflection coefficient is a function of the slope and roughness of the reflecting surface, and the steepness of the incident waves. Moraes (1970) found that the effect of surface roughness is greatest when the surface is steep. Domzig (1955) and Moraes emphasized the significance of the ratio of water depth to wave length in front of the structure. Domzig and Greslou and Mahe found that the reflection coefficient decreases with increasing wave steepness, and with the ratio of wave height to water depth, although Goda and Abe (1968) suggested that these observations are the result of measurement error. Other investigations of the reflection coefficient have been conducted by Dean (1945), Ursell (1947), Iribarren and Nogales y Olano (1949), Epstein and Tyrrell (1949), Schoemaker and Thijsse (1949), Cooper and Longuet-Higgins (1951), Laurent and Devimeux (1951), and Carry (1953). Because reflection is rarely perfect, the reflected waves are almost always smaller than the incident waves, although wave pressure theories are based upon either the assumption that reflection is perfect ($R_f = 1$), or that there is no reflection ($R_f = 0$). Perfect reflection is rare on natural structures in the field, where rock surfaces are rough and often fronted by rubble or pebble slopes.

For deep and shallow water clapotis, the pressure exerted on vertical structures is primarily hydrostatic, although there is also a slight hydrodynamic component. These pressures are small, but they last for several seconds while the water is rising and falling. Numerous theories have been proposed to account for the pressures exerted by non-breaking waves. The static–dynamic theories assume that wave motion is not disturbed by the presence of the structure (D'Auria 1890, Lira 1926, 1927, Richter 1930, Iribarren 1938, Broikos 1955), so that the pressures are those which would be exerted on a plane suddenly inserted at the wave crest (Hudson 1952). The pressure is the sum of a static component determined by the elevation of the water surface, and a dynamic component which is the result of the collision of the water particles with the structure. The static–dynamic theories have been criticized because they are not based upon any acceptable theory of wave motion in front of vertical structures, and because they do not provide a reasonable estimate of the actual pressures exerted by non-breaking waves. Molitor (1934) used an empirical approach to determine the pressures exerted by standing waves, using wave pressure data from the Great Lakes. His method depends upon the judicious selection of a number of coefficients, according to wind velocity and duration. Other models have been described by Wey (1920), Butawand (1926), Jacoby (1936), and Hansen (1940).

The best wave pressure theories consider the occurrence of standing waves or clapotis in front of vertical structures. Boussinesq (1877) and Saint-Venant and Flamant (1888) studied the formation of clapotis, but Bénézit (1923) was the first to consider their effect on vertical structures in deep water. Other theories have subsequently been devised to encompass shallow water conditions. These theories are derived from either classical rotational or irrotational wave theory (Sainflou 1928, 1935, Gourret 1935, Larras 1937a, Miche 1944, Biésel 1952, and Carry 1953). They are also based upon either the Lagrangian or Eulerian systems. In the Lagrangian system, particle motion is considered as a function of time, whereas in the Eulerian system, velocity, density, and pressure variations are considered in terms of the location within the water body. The Lagrangian system therefore provides a trajectory viewpoint of particle motion, and the Eulerian provides a streamline view of the flow field. The Eulerian system is preferable for the consideration of velocities and accelerations, but the Lagrangian is generally better for the amplitudes of orbital motion (Silvester 1974). Although the systems give similar results, particularly for areas near the bed, models based on the Lagrangian system provided a better measure of the distribution of pressures above the still water level (unless wave steepness is considerable) (Rundgren 1958).

Standing wave pressure theories satisfy the fundamental hydrodynamic equations to a certain degree of approximation. If all terms of the velocity or displacements of $n + 1$ or higher are omitted, then the equations may be described as being homogenous of the nth order of approximation. The applicability and limitations of the various orders of theory have been discussed by Tsuchiya and Yamaguchi (1970). In general, equations of the second order of approximation, including Miche's (1944) in the Lagrangian system, and Biésel's (1952) in the Eulerian system, provide a better fit to recorded pressure distributions than first order approximations. Higher order equations are usually much better, but they are more laborious to apply (Lappo and Zagryadskaya 1977, Fenton 1985). Goda and Kakizaki (1966), for example, developed a fourth order approximation which provided a better fit to recorded data than Nagai's (1969) third order approximation. Third and fourth order approximations are applicable to a much wider range of wave types and water depths than are approximations of lower order (Tsuchiya and Yamaguchi 1970). First and second order theories should only be used in situations where the water depth is more than five times the incident wave height (Silvester 1974). Despite these limitations, Sainflou's (1928) technique has been generally adopted, even though it is only partly of the second order of approximation. Sainflou's expression for the pressure exerted on a vertical structure is:

$$\frac{P}{w} = y_0 \pm H \sin \frac{2\pi t}{T} \left[\frac{\cosh \frac{2\pi}{L} (d + y_0)}{\cosh \frac{2\pi d}{L}} - \frac{\sinh \frac{2\pi}{L} (d + y_0)}{\sinh \frac{2\pi d}{L}} \right].$$

where P is the pressure intensity; w is the specific weight of water (ρg); Y_0 is the ordinate of a water particle at rest; H is wave height; t is the time elapsed by a particle in wave motion; T is the wave period; L is the wave length; and d is water depth measured from the still water level. This expression provides estimates for the pressures exerted by clapotis in the crest and trough positions at the structure (Hudson 1952). In the crest position, the expression describes a pressure curve with a zero value at the crest ($H + h_0$, where h_0 is the height of a surface particle's orbital axis above the still water level) and a pressure at the bottom ($y_0 = -d$) of:

$$P/w = d + [H/\cosh (2\pi d/L)].$$

In the trough position, the pressure curve has a zero value at $H - h_0$ below still water level and a pressure at the bottom of:

$$P/w = d - [H/\cosh (2\pi d/L)].$$

The pressure distribution can be approximated by straight lines, which makes it quite simple to calculate pressures at intermediate points on the structure (Hudson 1952).

The use of low order approximations is acceptable if the reflection coefficient is approximately one, but higher order equations must be used when the water is shallow, or when the waves are steep. Sainflou's expressions, for example, have been found to overestimate the pressures exerted by steep, non-breaking waves. Bruns (1951) compared 26 models of low order approximations, and found that only Lira's (1926, 1927), Sainflou's (1928), Richter's (1930), Gourret's (1935), and Miche's (1944) provide reasonable estimates of the pressures exerted by standing waves. Rundgren (1958) derived second order approximations in the Lagrangian and Eulerian systems. His expressions are capable of accommodating different values for the reflection coefficient, and are therefore applicable to situations in which partial or full clapotis conditions develop. Rundgren's modification of Miche's (1944) technique was preferred by the Coastal Engineering Research Center, CERC (1977), rather than the Sainflou model, which had been the choice of its predecessor, the Beach Erosion Board, BEB (1961). The Miche–Rundgren model indicates that when the wave crest is at the structure (Fig. 1.4*a*), pressures vary from zero at the crest to $wd + P_1$ at the bottom. The value of P_1 is given by:

$$P_1 = (1 + R_f)\,(wH_i)/2 \cosh 2\pi d/L,$$

where H_i and L are the height and length, respectively, of the incident wave at depth d; and the reflection coefficient R_f is equal to H_r/H_i, where H_r is the height of the reflected waves.

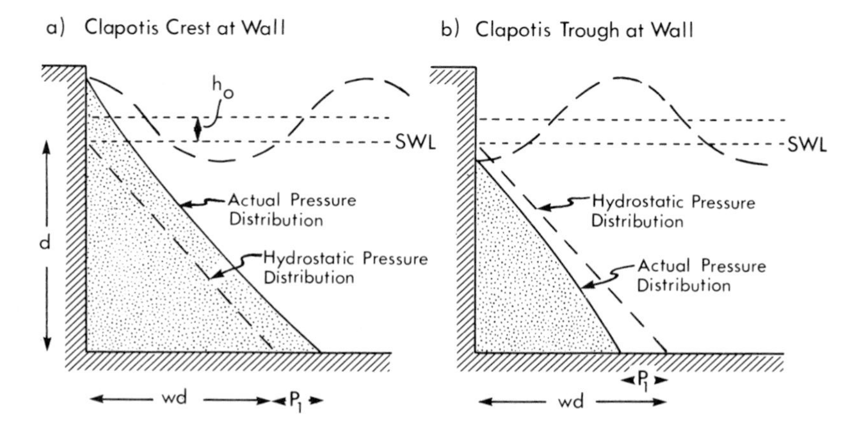

Figure 1.4. The Miche–Rundgren model for the pressure distribution of non-breaking waves on a vertical structure (CERC 1977).

When the trough is at the structure (Fig. 1.4*b*), the pressure increases from zero at the water surface to $wd - P_1$ at the bottom. The CERC have graphed the pressures exerted at intermediate points on a structure for reflection coefficients of 0.9 and 1.0. Although the Miche–Rundgren technique has been widely used in recent years, it is inaccurate when the relative depth (d/L) is small, and should not be used when it is less than 0.132 (Shtencel 1972).

The elevation of the maximum wave pressure is of considerable geomorphological importance. Most models suggest that the maximum wave pressures (as opposed to the hydrostatic pressures, which are greatest at the base) are exerted at the still water level (Lira 1927, Sainflou 1928, Richter 1930, Gourret 1935, Iribarren 1938, Miche 1944, Biésel 1952), although Molitor (1934) considered it to be higher at the elevation of a surface particle's orbital axis. This axis, which represents the mean level of the waves, lies above the still water level by an amount (h_0) which is proportional to the wave height (H) and steepness. Several equations have been used to determine this difference in elevation:

$h_0 = \pi H^2/4L$ (Bénézit 1923),
$h_0 = (\pi H^2/4L)\coth 2\pi d/L$ (Saint-Venant and Flamant 1888, Iribarren 1938)
$h_0 = (\pi H^2/L)\coth 2\pi d/L$ (Sainflou 1928).

Coen-Cagli (1935, 1936) argued that maximum pressures are exerted slightly below the still water level. He considered that the total pressure is greater than that of the clapotis, and roughly equal to wH.

Many workers have found that waves slightly larger than those which produce the typical standing wave pressure–time curves give double peaked or humped curves (Fig. 1.5). Double humped curves are transitional forms between the simple curves of small amplitude standing waves, and the shock pressures associated with large breaking waves. Double humps first occur near the bed as wave steepness increases or as the depth ratio decreases. Greater extremes are needed before double humped pressure curves are recorded at the mean water level. Their occurrence can be predicted using the techniques of Nagai (1969) and Goda (1970).

The effect of oblique, non-breaking waves on vertical structures is not completely understood. Larras (1937b) determined experimentally and theoretically that pressures do not vary appreciably with the angle of incidence of the waves. Kuznetsov (1970), however, reported that Russian workers have found that the forces exerted by oblique, non-breaking waves can greatly exceed those produced by normally approaching waves. The difference in pressure depends upon the incident angle, the wave steepness, and the relative depth (d/L). When waves are reflected obliquely, a

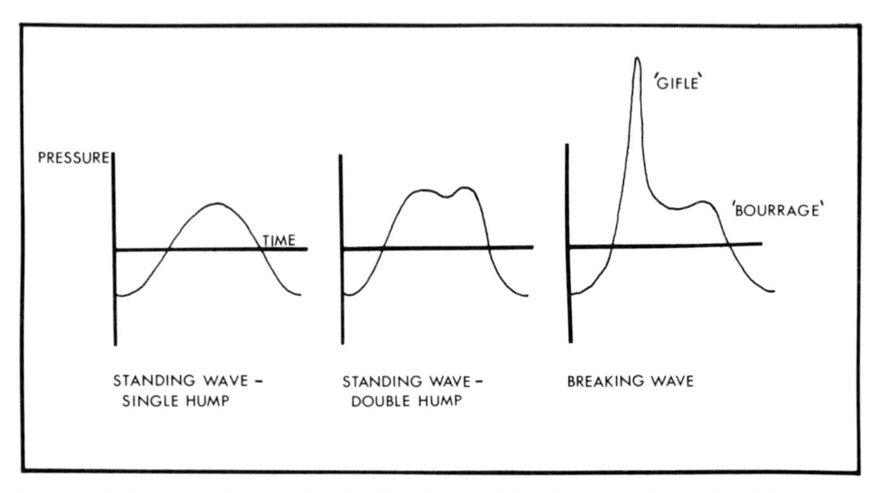

Figure 1.5. Pressure–time graphs showing the transition from standing to breaking waves (Horikawa 1978).

diamond-shaped surface pattern, the *clapotis gaufre* of French workers, develops in front of the structure. These short-crested waves may be important erosive agents in shallow water.

Large storm waves do not form clapotis, but retain the essential characteristics of unobstructed waves. When the sea is stormy, or when short-period waves are superimposed upon a longer swell, waves break in fairly deep water. Furthermore, natural structures such as sea cliffs rarely provide the smooth, vertical surfaces required for perfect reflection, nor are the waves usually exactly parallel to the cliff. Small deviations from the vertical are not significant, however, because pressures exerted by clapotis are similar on vertical and sloping surfaces, although the reach and reflection are somewhat modified (Silvester 1974). As the junction between marine cliffs and shore platforms is usually close to the high tidal level (Wright 1970, Trenhaile 1972), however, only rarely will the water be deep enough for standing waves to develop in front of cliffs. Full clapotis are most likely to form before cliffs in fairly shallow water where the fetch is small, or in more exposed areas where the cliffs plunge directly into deep water. Although the rise and fall of standing waves can facilitate hydraulic quarrying and the formation of caves and notches through the compression of air in rock clefts, effective mechanical erosion of cliffs which were rapidly drowned during Pleistocene marine transgressions has been inhibited by the reflection of the incoming waves (Chapter 8). This has prevented erosional features developing on many plunging cliff faces, accounting in part for the lack of evidence for some palaeosea-levels on rock coasts (Flemming 1965, Cotton 1967).

Breaking Waves

There exists at present no theory capable of successfully describing the motion of water particles in a breaking wave (Cowell 1980, Longuet-Higgins 1980). The relationship between breaking wave height and depth, and a variety of other factors, is presently based upon empirical and theoretical evidence. According to a number of workers, progressive waves moving over a gently sloping bottom usually break when the ratio of wave height to water depth falls between 0.71 and 0.78 (McCowan 1891, Russell 1939, Keulegan and Patterson 1940, Munk 1949). Miche (1944) determined the point at which a wave moving into shallow water becomes unstable. This occurs when:

$$(H/L)_b = 0.142 \tanh 2\pi d_b/L,$$

where subscript b refers to wave characteristics at the breakpoint. Danel (1952) subsequently proposed that the value of the constant is closer to 0.12. The depth at which waves break, however, is also influenced by the slope of the bottom. Collins (1970) found that:

$$H_b/d_b = 0.72 + 5.6 \tan \alpha,$$

where α is the slope of the bottom. Graphs have been provided which enable breaking wave height and depth to be estimated (Goda 1970, CERC 1977, Horikawa 1978). Collins' and Goda's data indicate that the ratio H_b/d_b increases as the bottom gradient decreases.

The point at which a wave breaks is therefore determined by its deep water wave characteristics, the submarine topography, and the water depth, which varies according to tidal and weather conditions. Occasionally, a structure such as a sea cliff stands in water which is at the breaking depth of the incident waves. In these cases, waves break directly onto the structure. Stevenson (1874) first provided data showing that pressures up to 316 kPa are exerted by breaking waves on vertical surfaces. Other data were provided by Gaillard (1904) and Hiroi (1919, 1920). Hiroi designed a model to consider the pressures exerted on a vertical structure by breaking waves. Dynamic pressure is exerted as a result of the fall velocity and the orbital crest velocity of a water particle as it falls onto the still water surface at a 45° angle from the breaking wave crest. Hiroi was concerned with the pressures which may shear a vertical breakwater from its base. He assumed, therefore, that the pressure acts uniformly over the surface of the structure, from the bottom up to $1.25H$ (H is the progressive wave height) above still water level, or to the top of the structure if it is lower than this. Mean presure ($\bar{P}$) is given by:

$$\bar{P} = 1.5\rho g H.$$

Hiroi acknowledged that the maximum wave pressures, which were found to be up to 430 kPa in two harbours in Japan, are exerted near to the water surface, but he considered that it is the mean pressure acting on the whole of the exposed surface of the structure which determines its stability. Although his model describes a uniform pressure distribution which is incompatible with measurements made experimentally and in the field, it has been found to provide a reasonable estimate of the mean pressure exerted on structures by breaking waves (Nagai 1961, Horikawa 1978).

Field measurements at Dieppe and Genoa by French and Italian engineers (Luiggi 1922, Coen-Cagli 1935, 1936, Rouville *et al*. 1938) have shown that when a wave breaks against a structure, a high intensity, short-duration shock pressure related to the uprush of water may occur, followed by a longer period of lower pressure corresponding to the downrush of water; these periods have been termed the 'gifle' and 'bourrage', respectively (Larras 1937c) (Fig. 1.5). The bourrage pressure is largely determined by wave period or steepness, and is essentially independent of bottom slope (Larras 1937c, Kirkgoz 1983). Shock pressures are much greater than the pressures associated with clapotis conditions. At Dieppe, for example, shock pressures of up to 690 kPa have been recorded, although only about 3% of the recorded waves registered pressures greater than those attributable to hydrostatic factors (Rouville *et al*. 1938). Larras (1937b) provided the first data on the duration of shock pressures. At Dieppe, pressures greater than 290 MPa lasted only 0.01 seconds, and experimental work has generally found that durations of only a few hundredths to a thousandth of a second are most typical (Bagnold 1939, Denny 1951, Rundgren 1958, Mitsuyasu 1966, Richert 1974).

Bagnold (1939) has shown experimentally that a number of critical requirements are necessary for the generation of shock pressures. A breaking wave can trap a pocket of air between itself and a vertical structure. This air cushion is increasingly compressed by the advancing wave front, until it is able to burst upwards. Bagnold found that shock pressures are only generated when the air cushion is very thin, when the wave front is flat and vertical, and when the waterline has had time to rise some distance up the structure before the wave crest makes contact with it (Fig. 1.6*a*). The presence of small, parasitic wavelets superimposed on the main wave can prevent the generation of high shock pressures. The prominences associated with these wavelets subdivide the thin air cushion into a number of smaller pockets. Because compression of these pockets is unlikely to occur simultaneously, high pressures generated by compression of one pocket are relieved by expansion into an adjacent pocket where compression has not yet proceeded as far (Fig. 1.6*b*). Denny (1951), for example, found that the presence of small ripples on the main wave, which were only one-tenth of its size, reduced the generated pressures by half.

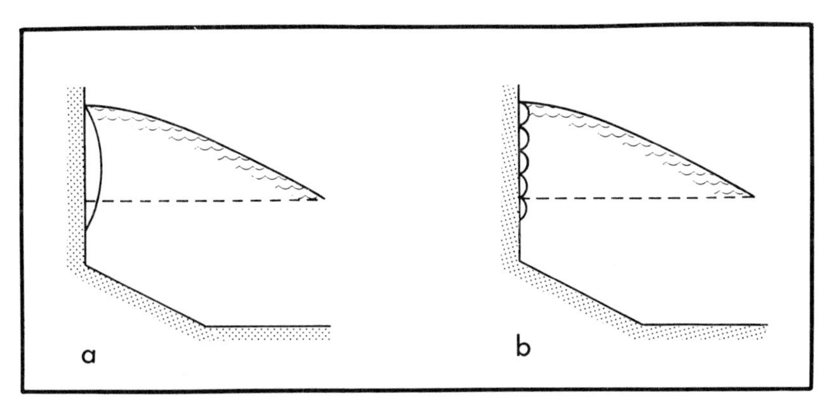

Figure 1.6. (*a*) Thin air cushion trapped by a breaking wave, and (*b*) a breaking wave with small wavelets (Bagnold 1939).

Bagnold derived the following expression for the peak shock pressures (P_{max}) for any consistent units:

$$P_{max} - P_0 = 2.7\rho U^2 k/B,$$

where U is the velocity of a water column, and k is its length; B is the initial thickness of the air cushion; ρ is the density of the water; and P_0 is the initial pressure of the air (atmospheric). The utility of this equation is improved by substituting for:

$$k = 0.2H,$$

where H is the wave height in metres from crest to trough. The peak shock pressure is therefore given by:

$$P_{max} - P_0 = 0.54\rho U^2 H/B.$$

Morison (1948) estimated the thickness of an air cushion to be about 6.1 mm for a particular model wave. Because of the impossibility of measuring the thickness of the air cushion in the field, however, Bagnold's equation is of theoretical rather than practical interest. Minikin's (1950) technique is therefore used in situations where the occurrence of strong shock pressures must be considered (Horikawa 1978). Minikin's model is based upon Bagnold's work on shock pressures, and on European wave pressure measurements at Dieppe and Genoa. His original equations were slightly modified by the BEB (1961) and the CERC (1977) to make them applicable to simple vertical structures. The total pressure exerted on a structure by breaking waves consists of a dynamic or shock pressure component, and a hydrostatic component. Dynamic pressures are at a maximum at the still water level, decreasing parabolically above and below this level. Dynamic pressures are zero at the breaker crest and trough.

Hydrostatic pressures are distributed in a triangular pattern, with a zero value at the wave crest, and a maximum value at the bottom of the structure (Fig. 1.7). The dynamic pressure is given by:

$$P_m = \pi \rho g \, d \, H_b \, (d_w + d)/L_w d_w$$

and

$$P_y = P_m[(H_b - 2y)H_b]^2,$$

where P_m is the maximum dynamic pressure at the still water level; P_y is the dynamic pressure at y above the still water level; d is the depth of the water at the structure; H_b is the breaking wave height; ρ is the water density; and d_w and L_w are water depth and length of the wave respectively, at a distance of one wavelength seawards of the structure.

Hydrostatic pressures are given by:

$$P_1 = \rho g H_b/2,$$
$$P_s = \rho g(d + H_b/2),$$

where P_1 is the static pressure at the still water level and P_s is the hydrostatic pressure at the base of the structure.

Minikin's technique should be used with caution when the bottom slopes in front of the structure are less than 1 : 20 (BEB 1961). Nagai (1961) found that the model provided estimates of shock pressures which were within ± 60% of the actual shock pressures generated by moderately sized waves. Very high shock pressures associated with large storm waves, however, are poorly predicted by this technique.

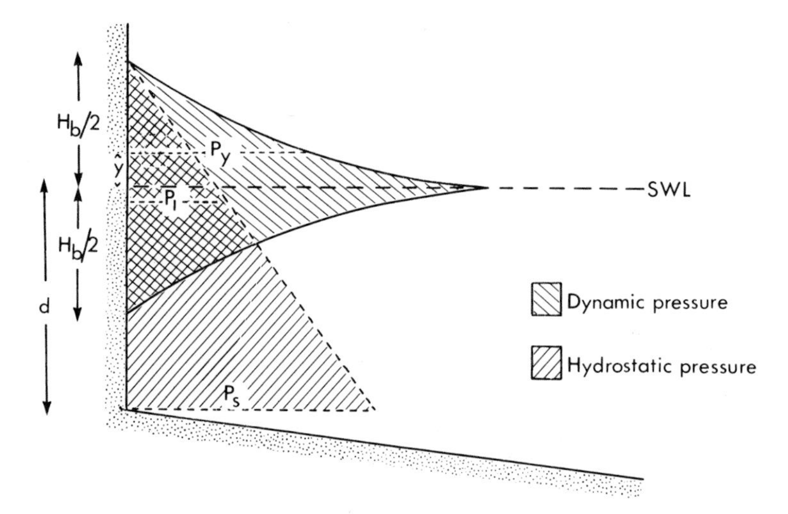

Figure 1.7. Minikin's breaking wave pressure model (Minikin 1950, BEB 1961, CERC 1977).

Bagnold's model suggests that there is a direct relationship between the maximum shock pressures and the height of the waves. Denny (1951), Ross (1954), and Rundgren (1958) provided further evidence in support of this hypothesis, but the exact nature of the relationship has not been determined. Pressures up to 689 kPa were recorded from waves 4.6 m high at Dieppe, compared with pressures of 551.6 kPa and 145 kPa from waves 0.25 and 0.19 m in height, respectively (Rouville *et al.* 1938, Bagnold 1939, Ross 1955). Rundgren (1958) derived an expression for the relationship between maximum shock pressures and deep water wave height (H) and length (L), within the range $0.006 < H/L < 0.06$:

$$P_{max}/\rho gH = C_1 \ln (H/L) + C_2.$$

The values of C_1 and C_2 depend upon the location of the pressure sensors above the still water level. The expression shows that for constant wave steepness, maximum shock pressures increase with wave height. When wave height is constant, pressure increases with the wave length. The minimum height of waves which are capable of generating shock pressures has been found to increase with decreasing wave period (Richert 1974). Carr (1954) and Richert (1968, 1974) have shown that a narrower range of wave heights appear to be capable of generating shock pressures when the submarine slope is gentle than when it is steep. Efforts to determine the relationship between the maximum shock pressure intensity and duration, the nature of the waves, and boundary conditions have generally proved to be unsatisfactory. Morison *et al.* (1954) and Miller *et al.* (1974) found that pressures exerted by plunging breakers are greater than those associated with spilling breakers. This may be because the periodic pressure field of spilling breakers deviates from hydrostatic pressure at the breaking wave front and at the wave crest (Peregrine and Svendsen 1978, Stive 1980). It is not possible, therefore, to derive a single expression relating impact pressures to wave or breaker height (Miller *et al.* 1974). The use of different bottom gradients, ranging from 1 : 50 (Mitsuyasu 1966) to 1 : 2 (Nagai 1961), and different wave types, including solitary waves (Bagnold 1939, Denny 1951, Hayashi and Hattori 1958), consecutive breaking waves (Rundgren 1958, Mitsuyasu 1966), and breaking waves preceded by non-breaking waves (Ross 1955, Richert 1968) make it extremely difficult to compare and evaluate the often contradictory results of investigations in this field. Not only have a number of techniques been employed to measure shock pressures, but the equipment has often been unsuitable for the measurement of high frequency pressure oscillations. Even in controlled experimental environments, it is difficult to exactly reproduce such variables as the impact velocity and area, the amount of entrained air, and the ambient pressure, and to relate the results to field conditions (Lundgren 1969, Ackermann and Chen 1974, Mogridge and Jamieson 1980).

The occurrence of shock pressures requires such specific conditions that their magnitude and duration have increasingly been treated statistically, as random variables (Kamel 1970, Massel *et al*. 1978).

It has been argued that entrapped air is not essential for high shock pressures, and that maximum impact pressures are actually generated when a vertical structure is struck by a breaker with a perfectly vertical face (Ross 1954, Hayashi and Hattori 1958, Kirkgoz 1982). Mitsuyasu (1966) has suggested that shock pressures are generated where the concave portion of a breaking wave encloses an air pocket, whereas water hammer pressures occur where the wave front is vertical and planar. Several studies have demonstrated that the intensity of shock pressures increases as the thickness of the air cushion decreases (Richert 1968, Kamel 1970). Water hammer pressures may therefore provide an upper limit to the pressures generated by breaking waves. Recorded pressures, however, are always of an order of magnitude less than that predicted by the water hammer equation:

$$P_h = \rho UC,$$

where P_h is the water hammer pressure, U is the initial velocity of the water, and C is the celerity of sound in water. Ackermann and Chen (1974) showed that water hammer pressures are not attained even when the ambient pressure of the entrapped air is equal to the vapour pressure of the water. Breaking wave pressures probably can never approach theoretical water hammer pressures because of the inevitable pressure of entrained air. The velocity of sound in water which contains only 1% of air by volume, for example, is 0.18 km sec^{-1}, compared with 1.22 km sec^{-1} in water which contains no air (Silvester 1974).

Several workers have found that the maximum pressures exerted by breaking waves occur within the upper four-tenths of the wave height (Bagnold 1939, Denny 1951, Ross 1955, Weggel and Maxwell 1970). Others, however, found that maximum pressures are usually at the still water level (Morison 1948, Mitsuyasu 1963, 1966, Minikin 1950, Nagai 1961), or even below (Richert 1968, 1974). This partly reflects different research designs, as the distribution of breaking wave pressures is affected by the characteristics of the breaking wave, and by the shape of the structure it breaks against (Nagai 1961). Nevertheless, most model and full-scale field investigations have shown that the pressures exerted by breaking waves are greatest at or slightly above the still water level, rapidly declining above and below the level of maximum intensity.

It is difficult to determine whether very high shock pressures can be generated by breaking waves on rock coasts. Shock pressures have proven difficult to produce under controlled laboratory conditions, but in the field the number of complicating factors is greater. Very few waves break in just the right position to generate shock pressures. Of those which do break against a cliff, only a small proportion are likely to present a smooth, vertical water surface parallel to the cliff face. The characteristics of the cliff are also important. Miller *et al*. (1974) failed to record any high shock pressures in the field which could be attributed to the compression of entrapped air. They concluded that the confining sidewalls of laboratory flumes, or a natural air pocket provided by the configuration of the surface of the structure, are prerequisites for the generation of shock pressures. Although clefts in a rock surface may facilitate shock pressure generation, a rough, impermeable surface also permits air to be trapped between itself and the advancing wave front, preventing the generation of very high water hammer pressures. Nevertheless, considering the millions of waves which break on storm wave coasts each year, it is probable that high shock pressures are occasionally produced in the field. Goda's (1970) data suggest that high shock pressures are most likely to be associated with quite steep bottom slopes, possibly of the order of about 1 : 10 (Richert 1968, Kirkgoz 1982); waves which are low in deep water; and cliff-foot water depths of less than about twice the deep water wave height. Despite the rather specific conditions necessary for the generation of shock pressures it is unwise to dismiss their possible contribution to the development of rock coasts, given that the dislodgement of even a single joint block by shock pressures once per century would account for the slow erosion recorded in some areas.

Broken Waves

There have been several attempts to investigate wave transformation in the surf zone (Iverson 1951, Horikawa and Kuo 1966, Hotta *et al*. 1982, Yasuda *et al*. 1982), but the predictive equations are still not entirely satisfactory. It is therefore difficult to determine the relationship between wave parameters and the pressures they exert in the surf zone. When a wave breaks, a bore-like wave front of a random, turbulent nature may persist up to the shoreline, if the bottom slope is less than 0.035 (2°). If the bottom slope is less than 0.022 (1.25°), however, the bore may re-form into a non-breaking wave, assuming that there is enough distance between the breakpoint and the shoreline (Nakamura *et al*. 1966). The energy and the forces exerted by broken waves are much greater than those of unbroken oscillatory waves, because the orbital motion is replaced by the shoreward movement of the whole water mass (Zenkovitch 1967). Because of bottom friction and particularly turbulence, however, wave energy rapidly declines in the surf zone in proportion to the distance travelled from the breakpoint (Sawaragi and Iwata 1974, Kirkgoz 1981). Water depth and the shape of the bottom determine the position of the breakpoint, the condition of the breaking waves, the energy flux, and the rate of attenuation of wave height in the surf zone (Nakamura *et al.* 1966, Svendsen *et al.* 1978). Wave height is greatest when the bed slopes are steep (Horikawa and Kuo 1966). Petrashen (1956—cited by Zenkovitch 1967) found that for given wave parameters, wave force in the surf zone is greatest when the bottom gradients are between 11.3 and 16.7°. These factors determine the ability of a wave to perform work, and the way in which its energy is distributed within the surf zone.

Morison *et al.* (1954) and Miller *et al.* (1974) measured the forces exerted by breaking and broken waves in the field. Their data indicate that the greatest pressures are generated by the bores of plunging breakers, followed in turn by the bores of spilling breakers, plunging breakers, spilling breakers, and finally by near breaking waves. The turbulent upper portion of broken wave bores can contain 20 to 30% air bubbles by volume (Fuhrboter 1970, Miller 1972). Pressures exerted by bores are therefore greatest near the base of the water body. Miller *et al.* (1974) have suggested that the high pressures recorded at Dieppe (Rouville *et al.* 1938) were generated by broken waves, rather than by breaking wave shocks.

The characteristics of a breaking wave are determined by the deep water wave steepness and the slope of the bottom. The parameters which define the occurrence of each type of breaker have been determined experimentally (Galvin 1968, Horikawa 1978, p. 54). Although the results of these experiments differ in detail, they indicate that spilling breakers usually occur where bottom gradients are low (0 to about 4°, say). Spilling breakers also tend to be associated with steep waves in deep water (Iverson 1951). Steep waves plunge only if bottom gradients are fairly steep. This suggests that the erosion of steep rock slopes by plunging breakers, which dissipate a large proportion of their energy near the breakpoint (Galvin 1972), is more effective than the erosion of low rock slopes, such as shore platforms, by spilling breakers and bores.

In Japan, the Hiroi (1919) equation has been used to estimate the pressures exerted by broken waves. The BEB (1961) has derived expressions for the pressures exerted by broken waves. They are based upon simplifications of wave behaviour, however, and probably provide rather conservative estimates. When a wave breaks, intense turbulence is created, and the turbulent water rises up above the still water level. This divides the surf zone into two portions, one lying landward and the other seaward of the waterline. The total pressure is the sum of the static and the dynamic components. If the structure is between the breakpoint and the shoreline, then:

$$P_m = \rho g d_b/2,$$
$$P_s = \rho g(d + h_c),$$

where P_m is the dynamic pressure at a height h_c above still water level; P_s is the static pressure at the base of the structure; d_b is breaking water depth; d is the water depth at the structure; and h_c is the height of that portion of the breaking wave above the still water level (Fig. 1.8*a*).

If the structure is shorewards of the waterline, the corresponding expressions are:

$$P_m = (\rho g d_b/2)(1 - x_1/x_2)^2,$$

$$P_s = (\rho g h_c)(1 - x_1/x_2),$$

where x_1 is the distance from the shoreline to the structure, and x_2 is the distance from the shoreline to the limit of wave swash (Fig. 1.8*b*).

Homma and Horikawa (1965a,b) found the BEB models inadequate. They derived equations based upon solitary wave theory and the assumption that the height of the broken waves in the surf zone is given by $0.78d$, where d is the water depth. This approximation is justified if the bottom slope is greater than 1.9° (1 : 30). Their model describes a triangular distribution of dynamic pressures, with a maximum intensity at the still water level, and a zero value at the base of the structure and at a height h_d above the still water level (where $h_d = 1.2d$). The distribution of static pressure is also described by a triangle, with a maximum value at the base of the structure, and a zero value $1.2d$ above the still water level. A similar expression considers the pressures exerted by broken waves on a non-vertical structure (Fig. 1.9). The authors considered that their model provides a good estimate of the mean pressures exerted by broken waves, except where water depths are less than one-fifth of the offshore wave height.

The dynamic pressures exerted by oblique breaking or broken waves (P_a) on a vertical structure are less than those generated by waves approaching the structure normally (P). The CERC (1977) provided the following expression:

$$P_a = P \sin^2\alpha,$$

where α is the angle between the direction of the wave and the axis of the structure. If the structure slopes backwards at an angle θ to the horizontal ($\theta \geqq 90°$), the pressure (P_c) is given by:

$$P_c = P_b \sin^2\theta,$$

where P_b is the pressure generated on a vertical structure.

Processes of Mechanical Wave Erosion

Most models which estimate the pressures exerted by standing, breaking, or broken waves predict that the maximum pressures occur at or slightly

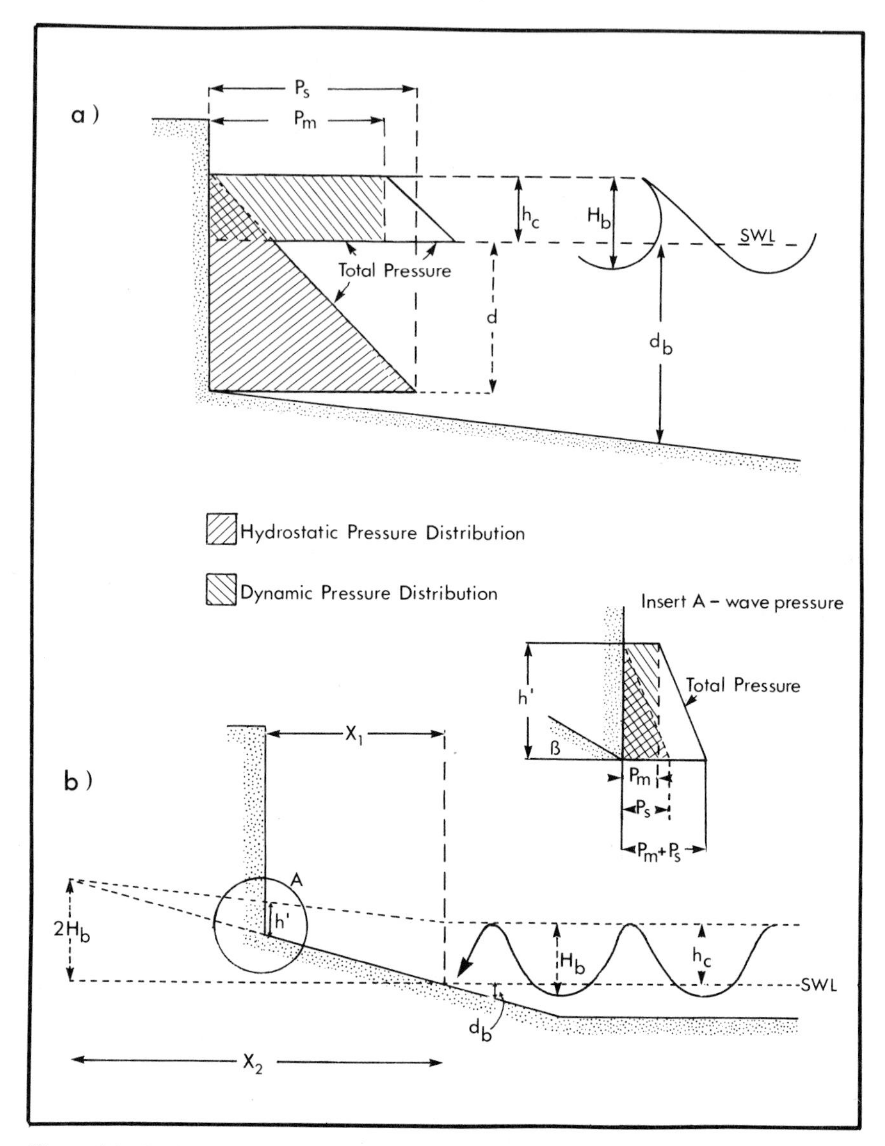

Figure 1.8. Broken wave pressure (*a*) when the structure is seawards of the stillwater line; and (*b*) when the structure is landward of the stillwater line (BEB 1961, CERC 1977).

above the still water level. On a one-to-one basis, breaking waves may be more effective erosional agents than broken waves, although because of the critical conditions which are necessary, only a small proportion of the waves actually break against steep, natural structures. The erosional efficacy of broken waves at or close to the still water level is determined by the

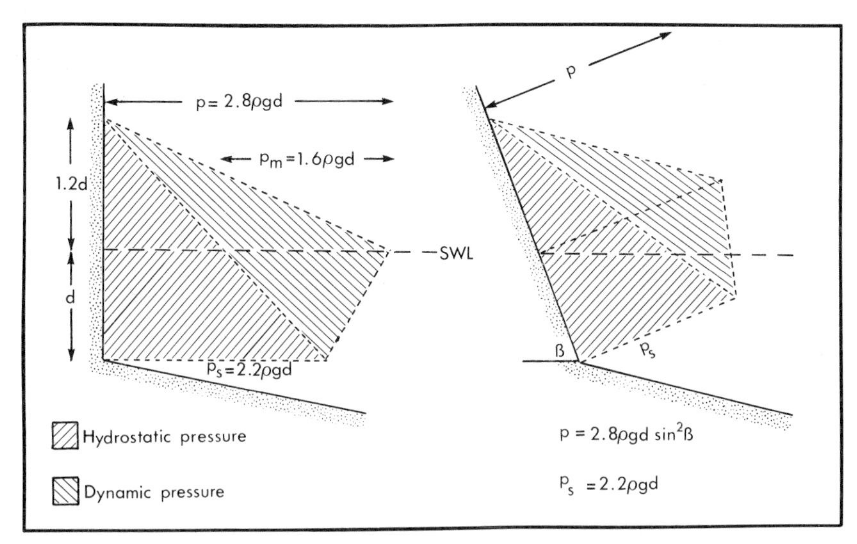

Figure 1.9. The distribution of broken wave pressure on a vertical and non-vertical structure (Homma and Horikawa 1965a, b).

local depth and the bed slope in the surf zone, which determine the energy flux and wave attenuation after breaking.

While mechanical wave erosion is accomplished by a number of processes, their relative importance has usually been inferred from morphological evidence, which is often ambiguous: the use of the microerosion meter (MEM) facilitates the measurement of erosion rates, but the processes responsible must still be inferred from the data (Kirk 1977, Robinson 1977a,b). Furthermore, the technique is incapable of considering the quarrying of large rock fragments, and like other methods has difficulty in assessing the role of high magnitude, low frequency, erosional events. Consequently, despite the lack of direct measurement, the presence of large, angular debris, fresh rock scars, and other indirect evidence has convinced many workers that wave quarrying is the dominant erosive mechanism in many areas (Bartrum 1938, Everard *et al.* 1964, Trenhaile 1969, 1972, 1978, Sunamura 1978a).

The most effective wave quarrying processes appear to be the most restricted in extent. Quarrying caused by the shock pressures of breaking waves, water hammer, and air compression in joints requires the alternate presence of air and water, and is therefore restricted to a rather narrow range extending from just below the still water level up to the wave crest (Sanders 1968a) (Fig. 1.10). Theory, and the presence of fragile solutional features on vertical rock faces in some wave exposed areas, however, suggests that the role of water hammer is quite slight (Ackermann and

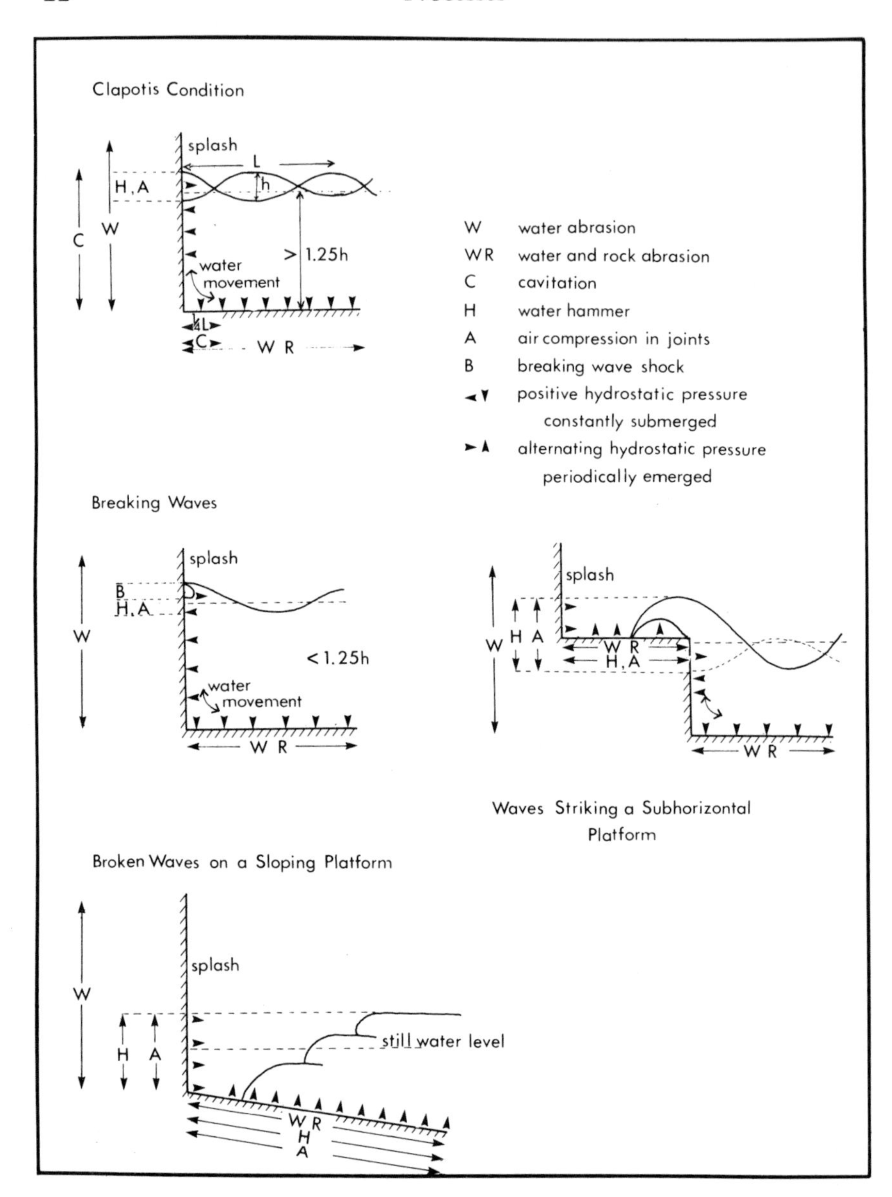

Figure 1.10. Mechanical wave erosion processes and their distribution on rock coasts (Sanders 1968a).

Chen 1974, Trenhaile 1969). Bagnold (1939) argued that under natural conditions, true water hammer (defined as the impact between a body of water and a solid) can never occur, as there must always be an intervening cushion of air. Even if a continuous air pocket is not trapped between the wave front and the cliff, the presence of air in the violently agitated water near the breakpoint and in the surf zone ensures that some air is always interposed between the rock and the water. Bagnold made the interesting suggestion that as the foam-forming properties of water are sensitive to its mineral and organic content, water hammer and shock pressures should vary geographically. As discussed previously, wave shock requires critical conditions which must severely restrict its occurrence and intensity. The high pressures which they generate, however, may more than compensate for their low frequency, and it is possible that they play an important role in the long-term development of rock coasts. Joint blocks and other rock fragments can be dislodged as a result of the compression of pockets of air trapped along joints, bedding planes, erosional clefts, and other rock cavities. Air compression caused by waves breaking against a rock face, or by the inrushing bore of a broken wave, is followed by the sudden release of this pressure as the water recedes. The process is particularly effective if sand is washed into the rock fractures, helping to keep them open and making them more vulnerable to wave attack (Robinson 1977a).

The compression of air trapped in rock clefts is probably the most effective process of mechanical wave action, although it is also one of the least understood. Robinson (1977a) found that wave quarrying at the cliff foot in northern England is most pronounced in winter, when wave energy is greatest. Densely jointed rocks appear to be most susceptible to this process, particularly if they are thinly bedded. The more rapid erosion of shale beds often provides air prockets along the bedding planes of resistant strata. High pressures can only be generated if the wave front or bore is parallel to the rock face, or if for any other reason the pressure cannot be relieved laterally into areas of lower pressure.

Bagnold (1939) used a simple cylindrical piston-and-cup model to derive an expression for the generation of shock pressures. His model is therefore applicable to a situation in which an air cushion is compressed in a rock cleft by an inrushing mass of water. Bagnold assumed that the pressures fall within the range of about 0.20–1 MPa, and that compression occurs adiabatically. Mitsuyasu (1966) developed a similar model to describe the infinitesimal adiabatic compression of an air cushion (Fig. 1.11), in which:

$$P - P_0 = \rho U k \sigma \sin \sigma t + \rho U k \sigma ((\gamma + 1)/2) (a_1/B) \sin^2 \sigma t,$$

where P is the generated pressure; P_0 is air initially at atmospheric pressure; B is the original thickness of the air cushion; U is the initial velocity of the water; k is the length of the water column responsible for air compression; t is the elapsed time; and γ is the compression index, or the ratio between the specific heat at constant pressure and that at constant volume. The compression index is equal to 1.4 if compression is adiabatic. The values of σ and a_1 are given by $(P_0\gamma/\rho k B)^{1/2}$ and $(\rho k U^2 B/P_0\gamma)^{1/2}$, respectively. Mitsuyasu considered that a fairly good approximation is given by the first term on the right-hand side of the equation. He also

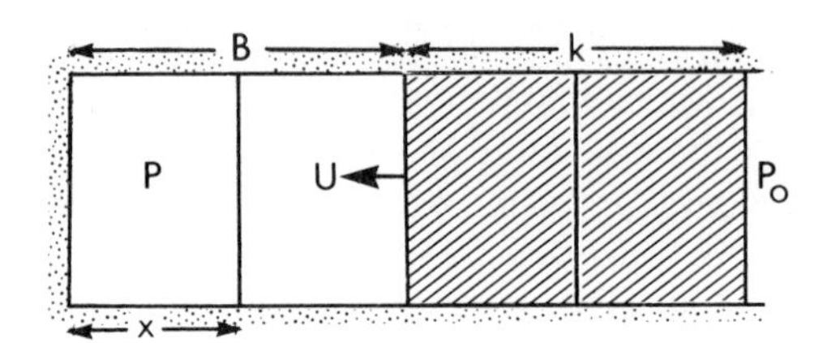

Figure 1.11. Mitsuyasu's (1966) air compression model.

provided an equation for the finite compression of an air cushion by a large mass of water:

$$P_{max-min}/P_0 = 1 \pm 1.18(Z)^{1/2} + 1.2(Z),$$

where Z is given by $(\rho k U^2/P_0 B)$. The +ve sign refers to the maximum pressure and the −ve to the minimum pressure. Mitsuyasu has also considered the situation in which there is leakage of the compressed air. This occurs through joints in rocks, or by lateral expansion into areas of lower pressure along the cleft itself.

Whether air compressions is adiabatic, isothermal, or a combination of the two depends upon a number of factors, such as the pressure and its duration, the shape and acceleration of the air/water boundary, the presence of air bubbles, and the irreversibility of the process (Ramkema 1978). Bagnold (1939) thought that the compression index would vary during compression, but that rapid cooling might bring about isothermal compression towards the end of the stroke. The matter, however, has not been resolved. Because of large air volumes and higher pressures, it is possible that the process in nature may be more adiabatic than in model experiments, but on the other hand, because of the larger air/water surface, the longer characteristic period, and the state of the water surface, the process in nature could be more isothermal (Ramkema 1978). Ramkema has developed models for adiabatic and isothermal compression. For linear compression:

$$P_{max} = P_0[1 + (2\gamma S)^{1/2}],$$

where γ is 1.4 for adiabatic and 1.0 for isothermal compression; and S is defined as the impact number, equal to $\rho U^2 k/2P_0 B$. He has also graphed the relationship between peak pressures and the impact number for linear and non-linear, adiabatic, and isothermal compression.

Bagnold's, Mitsuyasu's and Ramkema's models suggest that the pressures generated by the compression of air cushions increase with the momentum of the impinging water mass, and decrease with the thickness of the cushion. The alternate presence of air and water, whether or not it is accompanied by the generation of high shock pressures, induces a rocking motion in the undermined joint blocks, which can eventually dislodge them (Trenhaile 1972). Pitty (1971, p. 174) has noted that suction associated with turbulent eddies can cause quite large blocks to be lifted from the bottom. He suggested that blocks could even by quarried from rocks which have been previously weathered by chemical or physical processes, or by pressure release. The significance of this process cannot be determined at this time.

Several other processes may be locally important, although they are not as intimately associated with the water surface as are the processes responsible for wave quarrying (Fig. 1.10). They include cavitation, hydrostatic pressure fluctuations, and abrasion. Cavitation can occur when water velocities near the bottom are sufficient to cause a significant drop in pressure. If the vapour pressure is attained, the rock surface may be damaged by the rapid formation and collapse of vapour pockets. Cavitation requires high velocities and water which is not aerated. The most suitable bottom conditions may be at a distance about one-quarter of the length of the wave in the clapotis condition from the reflecting wall (Sanders 1968), where water motion is twice as great as the incident wave velocity (Sainflou 1928).

Abrasion is the result of the sweeping, rolling, or dragging of rocks and sand across gently sloping rock surfaces, or the throwing of coarse material against steep surfaces. Abrasion of fairly homogeneous strata generally produces much smoother surfaces than those associated with quarrying. Grooves can develop where abrasives are concentrated along converging or parallel joint planes (Swinnerton 1927). In northern England, abrasion is usually more active in the stormy winters than in summers on the shore platform ramp, although the difference is slight, or even reversed in places, depending upon the local character and behaviour of the beaches (Robinson 1977b). The effect of abrasion is limited by the height to which significant quantities of beach material can be carried by waves. At the cliff foot in northeastern Yorkshire, England, intense erosion associated with a sand and small-pebble beach occurred in a zone extending up to a point about 10 cm above the beach surface. Maximum erosion was at a depth of about 14.5 cm below this surface, possibly because of the turbulence created by the waves striking the cliff foot (Robinson 1977a). A further difficulty is that of sufficiently agitating beach particles at the base of thick accumulations. In Yorkshire, on the platform-ramp, all waves are able to agitate particles at the base of deposits less than 5 cm thick, but only the largest waves can move particles at depths of more than 13.5 cm (Robinson 1977b). Abrasion can be retarded by large storm waves, however, because more particles are taken in saltation or suspension, providing less contact with the bedrock. Thick accumulations of abrasive material can also cause a reduction in the energy of the waves which cross it, providing a further reduction in the erosive efficacy (Sunamura 1976). The lack of suitable abrasive material, and the surface irregularity of many rocky foreshores can relegate corrasion to a secondary role in the lower intertidal zone, where its effect is often restricted to structural lines of weakness.

Potholing is a gyratory form of abrasion, which operates in a great variety of rock types and coastal environments. Little work has been conducted on marine potholes, although Dionne (1964) has provided a valuable review of the available literature. Potholes are approximately cylindrical depressions, produced where the water rotates coarse material in the surf or beaker zone. More irregular forms are the result of structural control or the coalescence of several depressions. The abrasives usually consist of rocks or large pebbles, although they can occasionally be sand or gravel. Erosion is likely to be particularly rapid where the abrasives consist of concretions or other material more resistant than the host rock (Henkrel 1906, Wentworth 1944). Frequently, only the lower portions of the depressions are polished, the upper surfaces, being above the level of contemporary abrasion remain rough and are covered in algae and lichen (Dionne 1964, Bird 1970a). It has been suggested that potholes can be produced by rapidly flowing water without the assistance of rock tools (Alexander 1932), but the temporary absence of rocks from a pothole is not necessarily evidence of this possibility. The initial entrapment of abrasives occurs at the foot of scarps, or where there are already structural or erosional depressions. In calcareous rocks, more widespread pothole-like features develop as a result of the inheritance or corrosional hollows. Steep sides, sometimes with overhanging lips, and a width-to-depth ratio of 1 : 1–1 : 1.5, may serve to distinguish marine potholes from the shallow, cupped, or flared depressions produced by corrosion. Dionne has suggested that as this ratio is rarely exceeded, it would appear that potholes develop towards a state of equilibrium, possibly related to the progressive weakening of the gyratory flow as the depression deepens and sediment accumulates in the bottom. Potholing and abrasional damage caused by the throwing of debris against steep rock surfaces are probably most significant in cool, storm wave environments, where wave action is vigorous and pebbles and other coarse beach materials are abundant. Nevertheless, it is likely that in the initial stages of pothole development, weak waves, which do not wash the abrasives from the shallow depressions, are more effective than vigorous waves. (Swinnerton 1927). Potholes tend to occur, however, in the upper portions of the intertidal zone, where the waves are strong. Dionne (1964) has calculated that in a semi-diurnal tidal environment, potholes are abraded for about 160 minutes each day. For a little more than half this time, they are subject to the most turbulent action of the breaking waves. Not all potholes are of recent origin, however; some have been partially inherited from ancient shorelines. In Victoria, Australia, for example, some abrasion potholes in calcareous aeolianite were formed from solution pipes, which were exhumed by the coastal erosion of a buried soil horizon (Bird 1970a).

Hydrostatic pressure on the bottom varies according to the height of the water column above. The pressure at a depth of about 10 m is twice that at the water surface. High frequency fluctuations associated with the position of wave crests and troughs are superimposed upon low frequency oscillations related to tides (Hoffmeister and Wentworth 1942). Pressure differences on the bottom attributable to waves are given by ρgH (that is, about $10.1H$ kPa, where H is wave height in metres). Steep waves provide the greatest variations in pressure with the highest frequency. Hydrostatic pressure variation induced by standing waves or clapotis may be important because the reflected wave is up to twice the height of the incident wave. Stem waves associated with Mach reflection can also be twice the incident wave height. This occurs when waves approach a reflecting surface at a low angle. The reflected waves do not leave the surface entirely but travel along it, steepening as they progress along its face (Berger and Kohlhase 1976). In the surf zone, where the rock surface is periodically covered and uncovered by the turbulent bore, small hydrostatic variations are proportional to the height of the bore. Whether these pressure variations are sufficient to cause deterioration of the rocks depends upon whether the strain exceeds the tolerance of the material. The permeability of the rock probably influences the depth to which changes in pore pressure are significant. It remains to be determined whether these low intensity, high frequency pressure pulsations can, for example, loosen joint blocks, or whether they can contribute to the destruction of fissile shales and mudstones.

Although we presently know so little about the operation and relative efficacy of mechanical wave processes, most studies show that the zone of maximum pressure, and the area in which the processes are apparently most effective, are essentially coincident, at or near the still water level. In a tidal environment, this zone fluctuates within the tidal range. The distribution of the still water zone within the tidal range is therefore of fundamental importance to the development of rock coasts.

Tidal duration

In 1880, Dana suggested that there is 'a level of greatest wear', a little above the mid-tidal level. So (1965) introduced, but failed to substantiate, the interesting hypothesis that the effect of storm waves is greatest at the mid-tidal level because they operate most frequently at that elevation. The tide determines where wave quarrying, by processes which are most active at about the still water level, is most effective. Waves reach the

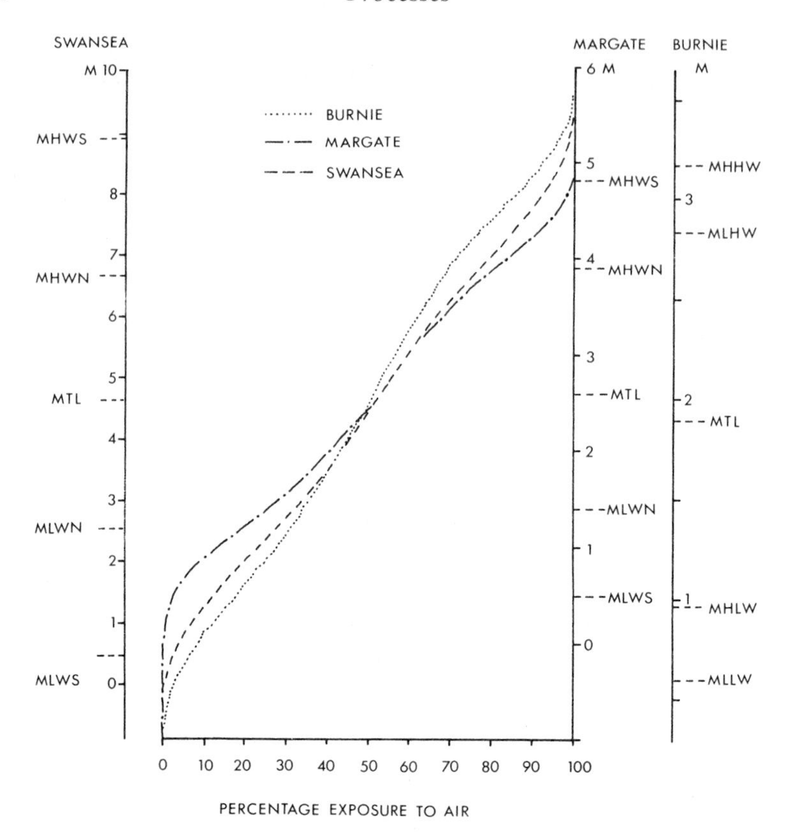

Figure 1.12. Tidal emersion curves for Margate and Swansea in southern Britain, and Burnie in Tasmania (Carr and Graff 1982).

shoreline at elevations determined by the tidal level. Although waves, and particularly breaking or broken storm waves, perform the erosive work, it is the tidally controlled distribution of wave energy which determines where this work is performed (Trenhaile 1978, Trenhaile and Layzell 1981). Vigorous storm waves can approach a coast at any tidal level, but over the long periods of time commensurate with the development of rock coasts, the 'level of greatest wear' must be closely associated with the elevation most frequently occupied by the still water level.

The total time per year in which a particular intertidal level is occupied by the still water level is the sum of two components: the time during which the elevation coincides with a tidal extreme (high or low tide); and the time during which the elevation occupies an intermediate point at the ebb or flow stage of a tidal cycle. At high and low tide, the still water level is almost stationary. The height of these tidal extremes, however, varies from

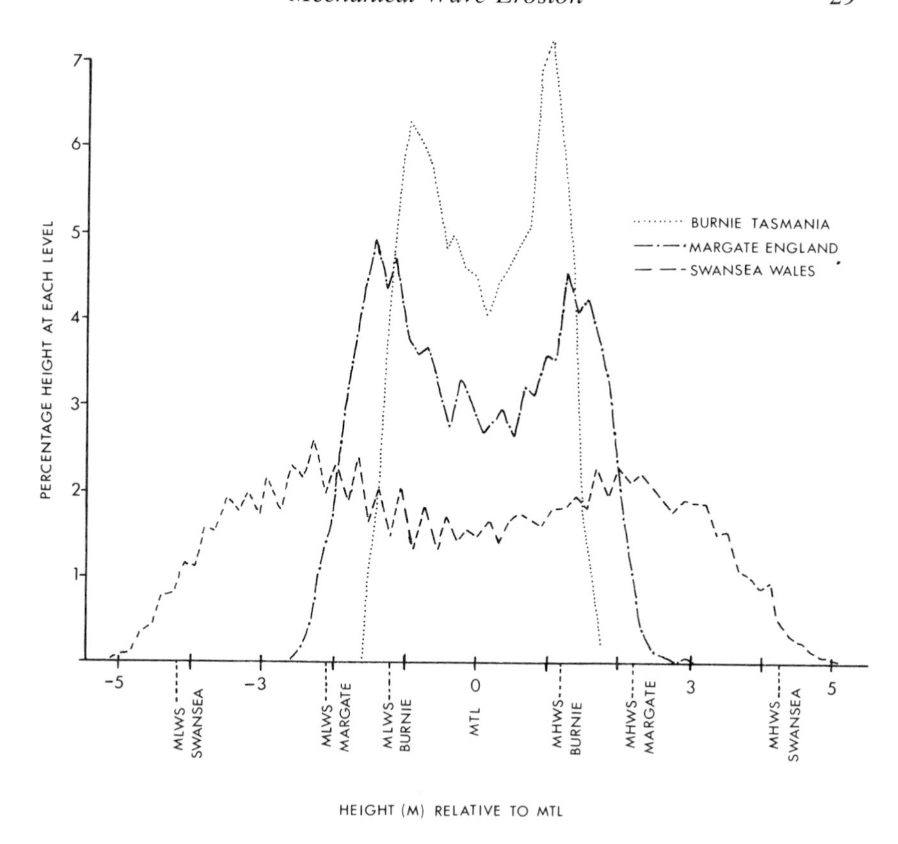

Figure 1.13. Tidal duration curves for the same areas as in Figure 12 (Carr and Graff 1982).

one tidal cycle to the next. Elevations which fall between the spring and neap tidal extremes are exposed at high tide and submerged at low tide for numerous tidal cycles of smaller amplitude. The distribution of the water level can be obtained from the tidal records for the major ports (Chastain 1976, Carr and Graf 1982) (Figs. 1.12 and 1.13). Still water level is most frequently at, or close to, the neap high and low tidal levels, declining by about one-third towards the mid-tidal level, and precipitously towards the levels of the spring tidal extremes. The decline at the mid-tidal level relative to the neaps reflects the lack of tidal extremes at this elevation, whereas the neap tidal levels coincide with the extremes of some cycles, and they experience all the ebb and flow stages of cycles of greater amplitude.

The amplitude of the tidal level distribution (tidal duration distribution) is inversely proportional to the tidal range. This demonstrates that the work of wave processes, which are intimately associated with wave action about the waterline, is particularly concentrated within microtidal environments. The level of most effective mechanical wave erosion is not necessarily between the neap tidal levels, as the largest storm waves frequently operate when sea level is raised above the high tidal level. The discrepancy between the levels of greatest erosion and the maximum tidal frequency may not be great in macrotidal environments, relative to the range of elevations over which the waves operate. In micro- and even mesotidal areas, however, the difference, although similar in absolute terms to macrotidal regions, could be sufficient to elevate the level of maximum erosion to the high tidal level, or even above. This difference may be exacerbated by more vigorous wave action at the high than at the low tidal level, related (depending upon the submarine topography), to deeper water in the nearshore and breaker zones. The resistance of the rocks acts as a filter to wave action. Where the rocks are resistant, only the more vigorous storm waves, which act at the higher elevations, are able to accomplish significant amounts of erosion. The difference between the level of greatest erosion and the most frequent still water level may therefore increase with the resistance of the rock.

2

Chemical and Salt Weathering

Coastal environments are particularly suitable for many chemical and physical weathering processes (Fookes and Poole 1981, Dibb *et al*. 1983). Spray, splash, and tides subject coastal cliffs and intertidal rock platforms to alternate wetting and drying, salt crystallization, chemical reactions, and other weathering processes which are the result of, or are exacerbated by, the presence of salts. These processes play an important and in some cases a dominant role in the development of rock coasts in warm climates, particularly where they are sheltered from vigorous wave action. Their role in cooler regions is less well documented, but because of mechanical wave action and frost in these areas, chemical and salt weathering are probably the major erosional processes only in sheltered locations, or where the local rock is particularly susceptible to these processes.

For convenience, chemical and salt weathering will be considered separately, although they generally operate together in coastal regions, and the results are often difficult to distinguish in the field.

Chemical Weathering

Several texts are devoted to the subject of chemical weathering (Reiche 1950, Keller 1957, Loughnan 1969, Carroll 1970), and a number of other useful summaries have been published (Ollier 1969, Chorley 1969, Evans 1970, Winkler 1973, Brunsden 1979). It is not intended to duplicate these sources, although some discussion of the processes of chemical weathering and the factors which determine their efficiency is necessary to appreciate their role in the coastal environment.

Le Chatelier's principle, which essentially states that a system in equilibrium will adjust to restore equilibrium if any force is applied, is the basic tenet of chemical weathering. Many rocks formed in the Earth's crust under conditions of high temperature and pressure are now exposed to atmospheric, hydrospheric, and biologic agencies on or near the Earth's surface. Weathering is the result of the adjustment or readjustment of rock minerals to more stable mineral phases under the prevailing conditions. The main chemical changes involve the removal of the more soluble components of the minerals, and the addition of hydroxyl groups and atmospheric oxygen and carbon dioxide (Loughnan 1969).

Water assumes a dominant role in chemical weathering. The attraction of dipolar water molecules to ions weakens the ionic bonds between the atoms of other molecules. This may cause them to leave the crystal surface and enter into solution, suspended by the water molecules which are oriented around them. The separation of water molecules into H^+ and OH^- ions facilitates chemical reactions. The negative log of the hydrogen ion concentration is the pH, which describes the acidity of solutions. The hydrogen ion is chemically very active because of its small size, which permits it to penetrate crystal structures, and its high charge-to-radius ratio, which is greater than that of any other ion. Hydrogen ions readily replace other cations to form new compounds, and they may remove OH^- ions from a system by combining with the hydroxyls to form water.

Chemical Reactions

Rock weathering is usually the result of a number of chemical reactions working together. These include:

(*a*) *Hydrolysis and Ion Exchange*. Hydrolysis, which is the dominant primary weathering process of igneous rocks, is the reaction between the H^+ and OH^- ions of water and mineral ions. In the silicates, the H^+ ions replace metal cations, and the OH^- ions combine with these cations to form soluble products. Hydrolysis, therefore, is facilitated by the repeated leaching away of the soluble products, rather than by saturation with immobile water (Keller 1957). Although silicates are decomposed largely by hydrolysis, ion exchange, involving the substitution of other cations for those of a mineral, is also important. The common metal cations (Na^+, K^+, Ca^{2+}, Mg^{2+}) are readily exchangeable. Rock weathering can result from exchange of cations in solution with the loosely held exchangeable ions of a mineral, or in the case of contact exchange, direct exchange of the cations of one mineral with another (Carroll 1970).

(*b*) *Oxidation*. Oxidation occurs when a substance, or one of its atoms, loses an electron and takes on a positive charge. It usually involves the effect of oxygen on the weathered material, although oxygen need not be involved. Oxidation probably occurs entirely through the presence of water containing dissolved oxygen, which can be present in quantities ranging from a surface film to complete immersion. Generally, however, oxidation occurs in the aerated zone above the level of permanent water saturation. One of the most readily oxidized elements is iron, and oxidation is often evinced by the presence of red and yellow iron oxide and hydroxide stains. Bacteria can also cause organic oxidation. Reduction, which is the opposite of oxidation, usually occurs in waterlogged conditions, where the oxygen content is low and anaerobic bacteria flourish. Organic matter is such an important reducing agent, however, that where it is abundant, it can provide reducing conditions above the water table (Loughnan 1969).

The stability of an element in an oxidation state is a function of the change in energy which would be required to add or subtract an electron. The redox potential of oxidation (Eh) has been defined as the electron-escaping tendency of a reversible oxidation–reduction system (Degens 1965). It varies according to the concentration of the reacting substances, and if H^+ or OH^- are involved, with the pH of the solution. Since the Eh decreases as the pH increases, oxidation occurs most readily in alkaline environments. The potential for oxidation is low in igneous and metamorphic rocks, and in black and grey shales containing pyrite, and is high in sandstones, limestones, and red shales (Carroll 1970). Oxidation is particularly important for weathering clays in sedimentary rocks, through reactions involving the Fe^{2+} and Mg^{2+} ions (Chorley 1969).

(*c*) *Carbonation*. Carbonation is the reaction between carbonate or bicarbonate ions and minerals. The solution of carbon dioxide in water forms carbonic acid. Despite its weak acidity, carbonic acid assists base exchange and the solution of carbonates. Calcium and magnesium, and to a lesser extent iron, are susceptible to weathering by carbonation. Although carbonates generally are not the end products of weathering, their formation constitutes an important step in the breakdown of some minerals such as the feldspars (Keller 1957, Ollier 1969).

(*d*) *Hydration*. Hydration is the absorption of water by a mineral in such a way that the water molecules are only very loosely attached to its structure. It prepares surfaces for, and often accompanies, hydrolysis, oxidation, and carbonation, and expedites ion exchange. The contribution of hydration to physical weathering associated with the pressures induced by swelling will be discussed later.

(*e*) *Chelation*. Decomposing organic matter is rich in chelating agents: ions which are able to remove or mobilize metallic ions from otherwise insoluble materials. They therefore play an important role in the breakdown of rocks in contact with decaying roots or plant matter.

(*f*) *Solution*. Winkler (1973) made the distinction between true solution, which is the complete dissociation of a mineral in a solvent (as in the case of carbonates and sulphates), and the dissociation of silicates by hydrolysis with residual clay. The rate of solution is affected by the degree of saturation and by the motion of the solvent, and by the solubility of the salt. Mineral solubility can be affected either by temperature, as for example with gypsum, or by the temperature and the pH of the solvent, as for the carbonates. The special problem of the solution of coastal limestones will be considered later (Chapter 3).

The Weathering of Rocks

Most of the Earth's rock-forming minerals are silicates. Their basic structure is the silica tetrahedron (SiO_4), in which one ion of silica is bonded covalently with four surrounding oxygen ions, sharing one electron with each. Silica tetrahedra can occur singly or in chains and sheets. The structure is very strong, but substitution of other metal cations in the rock magma, such as iron (Fe^{2+} or Fe^{3+}), magnesium (Mg^{2+}), aluminium (Al^{3+}), calcium (Ca^{2+}), sodium (Na^{+}), or potassium (K^{+}) produces weaknesses in the crystalline structure. Because their smaller electrical charges can be neutralized by water, the resistance of the mineral to weathering declines with the degree to which silica is replaced by these other cations. Most of the main igneous rock minerals belong to either the ferromagnesium group, in which Fe^{2+} and Mg^{2+} occur in the silica tetrahedra, or to the plagioclase feldspars, in which Al^{3+} and other cations replace some of the Si^{4+} ions. The resistance to weathering of the ferromagnesium minerals increases with the silica:oxygen ratio, and in the plagioclase feldspars with decreasing Al^{3+} substitution. In general, the order in which minerals crystallize out in a cooling melt determines their relative susceptibility to weathering. Minerals which form at high temperatures are the most susceptible, because the substitution of Al^{3+} and other ions in the silica tetrahedra is most easily accomplished at high temperatures. The scale of weathering susceptibilities (Polynov 1937, Goldich 1938) (Table 2.1) is very similar to the order of crystallization known as the Bowen Series.

Silicates mainly weather by hydrolysis, but ionic exchange with other cations and the oxidation of iron-bearing minerals are also important. Hydrogen ions penetrate the silicate surface and break down its structure, releasing cations and silica. Metal alkalies are leached away, and reconstruction of the residue can produce new minerals, leading to the formation of clays (Loughnan 1969).

Table 2.1. The order of Crystallization from a Silicate Melt (after Thomas 1974)

olivine 1 : 4				
	pyroxene (augite) 1 : 3			
		hornblende 4 : 11		
			biotite 2 : 5	
				othoclase–quartz muscovite 3 : 8
	calcic-plagioclase 1 : 4		sodic-plagioclase 3 : 8	
least stable	→	→	→	more stable

Note

Ratios refer to Si : O_2.

The rate at which minerals weather is determined by many factors other than their composition. Crystal size, shape, and degree of perfection influence the rate of decay, as does the access afforded to the weathering agent, and the rate at which the weathered products are evacuated from the system. There have been many attempts to determine the relative durability of the common minerals to weathering processes (Goldich 1938, Fairbairn 1943, Gruner 1950). Ollier (1969) has provided a weathering series which is based upon the results of many of these studies. In order of decreasing resistance, Ollier's series is: zircon, tourmaline, monazite, garnet, biotite, apatite, ilmenite, magnetite, staurolite, kyanite, epidote, hornblende, andalusite, topaz, sphene, zoisite, augite, sillimanite, hypersthene, diopside, actinolite, and olivine. Jackson *et al.* (1948) have produced a rather different series for minerals in very fine-grained rocks. Their series, in order of decreasing resistance is: anastase, hematite, gibbsite, kaolinite, montmorillonite, hydrous mica intermediate, illite, quartz, albite, biotite, olivine, calcite, and gypsum. Curtis (1976a) has discussed the criteria which have been used to classify the durability of rock minerals to chemical weathering. He considered that chemical composition is a poor indication, and suggested that the persistence of a mineral is related to the total energy released by its breakdown into weathering products.

It is even more difficult to rank the susceptibilities of rocks to weathering than it is for minerals. Rocks vary according to their texture, strength of bonding, degree of fracture, permeability, porosity, saturation coefficient, and water absorption. Rock dip, bedding planes, joints, and faults influence the movement of water through a rock mass and the evacuation of the soluble products. The presence of small-scale cracks in rocks is particularly important in determining the efficiency of chemical weathering (Whalley *et al.* 1982). Coarse-grained rocks are permeable and have large capillary spaces, but fine grains expose a greater surface area to weathering agents. Coarse-grained igneous rocks, however, usually weather more rapidly than fine-grained rocks (Birkeland 1974). Several attempts have been made to rank the susceptibility of various rock types, although the series can only be useful at the most general level. Rougerie (1960) provided a series ranging from the most stable (pegmatites and porphyritic granites) to the least stable (schists, microgabbros, microdiorites, pyroxenites, microgranites, dolerites, amphibolites and green rocks). He acknowledged, however, that his series is not consistent with the development of relief in the field. Birkeland (1974, p. 141) has also proposed a rock stability series which, from the more to the less stable, is: quartzite, chert, granite, basalt, sandstone, siltstone, dolomite, limestone.

Although generalizations are difficult, some comments on the weathering characteristics of common rock types may be justified. In granites, weathering is often concentrated along the joint planes, leaving the joint blocks as corestones of unweathered material. Because of large crystals and the presence of fractured grains, biotite weathers fairly quickly in granites and leads to its rapid breakdown (Eggler *et al*. 1969). Quartz is usually unaltered, but the micas are converted into clays and the feldspars into kaolinite. Joint weathering produces spheroidal forms in basalts, and the minerals are converted to clays and iron oxides. Basalts lack biotite, however, and weathering usually proceeds inwards, grain by grain, producing a sharp division between weathered and unweathered rock (Birkeland 1974). Dolerite weathers in much the same way, but rhyolites are generally more resistant. Andesites usually occupy an intermediate position between basalt and rhyolite in terms of resistance to weathering (Ollier 1969).

Metamorphic rocks are usually susceptible to chemical weathering because of the formation of secondary high-temperature minerals; and because of bonding and schistosity, which facilitate penetration of the rock by water. Gneiss weathers along the bands of the most susceptible minerals, and schist is vulnerable because of its marked fissility, despite the presence of some very resistant minerals. The low porosity and the chemical composition of quartzite makes it particularly resistant. Amphibolites, which consist almost entirely of hornblende, contain cleavage planes which provide access to water. The rock tends to weather in a similar way to basalts.

Sedimentary rocks consist of material which has already been subjected to weathering, and they generally contain a lower proportion of silicate minerals than igneous rocks. In igneous rocks, hydrolysis is the main primary weathering process, but in most sedimentary rocks oxidation, hydration, and solution are of greater significance. The way in which sedimentary rocks weather depends upon the nature of the rock clasts, the amount and the type of clay they contain, and the type of rock cement. The quartz grains in sandstones are extremely resistant, and weathering is dependent on the breakdown of the cement. This may be iron, calcite and gypsum, or silica which is very resistant. The high porosity of sandstones provides easy access to water. The presence of shales in greywackes, and shales and feldspars in arkose, determines their susceptibility to chemical weathering. Shales, mudstones, and marls contain clays with mica and possibly other minerals. Metal cations in the open crystal lattice control the hydration and oxidation of shales. Joints and bedding and cleavage planes provide access to water, particularly if the beds dip steeply. Adsorbed water in these argillaceous sediments facilitates hydrolysis and other forms of chemical attack on the internal capillary surfaces (Hudec 1978a). Solution along joints and bedding planes is the main agent of chemical weathering in limestones, but in porous chalk and dune limestones solution is more evenly distributed through the rock mass.

The plane along which two lithological units meet is often the site of accelerated weathering. This is because of the presence of leachates from the rock above, and often because of the presence of water along the boundary. The occurrence of one rock type may therefore influence the weathering of an adjacent rock (Ollier 1969).

Climate and Weathering

The main factor which determines the efficacy of chemical weathering is the amount of water available for chemical reactions, and more crucially, for the removal of soluble products. If the weathered products remain in the system, equilibrium conditions may be attained, which will inhibit further weathering. Precipitation is most effective in penetrating to the weathering zone if it is evenly distributed, occurring as gentle but persistent showers. Its effect is much less if it is seasonally restricted, and occurs as brief violent showers (Loughnan 1969). Some reactions, however, are assisted by seasonal changes. In savanna regions, solutions move towards the rock surface in the dry season, but in the humid tropics weathering is continuous and unvarying (Ollier 1969). Most chemical reactions are also accelerated by high temperatures, although the influence of temperature may be partly countered by increased rates of evaporation and reductions in the solubility of oxygen and carbon dioxide in water.

Because of the complex relationship between climate and weathering (Trudgill 1976a), models based on climatic criteria are useful only in identifying gross weathering patterns at the global scale. Peltier (1950), for example, considered that chemical weathering is greatest in hot, wet climates, where the mean annual rainfall is between about 140 and 230 cm, and mean annual temperatures are between about 7 and 28°C. Strakhov (1967) suggested that the rate of weathering increases by a factor of 2 to 2.5 with every 10°C increase in temperature. He proposed that the deepest weathering zones are in the tropical forest regions, and the shallowest in deserts and semi-deserts. He believed that weathering rates in moist tropical areas are 20 to 40 times those in areas with moderate climates and less precipitation. Organic factors may make a major contribution to this: they play an important role in chemical weathering in all areas, but may be particularly significant in the humid tropics. It has been estimated, for example, that the presence of organic acids causes a 10-fold increase in the weathering rate (Huang and Kiang 1972, Huang and Keller 1973). In cold areas, the lack of liquid water may be more important in explaining the relatively slow rates of weathering than the low temperatures *per se* (Curtis 1976b).

Salt Weathering

Salt weathering is important in hot and cold deserts, on coasts, and possibly even on the surface of Mars (Malin 1974). Considerable quantities of salt are deposited in areas near to the coast (Chesselet *et al*. 1972, Wada and Kokubu 1973, Martens *et al*. 1973, Meszaros and Vissy 1974, Hoffman and Duce 1974). The particles are mainly NaCl, but sulphates, carbonates, potassium, calcium, magnesium and other salts, and organic matter are present in smaller quantities (Andreasen *et al*. 1978) (Table 2.2). The chemical composition of droplets relative to the sodium ion, however, varies with the wind speed and with a variety of other factors, and is therefore not necessarily the same as sea water. The chief source of these particles is the breaking of numerous bubbles as they reach the surface of the sea, but spray from breaking waves may be a secondary factor (Woodcock 1953, Wu 1981). There can be as many as 1×10^8 salt particles in a cubic metre of air over oceans, although 1×10^6 is more common (Cadle 1966, Rasool 1973). Near to the coast, a cubic metre of air can contain 9,500 particles on average (Junge 1963). This is much less than in urban areas, but much more than in geographically isolated regions. In coastal areas, the dissolved solids in rainwater range in concentration up to several tens of mg dm^{-3} (Freeze and Cherry 1979, p. 238). Estimates of salt fallout along coasts generally range between 28 and 337 kg ha^{-1} (25–300 lb $acre^{-1}$) per year. It may be less in coastal areas where the winds are offshore (Brierly 1965), but extremes of up to 4,500 kg ha^{-1} (4,000 lb $acre^{-1}$) per year can also occur (Rasool 1973). It has been estimated that more than 337 kg ha^{-1} (300 lb $acre^{-1}$) of salt can be deposited in a two-month period near Sydney, Australia, for example (Walker 1960). Brierly (1965) has provided other data on the fallout of sea salt around the world.

The nature of the salts and the rock, and the characteristics of the environment, determine the efficacy of the salt weathering processes. Experimental work and field observations over the last 150 years have

Table 2.2 The Ions and Salts of Sea Water (after Mottershead 1982a)

Ions in solution in sea water		Salts crystallizing out of sea water	
Ion	(g/kg)	Salt	(g/kg)
Cl^-	18.98	NaCl	27.21
Br^-	0.065	$MgCl_2$	3.81
SO_4^{2+}	2.65	$MgSo_4$	1.66
HCO_3^-	0.14	$CaSO_4$	1.26
Mg^{2+}	1.27	K_2SO_4	0.86
Ca^{2+}	0.40	$CaCO_3$	0.12
K^+	0.38	$MgBr_2$	0.08
Na^+	10.56		

provided some indication of the efficacy of various salts, and the susceptibility of different rock types (for example, Birot 1954, Pedro 1957, Tricart 1960; an excellent summary of this work is found in Evans 1970). The experimental conditions, however, generally bear little similarity to natural environments, and they do not allow the various processes of salt weathering to be distinguished. Although it is difficult to make reliable generalizations, many studies have emphasized the disruptive effects of the hydration and crystallization of the sodium and magnesium sulphates (Bakker *et al.* 1968, Goudie *et al.* 1970, Goudie 1977, Cooke 1979). The susceptibility of various rock types to salt weathering appears to be strongly related to such factors as their porosity, microporosity, water absorption capacity, and saturation coefficient. These are factors which determine the rate at which solutions can penetrate rocks, the amount of solution which can be contained, and the extent to which the smaller pores can feed crystal growth in the larger pores (Schaffer 1955, Tricart 1960, Minty 1965, Goudie *et al.* 1970, Goudie 1974, Cooke 1979). Consequently, many studies have found that sandstones are particularly vulnerable to salt weathering, whereas igneous and metamorphic rocks of low porosity are among the more resistant rock types. Coarse-grained rocks are generally more susceptible than fine-grained rocks. The environmental factors which affect salt weathering have not been fully considered. An environment which provides a ready supply of salts and protects them from depletion by wind or rain, appears to be a fundamental requirement for effective salt weathering. Its effectiveness in the humid and subhumid tropics may therefore be limited by high humidity and leaching (Thomas 1974). Nevertheless, salt crystallization is considered to be the most important reason for the weathering of building stone in Britain (Honeyborne 1965). Further work is necessary to determine the possible effect of salt solutions on frost action. Experimental work suggests that the combination of frost and salt weathering can in some cases provide a particularly potent mechanism for weathering (Goudie 1974, Williams and Robinson 1981), although it can inhibit frost shattering in others (McGreevy 1982) (see Chapter 5).

The presence of salts in the capillaries of coastal rocks facilitates several processes of mechanical weathering. The main mechanisms include (Cooke and Smalley 1968):

1. volume changes induced by hydration;
2. expansion of salt crystals caused by changes in temperature; and
3. crystal growth from solution.

Hydration

The absorption of water by hydration causes crystals to swell and to exert pressures against the constraining walls of rock capillaries. Mortensen

(1933) first recognized its significance in deserts, and he attempted to calculate the pressures which can be generated. Winkler and Wilhelm (1970) derived an expression which they used to calculate the hydration pressures exerted by some common salts at a variety of temperatures and relative humidities. The expression is:

$$P = \frac{(n.\ R.\ T)}{Vh - Va} \times 2.3 \log \frac{Pw}{Pw'},$$

where P is the hydration pressure in atmospheres; n is the number of moles of water gained during hydration to the next higher hydrate; R is the gas constant; T is the absolute temperature (^{0}K); Vh is the volume of the hydrate (molecular weight of the hydrated salt/density in g cm^{-3}); Va is the volume of the original salt (molecular weight of the original salt/density in g cm^{-3}); Pw is the vapour pressure of water in the atmosphere (in millimetres of mercury) at the given temperature; and Pw' is the vapour pressure of hydrated salt (in millimetres of mercury) at the given temperature.

The greatest pressures usually occur at low temperatures and high relative humidities (Winkler 1973). Salt crystals, which are low in the water of crystallization, are likely to form during the day when temperatures are high. At lower night-time temperatures, water is absorbed from the atmosphere, and higher hydrates may be formed (Cooke and Doornkamp 1974). The alternate hydration and dehydration of entrapped salts can occur many times in a single day, and the process is probably most effective where changes in temperature and humidity cause the hydration thresholds to be crossed most often (Cooke 1979). The most potent salts are those which expand rapidly and by a significant amount when hydrated. Carbonates and sulphates are particularly effective, but the process cannot account for the very common situation in which sodium chloride constitutes the main disruptive salt (Evans 1970, Chapman 1980).

Thermal Expansion

The coefficient of thermal expansion of many common salts, such as sodium, potassium, barium, magnesium, and calcium, is higher than for many rocks. Halite (NaCl) for example, expands by 0.5% when the temperature rises from near the freezing point to 60°C, whereas granites expand by only 0.2% (Cooke and Smalley 1968). Near the rock surface, the expansion of entrapped salts causes pressures on the walls of the rock capillaries, which can result in granular disintegration or splitting. Thermal expansion also propagates the fractures in which salts have crystallized (Cooke and Smalley 1968), and is probably most effective where there are high diurnal ranges in temperature (Cooke and Doornkamp 1974). It has been proposed that the heat expansion of salts is probably the major erosive mechanism above the high tidal zone on the sunny, south-facing sandstone cliffs of Oregon (Johannessen *et al.* 1982).

Crystallization

The crystallization of salts from solution can also exert disruptive pressures within rocks. It is difficult to account for the continued growth of a salt crystal, however, when it becomes confined by the walls of a rock capillary. Correns (1949) suggested that continued growth can occur against a confining pressure when there is a film of solution at the salt/rock interface. When the sum of the interfacial tensions at the solution/rock and salt/solution interfaces is less than that at the salt/rock surface, the solution is able to penetrate between the salt and the rock. Correns provided an expression for the pressure exerted by crystallization:

$$P = \frac{R \times T}{Vs} \times \ln \frac{C}{Cs},$$

where: P is the crystallization pressure, in atmospheres; R is the gas constant; T is the absolute temperature (^{0}K); Vs is the molecular volume of the solid salt (dm^3 mole^{-1}); C is the actual concentration of the solute during crystallization under pressure P; and Cs is the concentration of the solute at saturation. C/Cs, therefore, is the degree of supersaturation.

Everett (1961) and Wellman and Wilson (1965, 1968) have suggested that growing crystals are able to press against the confining walls of rock capillaries because they are generally unable to expand into the smaller unfilled capillaries. Large crystals forming in large capillaries grow at the expense of the small crystals in the small capillaries. This is because the chemical free energy of a solid increases with its surface area. When the larger capillaries are completely filled by salt crystals, they are unlikely to grow into the smaller unfilled capillaries. Extension into the small capillaries would require a considerable increase in surface area relative to the small volumetric increase, resulting in a large increase in the chemical potential of the crystal extensions. Pressures must therefore increase as the crystals continue to grow in the large capillaries, until either fracture occurs or the chemical potential is raised to that of the crystal extensions, thereby permitting them to grow down the capillaries. The smaller the unfilled capillaries, therefore, the greater are the pressures which can be exerted by crystal growth in the large capillaries.

The work done during crystal growth on one face of a crystal equals the pressure difference ($Ps - Pl$) between solid (crystal) and liquid, multiplied by the increase in volume (dV). This must equal σdA, where σ is the surface tension between a crystal and the saturated solution (Wellman and Wilson 1965). That is:

$$Ps - Pl = \sigma \mathrm{d}A/\mathrm{d}V.$$

Wellman and Wilson (1968) proposed that rock fracture will occur if $\sigma(1/r - 1/R)$ exceeds the tensile strength of the rock, where r and R are the radii of the small and the large capillaries, respectively. This suggests that rocks with large capillaries separated by areas of micropores are most susceptible to disruption by the crystallization of salts.

Weyl (1959) derived a general model for the crystallization of rock minerals. The forces of crystallization are the result of the deposition of matter in the interface between mineral grains. If the solution is supersaturated, precipitation in the area of contact will occur if the ratio of the supersaturation to the stress coefficient of solubility is greater than the average normal stress between the grains. Pressure will occur, however, if the average normal stress across the solution film is greater than this ratio. Weyl provided expressions which permit the forces of crystallization to be calculated. They show that considerable pressure can be generated even if crystals do not completely fill rock capillaries. Evans (1970) considered Weyl's model is the only acceptable explanation for the forces of crystallization.

A state of supersaturation is essential for crystallization. With high concentrations, crystallization is rapid and the crystals are larger than those which form from lower concentrations. Mullin (1961, pp. 27–8) has considered the relationship between crystallization and the temperature and concentration of solutions. A solution can become saturated as a result of a drop in temperature, or at constant temperature, by evaporation of the solution. Alternatively, rainwater can dilute the concentration. Supersaturation and crystallization may also be the result of changes in temperature and concentration levels. Supersaturation is probably induced most frequently by evaporation rather than by cooling (Cooke and Doornkamp 1974). In the coastal zone, strong winds facilitate salt weathering, even when rainfall is quite high, but several days of strong evaporation without rainfall or spray are necessary to crystallize solutions within rock capillaries (Evans 1970). A drop in temperature, however, affects a much larger volume of salt per unit time than does evaporation, which is a gradual process (Kwaad 1970).

Winkler and Singer (1972) used Correns' equation to calculate the crystallization pressures of some common salts. The pressures induced by halite are particularly high in relation to those of the other salts, and they rapidly increase with the degree of saturation. Whereas the solubility of most salts declines with temperature, leading to crystallization, the solubility of halite decreases only very slightly. In general, experiments have suggested that sulphates are more effective in damaging rocks by crystallization than carbonates, which are in turn more effective than chlorides. The efficiency of some salts, however, is affected by the presence of others. Honeyborne (1965), for example, found that a mixture of NaCl and $CaSO_4$ causes more damage to stone than either of the salts in isolation. The effect of interactions is poorly understood, but they may play a significant role in salt weathering involving complex seawater solutions.

Salt weathering plays an important role in the disaggregation of rocks in the coastal zone. Dunn (1915) and Coleman *et al.* (1966) have described the effects of salt crystallization on high tidal flats in the Northern Territories and in Queensland, Australia, respectively. In the Northern Territories, quartzites are rounded *in situ*, and shales are disintegrated by the crystallization of salts which are precipitated by sea spray. In Queensland, the rocks are thoroughly wetted by high spring tides. Rapid evaporation dries the rock, and the salts crystallize after the tides have returned to more normal levels. The Eh and pH reach their maximum values just before the rocks have completely dried. Deliquescence of salts and nightly dewfall maintain a thin film of water around the rocks, which assists chemical weathering.

Jutson (1918) discussed the effect of salt weathering on the granite cliffs around some saline lakes in Western Australia. Recession of these cliffs by salt attack maintains their steepness and produces a level rock floor at their base. In southern Italy, rapid recession of flysch cliffs is accomplished by saline spray, rather than by weak wave action (Baggioni 1975). Tricart (1962) considered that basins 4 to 5 m deep in dolerites at Dakar, Senegal, were produced by salt crystallization. At Mamba Point in Liberia, however, salt crystallization is inhibited by higher humidity and by closely spaced joints. Similarly, troughs have developed in the wet climate of southeastern Brazil because of local winds and locally low humidity, but not in the humid environment further north at Ilheus (Tricart 1959, 1972). Tricart (1972) noted that high evaporation is necessary for the development of spray corrosion forms, including salt spray benches. They are therefore typical of climates with long dry seasons, or high isolation. He considered that they are generally poorly developed in the wet tropics where evaporation is low, but are prominent features of other areas of the tropics, arid and semi-arid climates, and also Mediterranean regions.

In southwestern Britain, the development of a number of microtopographical forms in greenschist in the immediate supratidal zone has been attributed to the crystallization of sodium chloride in the rock pores. Laboratory experimentation, calculation of the pressure of crystal growth using Correns' equation, and consideration of the tensile strength of the rock, suggest that a concentration of the saline solution only 1.5% above normal saturation concentration is sufficient to disrupt the rock. Erosion of the bedrock surface in this area is occurring at about 0.6 mm yr^{-1} (Mottershead 1982a,b). Salt weathering can also be important in dry polar climates, where it contributes to the formation of such features as tafoni and honeycombs, and it may even play a role in the formation of shore platforms in high latitudes (Loney 1964, p. 59). Nevertheless, intertidal and spray salt fretting usually becomes less significant in the higher latitudes, as temperature and evaporation decrease (Guilcher *et al.* 1962, Guilcher and Bodere 1975).

Tafoni

Tafoni (sing. tafone, French *taffoni*(s)) are a form of cavernous weathering (Fig. 2.1). They are spherical or elliptical hollows with broad, rounded interiors. They range from a decimetre up to several metres in depth and diameter, the larger forms generally having a thin overhang or lip. Tafoni is a 'Mediterranean' term: a language supposed, on the basis of Greek and Latin words which are of non-Indo-European origin, to be pre-Latin (Devoto and Oli 1971). 'Tafo' is an Italian root, from the Greek, meaning tomb or sepulchre (as in epitaph, cenotaph) (Battisti and Alessio 1957).

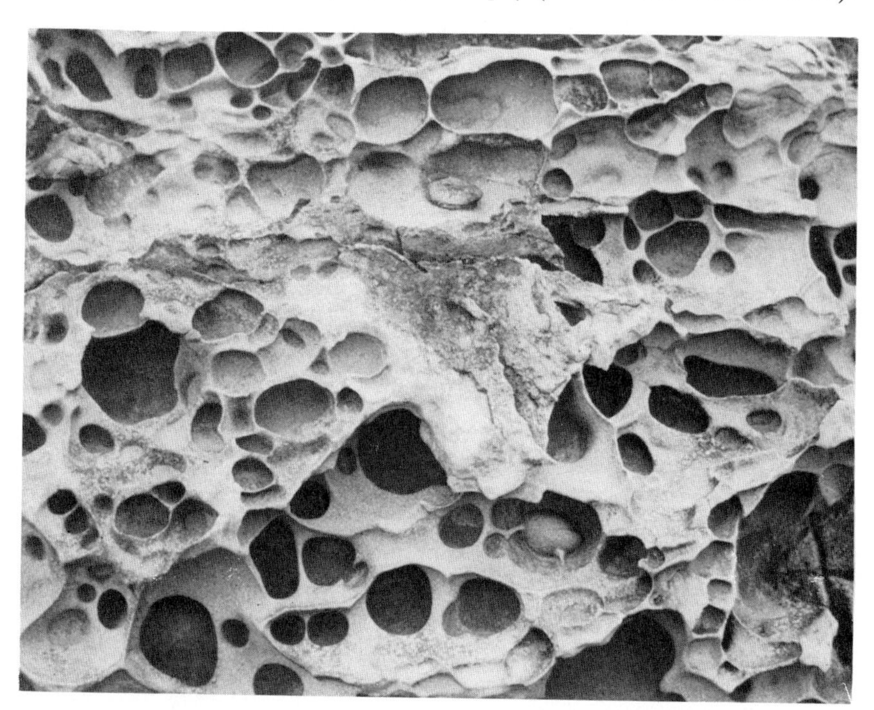

Figure 2.1. Small tafoni up to 15 cm in diameter in sandstones on the northeastern coast of South Island, New Zealand.

Although the term 'tafone' is exclusively geological, Wilhelmy (1964) noted that 'tafonare' is used in Corsica to mean 'to perforate'. Dury (1965, cited by Ollier 1969, p. 191) has claimed that the term means 'windows', and should therefore only be used in situations where a weathering hollow has broken through the roof of a rock outcrop or boulder; his proposal, however, has received little support.

Tafoni usually occur in medium- and coarse-grained rocks, including granites, lavas, sandstones, conglomerates, gneisses, schists and limestones (see, for example, Kelletat 1980). They also occur in a variety of climatic environments, although they are most typical of Mediterranean regions and foggy, arid coastal zones, ranging from the equatorial to the polar (Martini 1978). They are usually small in polar regions, and in subhumid and humid climates they are generally restricted to the coastal zone, where strong sea winds produce microclimatic aridity (Jennings 1968).

Wilhelmy (1958) distinguished between sidewall and basal tafoni. Sidewall tafoni occur on the sides of vertical or almost vertical rock faces. Basal tafoni, which are generally larger than the sidewall forms (Dragovich 1969), develop on the undersides of boulders, at the base of cliffs, and at the foot of exfoliation sheets.

A number of mechanisms could be responsible for the form and occurrence of tafoni, although they appear to be initiated where moisture accumulates, along structural weaknesses, at the foot of scarps, and on the shaded undersides of rock slabs (Twidale 1982). The 'German school' (Evans 1970) considered that they are the result of temperature variations, corrosion, or corrasion by wind. Bartrum (1936) thought that the wind may help to remove the weathered products, and Rondeau (1961) suggested that it polishes the rock surfaces. Evteev (1960) proposed that in Antarctica it is capable of boring cells in rock, and Sekyra (1972) believed that the intensity and duration of wind erosion determines the frequency and size of cavernous forms. However, many other workers have found that tafoni do not have preferred orientations, which suggests that they are not initiated by the wind (Cailleux 1962, Calkin and Cailleux 1962, Prebble 1967, Dragovich 1969, Calkin and Nichols 1972, Smith 1978, Watts 1979).

Chemical weathering probably plays an important role in the development of tafoni, particularly in granites (Jennings 1968, Dragovich 1969). In sandstones near Sydney, Australia, their formation is partly the result of the seepage of water containing organic acids from decaying vegetation, which reacts with iron, silica, and released clays (Johnson 1974). It has been suggested that in deserts, hydration is primarily responsible for the disintegration of the walls of tafoni (Grenier 1968, Mabbutt 1977). Rondeau (1961), however, considered that tafoni are the result of hydrolysis associated with slowly evaporating water. Near Auckland, New Zealand, the formation of tafoni in andesitic rocks is aided by the removal of the calcium carbonate cement by percolating water and rain (Bartrum 1936). The removal of calcium and magnesium carbonates also facilitates tafoni development in limestones (Smith 1978). The interiors of developed tafoni are particularly suitable environments for chemical weathering. Tafoni basins are cooler, moister, and experience smaller ranges of temperature and relative humidity than do the rock faces outside (Cooke and Warren 1973). Martini (1978) considered that chemical weathering—including such processes as hydration, hydrolysis, cation exchange, and diffusion, which are induced by the microclimatological differences between rock basins and the outside walls—is essential for the development of tafoni.

The presence of salt crystals on the walls of coastal and desert tafoni suggests that salt crystallization also plays an important role in their formation (Klaer 1956, Rondeau 1961, Cooke and Warren 1973, Smith 1978, Bradley *et al.* 1978). In Corsica, tafoni are most numerous and freshest near the sea, where spray provides a ready supply of salts (Bourcart 1957, Rondeau 1961, Evans 1970). Tafoni and honeycombs are also most numerous at sea level near Naples, and they are absent at elevations between about 40 and 60 m (Baggioni 1975). The formation of small conical, cylindical, and cavernous pits in the greenschists of south Devon, England, has been attributed to the effects of salt crystallization (Mottershead 1982a). The crystallization of salts precipitated by sea fog is also important in the coastal Namib Desert (Goudie 1970, see Evans 1970), and it is thought that salt shattering plays a significant role in coastal Peru (Tricart 1969a, p. 107). Salt weathering has produced tafoni in basalts on Madeira (Guilcher and Bodere 1975), and in sandstones in Utah (Mustoe 1983). In Antarctica, where chemical weathering is generally considered to be slow (Kelly and Zumberge 1961, Tasch and Gafford 1969, Boyer 1975), fine saline debris or 'rock meal' has been found in the basins of tafoni (Wellman and Wilson 1965). Kelly and Zumberge believed that the breakup of rocks in Antarctica is accomplished by frost and by the crystallization of NaCl supplied by sea spray. Treves (1967) has also noted the presence of cavernous weathering forms in an area of coastal Antarctica where the rocks are covered in salt. In the Arctic, Watts (1979) found no evidence of chemical weathering associated with tafoni on Baffin Island in sheltered areas of the coast, and concluded that they are probably the result of salt crystallization. At Cape Evans in Antarctica, Cailleux (1962) found that tafoni are best developed on shaded rock faces, where the last remnants of water with strong salt concentrations are retained during periods of thaw. A tendency for tafoni to develop in shaded sites in deserts (Mabbutt 1977) may provide further evidence of the role of salt crystallization. In South Australia, total soluble salts, consisting mainly of halite and gypsum, are 2 to 13 times more abundant in the tafoni rock flakes than in massive granite (Bradley *et al.* 1978, 1979).

There is no conclusive evidence that salt weathering is essential for tafoni development. Salt crystals in tafoni may be the product, rather than the cause, of weathering. Johnson (1974) did not find salt crystals in tafoni near Sydney, and Dragovich (1969) only found them in discernable quantities in a few coastal tafoni in South Australia. Dragovich considered that salt crystallization assists the formation of tafoni in coastal areas, but is not essential for their formation. Cailleux (1962) noted that salt crystallization may produce tafoni in Antarctica, but he considered frost a more likely mechanism. In Corsica, however, tafoni become less common inland, away from the coast, despite an increase in the frequency of frosts. Furthermore, the occurrence of tafoni in areas which seldom experience frost demonstrates that it is not essential for their formation. Nevertheless, as Martini (1978) has recognized, frost action may be a contributory factor in the development of tafoni in some areas.

Some tafoni penetrate a case-hardened crust on the rock surface (Tricart 1969a, Watts 1979, Winkler 1979, Conca and Rossman 1982), whereas others are associated with core softening of the interiors of boulders (Conca and Rossman 1985). Indeed, some workers consider that the occurrence of a hardened crust and a weakened underlayer is a fundamental reason for the occurrence of tafoni in deserts (Wilhelmy 1964, Mabbutt 1977). These crusts develop because of evaporation which causes saline solutions to move towards the rock surface. Deposition of silica, iron and magnesium oxides, calcium carbonate, and other cementing agents produces resistant surface rinds which protect the chemically or physically rotted rock interior. In places where the rock is persistently moist, however, the rind is broken or fails to develop, and tafoni can form as a result of the excavation of the rotted interior (Jennings 1968). Wilhelmy attributed the occurrence of tafoni on shaded rock surfaces to the absence of a crust, which protects the rock surface in more exposed areas. Case hardening has most frequently been reported on sandstones, limestones, and volcanic tuffs, but it does not appear to be essential for the formation of tafoni in these or other types of rock (Smith 1978, Bradley *et al.* 1978, 1979). Dragovich (1969) found no examples of case hardening associated with tafoni in granites in South Australia. She thought that the lips or hoods around their mouths are the natural result of cavern backwearing, rather than the rims formed by a breached surface rind. Dragovich ascribed the occurrence of tafoni to structural, mineralogical, and topographical factors which produce local concentrations of moisture and aid its penetration into the rock mass. She proposed that where moisture concentrates in basal tafoni, the rock flakes because of the alternate expansion and contraction caused by the hydration of altered minerals (Dragovich 1967). Granular disintegration can occur in sidewall tafoni, where there are strong temperature fluctuations. Smith (1978), however, considered that flaking is the main form of disintegration in both types of tafoni in limestones, although granular disintegration is more common in granular and crystalline rocks. Martini (1978) has also noted that flaking appears to be more important than granular disintegration in most areas.

Thus, no single explanation will suffice to account for the form and occurrence of tafoni. Chemical weathering appears to be the dominant mechanism in most cases, with salt weathering in coastal areas, although frost action and other forms of physical weathering, wind, and bioerosional agencies may also be contributory factors. Tafoni are most common where there is a good deal of insolation, variable winds and moisture conditions, frequent dews, fogs or other sources of saline moisture, and frequent cycles of wetting and drying (Barbaza 1970, Martini 1978). Coastal regions satisfy most of these requirements, and tafoni are conspicuous features of many coastal regions; but they are generally less significant than a related form, the honeycomb.

Honeycombs

Honeycombs (stone lattice, stone lace, or aveoles) are small, closely spaced depressions up to a few centimetres in diameter and depth (Fig. 2.2).

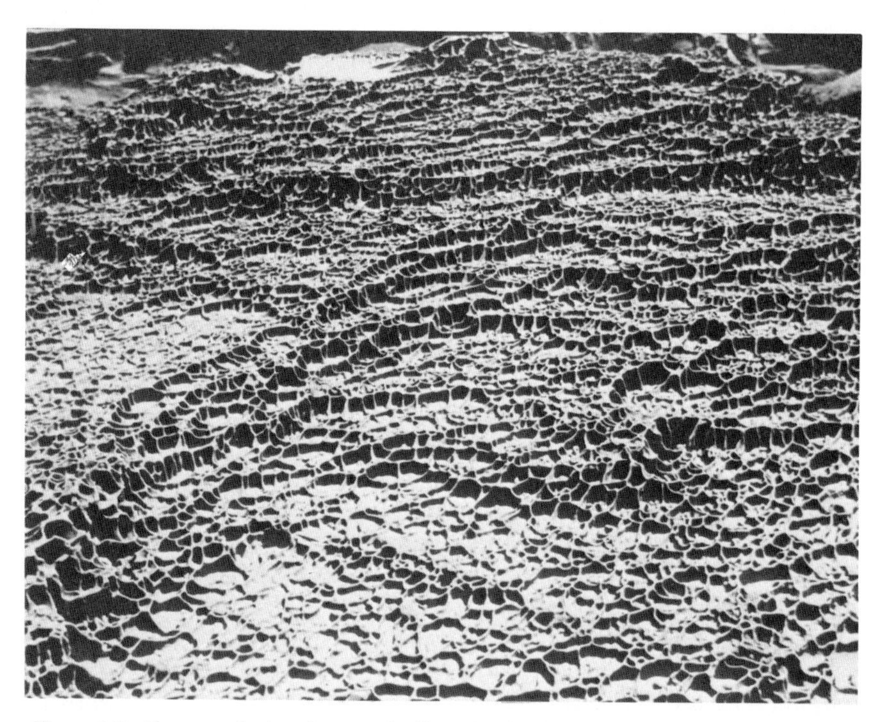

Figure 2.2. Honeycombs in arkose at Artillery Rocks, Otway Coast, Victoria, Australia.

However, the lack of a precise definition of honeycombs, tafoni, and other related forms makes it difficult to distinguish between honeycombs and small tafoni, and to determine the meaning of the terminology as it is used by different workers. The distinction is particularly difficult when honeycombs and tafoni occur in the same place. Bartrum (1936) considered tafoni to be fairly large depressions which are the product of differential rock weathering, whereas honeycombs are smaller features which develop in rock which is essentially homogeneous. Martini (1978) considered that honeycombs are a form of tafoni which develop under stress conditions, such as where there is too much moisture, too little salt, or shadowing of the whole rock outcrop. Under such conditions, tafoni weathering is inhibited by the lack of microclimatic differences between the inside and the outside of the developing tafoni. According to Martini, small fractures and minor heterogeneities in the rock then control the formation of honeycombs. Bartrum (1936) recognized that some honeycombs are controlled by dense rock jointing, but others develop in the absence of any joint control.

The best development of coastal honeycombs is usually in the zone which is above the level of normal wave action but periodically experiences heavy spray and splash during stormy weather. Honeycombs have been reported on steep or even vertical rock faces (Jutson 1954, Van Autenboer 1964, Gill 1973a, Kelletat 1980), but their best development is usually attained on low or subhorizontal slopes. Sandstones, including arkose and greywacke, are particularly suitable for honeycomb development, possibly because of their coarseness, loose cohesion, and in many cases their susceptibility to chemical weathering. Honeycombs have, however, been reported in volcanic and crystalline igneous rocks (Guilcher *et al.* 1962), and in all regions from the poles to the equator.

Competing views regarding the mechanisms responsible for honeycombing are similar to those which have been discussed for the origin of tafoni. As with tafoni in some areas, they may be of ancient antiquity. In Corsica for example, they have been attributed to the effects of frost action during the Pleistocene (Cailleux 1953). Cotton (1922, p. 258) and Gill *et al.* (1981) believed that honeycombs are the result of chemical weathering, but Bartrum (1936) proposed that salt crystallizaton caused by evaporation after a period of heavy spray may also be important. Nichols (1960) described honeycombs formed in gabbro in Antarctica, and concluded that the absence of staining in the honeycombs is evidence that the process of formation is largely physical. This view has been supported by further investigations in Antarctica (Van Autenboer 1964), and elsewhere (Bourcart 1957, Fairbridge 1968a, Guilcher and Bodere 1975). Microscopic examination of coastal honeycombs in arkosic sandstones in Washington State, USA, revealed little evidence of chemical decomposition, the loose grains appearing to be derived from salt crystallization and other physical processes (Mustoe 1982, 1983). Electron microscopic examination of sandstone honeycombs in Northern Ireland also suggests that salt crystallization could dislodge the quartz grains, although etching may be accomplished by chemical weathering (McGreevy 1985a). Salt does not appear to be a major factor in the formation of honeycombs in greywackes in Victoria, Australia, however, where halite crystals are absent in the recently formed cavities (Gill 1981).

The margins of developing honeycombs can be strengthened by the precipitation of iron oxides and other soluble products of chemical decomposition. In Spain and Tunisia, for example, honeycombs are associated with rocks which have an iron oxide cement (Rondeau 1965). Case hardening is particularly marked if the rock is densely jointed (Cotton 1922, Jutson 1954). Bartrum (1936) suggested that the whole surface of joint blocks is originally indurated, but the crust is thinnest in the centre of the upper and lower faces. This may be because the central portions remain moist longer than the extremities, and so receive less precipitate from solution. It is more probable that the joints are the channels along which the solutions flow, so that they become indurated as the solutions evaporate. Bartrum considered that capillary action also carries material which is leached from the joint block out to its indurated extremities. This strengthens the extremities, and as weathering loosens and removes the sand grains in the interior, honeycomb depressions form with prominent, indurated, rib-like walls. It has been proposed that the wind is the primary agent for removal of the loose grains (Cotton 1922, Bartrum 1936), but Jutson (1954) emphasized the effect of the impact of falling water from waves or spray.

An intricate honeycomb network is more difficult to explain in the absence of joint control. Palmer and Powers (1935) believed that honeycombs in Hawaian coastal lavas were initiated by the excavation of vesicles: small spherical cavities in the lava produced by gas bubbles. They considered that these hollows were enlarged by chemical action and by sand abrasion induced by breaking waves and spray. As Bartrum (1936) noted, the occurrence of very similar honeycombs in non-volcanic rocks belies any dependency upon the occurrence of vesicle-like weaknesses in the rock. Bartrum suggested that honeycombs, which are not controlled by joints, initially develop from hollows formed by salt crystallization and chemical weathering. Once formed, these hollows become receptacles for the collection of further spray water, thereby localizing the erosive processes. He did not, however, explain why each hollow develops where it does, nor how fairly regular honeycomb patterns develop from randomly located depressions. In the northwestern USA, the thin walls which separate adjacent honeycombs have been attributed to a microscopic green algal coating. The algae may help to keep the rock moist, or they may provide a physical barrier to the penetration of water. Rapid erosion or ecological factors, such as sunlight and humidity, can prevent algal coatings developing in the hollows (Mustoe 1982).

Jutson (1954) believed that where honeycombs are numerous, they play an important role in the lowering of coastal rock surfaces, both directly by their growth *per se*, and also indirectly, unless the walls are case hardened, by assisting the removal of rock by other processes. Honeycomb weathering can be quite rapid. Crystalline schists in a sea wall in the spray zone on the French Atlantic coast have been honeycombed at a maximum rate of about 1 mm yr^{-1} (Grisez 1960). In Northern Ireland, immature honeycombs have formed in a seawall built in the latter part of the last century (McGreevy 1985a). In the northwestern USA, arkose and metavolcanic greenstone used for constructional purposes have developed honeycombs within the last one hundred years (Mustoe 1982). Fresh blocks of greywacke placed in a seawall in Victoria, Australia, in 1943 and 1949 are now honeycombed (Gill 1981, Gill *et al.* 1981). This rate of development provides support for the contention that honeycombing is an important erosive mechanism in the splash and spray zones of some areas.

Water Layer Levelling

Bartrum and Turner (1928) first drew attention to the lowering, smoothing, and levelling of intertidal rock platforms by weathering processes apparently associated with pools of standing water. These processes operate on a wide variety of rock types, including tuffs, weathered basalts and granites, schists, and phyllites. Their activity appears to be greatest on fine-grained sedimentary rocks, such as sandstones, limestones, shales, siltstones, and mudstones. They may be particularly potent on rocks with a high proportion of silaceous material (Bowman 1970), high permeability, and low angles of dip (Davies 1972).

The processes, which are collectively referred to as 'water layer levelling', operate wherever spray and splash can accumulate. They can only operate effectively, however, up to the level reached by spray and splash, and down to the elevation at which the rocks are continuously washed by swash in normal weather (Wentworth 1938). Ongley (1940) described the formation of rock ledges by storm wave spray and splash 17 to 24 m above sea level in eastern New Zealand. Water layer levelling is most active at lower elevations, however, and Sanders (1968a) suggested that it is most effective in a zone extending from the high tidal level to about 1/3 m above. Bartrum (1935) believed that the processes operate down to an ultimate base level which is coincident with the floors of the irregular wave-cut depressions which are occupied by pools. This is because water layer levelling is induced by alternate wetting and drying, and must cease when the lowered rock surface remains permanently wet. Other early workers stressed that water layer levelling is controlled by local levels of saturation. Wentworth (1938) originally termed the mechanism 'water level weathering', but Hills (1949) preferred 'water layer weathering' to avoid the implication that it is related to a particular level, such as sea level (Johnson 1938). More recently, however, it has been suggested that the processes do operate down to a general base level, which is the elevation at which the rocks are permanently saturated with sea water (Bird and Dent 1966, Bird 1968, Sanders 1968a, Davies 1972, Takahashi 1977).

Water layer levelling is active around the margins of pools, where alternate wetting and drying occurs as the water is replenished by spray or splash and depleted through evaporation during quieter periods; or possibly where a capillary fringe can be continously maintained (Wentworth 1938). The processes cut into the divides between the rock pools, enlarging them and eventually causing them to merge. Most workers agree that the processes include alternate wetting and drying, salt crystallization, chemical weathering, and the movement of solutions through rock capillaries. The precise nature and relative importance of these mechanisms have not been determined. Bartrum (1935) recognized that salt crystallization may be significant, but he attached greater significance to other, unidentified processes, which are associated with wetting and drying. Davies (1972) also acknowledged the contribution of salt crystallization, but he believed that water layer levelling is largely chemical. Wentworth (1938) and Hoffmeister and Wentworth (1942) suggested that the effects of wetting and drying may be physically akin to the swelling of shales, which occurs when they are subject to alternating wet and dry conditions. Wentworth noted that surface tension phenomena and colloidal and dilation behaviours are probably also involved, but he felt that salts are washed away too rapidly for effective disruption by crystallization. Emery (1960) also proposed that it involves the loosening of rock grains by swelling, together with base exchange and hydration. Bowman (1970) provided further support for the role of swelling pressures induced by the alternate wetting and drying of clay minerals, but he also recognized the contribution made by chemical weathering and salt crystallization.

Case hardening often accompanies water layer levelling on platform surfaces. It usually occurs along joint planes, forming box work or frame-like patterns on shore platforms and other rock surfaces (Kieslinger 1959). Joints provide a system of channels for the movement of dissolved ions, depleting or impregnating the joint walls. In sedimentary rocks near Sydney, Bowman (1970) found that the chemical effects of pools appear to be associated with the typical cement weathering series:

carbonate and siderite $\longrightarrow$ limonite; or

pyrite and iron sulphates $\longrightarrow$ limonite.

Bowman considered that alternating wet and dry conditions are too unstable for the deposition and accumulation of upward-migrating mineralized solutions on the rock surface. They can accumulate and oxidize to limonite in the joints, however, which provide more stable, saturated environments.

Several types of weathering pits have been described in the coastal zone. If the impregnated joints are more resistant than the interiors of the joint blocks, a box or frame structure develops (Fig. 2.3); if the joints are weaker, the joint blocks may stand out as small plateaux. Iron and magnesium oxides, calcite, silica, and other substances can strengthen the joint planes. Case hardening frequently produces weathering pits which have raised rims as a result of their strengthening by secondary limonite

Figure 2.3. Frame structure associated with indurated joint planes on a shore platform on the Whangaparaoa Peninsula, north of Auckland, New Zealand.

and other precipitates (Wentworth 1938, Hills 1949, Guilcher *et al*. 1962). Bowman (1970) found the most pronounced forms occur towards the back of intertidal platforms, above the level of the normal high tides; they are poorly developed where they are exposed to wave attack at lower elevations, and in very exposed sites the joints blocks may even stand out in relief above the etched joint planes (Fig. 2.4); an almost identical series has been identified on Madagascar (Guilcher *et al*. 1962). Some pits

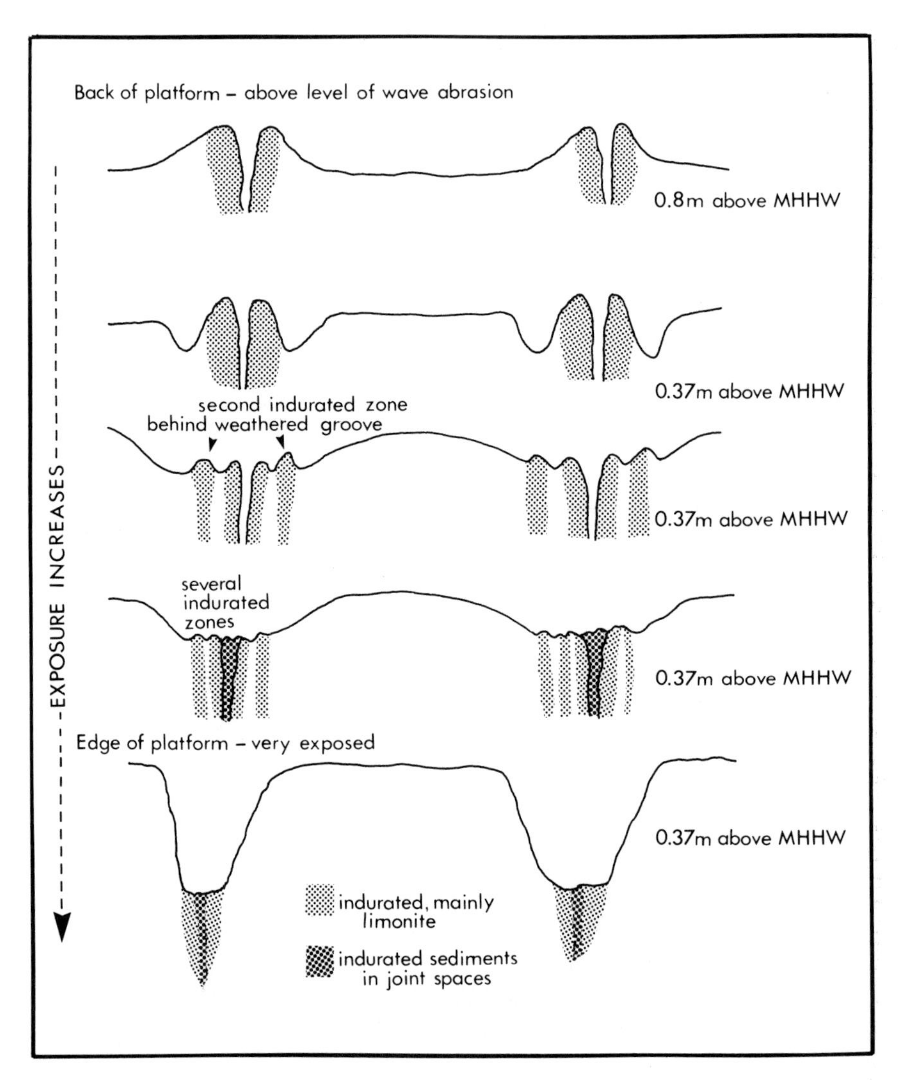

Figure 2.4. The distribution of weathering pits at Port Kembla, New South Wales (after Bowman 1970).

resemble miniature volcanoes with water-filled craters (Ollier 1969). These features are the result of small pools which keep the surrounding rock wet, while the general platform surface is reduced by wetting and drying. Jutson (1954) described 'saucers' or depressions which are found at the seaward margins of shore platforms, usually on the rampart. He attributed them to chemical weathering and induration of the joints, although he did not believe that water layer levelling is an important factor in the general development of coastal platforms. A similar view was expressed by Johnson (1938), who questioned whether a mechanism which is associated with isolated pools doesn't roughen rather than smooth rock surfaces. The mechanism may play a significant role in the modification of rock platforms in areas with high evaporation levels and diurnal or mixed tides (Davies 1972). It is unlikely to be important in cool, wet climates with low rates of evaporation. Nevertheless, small weathering pits which have developed in basic and intermediate rocks on the shores of fresh water lakes in southern Sweden are similar in several ways to pools in which water layer levelling is active. These pits develop in striae and other hollows close to the water level where there is alternate wetting and drying. Water in the hollows facilitates chemical weathering in summer and frost action in winter. This causes the pools to broaden and, because these processes are most active at the water level, an overhang often develops (Samuelsson and Werner 1978). Similar forms in granite and gneiss along lake and sea shores in Finland have been attributed to marine abrasion, frost and alternate wetting and drying (Aartolahti 1975, Uusinoka and Matti 1979).

The 'Old Hat'

Dana (1849) was the first to accord a major role to chemical weathering in the development of rock coasts in hot, wet climates. He suggested that marine cliffs are weathered down to the level at which the rocks are permanently saturated by sea water. Bell and Clarke (1909) first proposed that shore platforms develop at the saturation level, although the concept did not receive serious consideration until the later contributions of Bartrum (1916, 1926, 1938). Bartrum called these platforms 'Old Hat', following Hochstetter's use of the term to describe the form of Mill or Kaiaraara Island, in Russell, northern New Zealand (see Bartrum 1916). Bartrum cautioned that these platforms develop only in impermeable rocks which are free of joints, and in sheltered locations where wave action is feeble (Fig. 2.5). He believed that weak waves simply wash away the fine weathered materials, but other workers have suggested that wave abrasion plays an important role in the excavation of these platforms. (The implications of this assumption are discussed in Chapter 9.) Bartrum considered

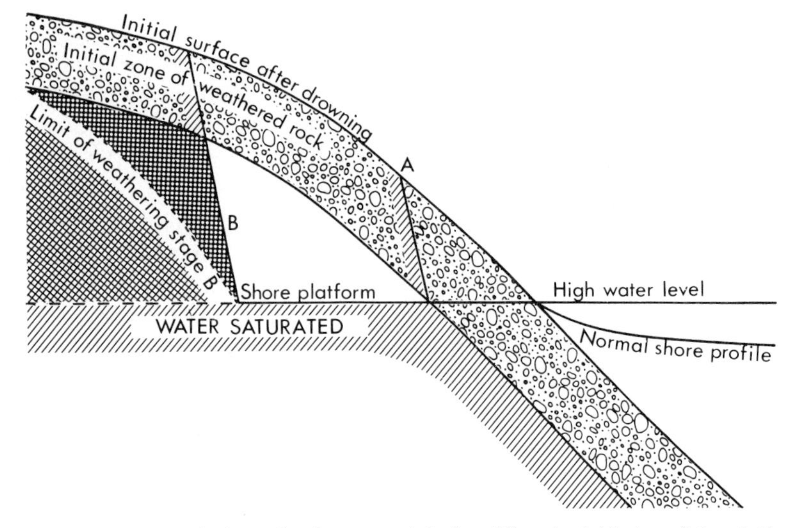

Figure 2.5. 'Old Hat' platform development. *A* is the cliff at the initiation of the platforms, and *B* is the cliff at a much later stage (after Bartrum 1916 *et seq.*).

that Old Hat platforms develop a little below the high tidal level, but Fairbridge (1952) argued that the saturation level, and therefore the platforms which it controls, are close to the low tidal level. Several other workers have claimed that shore platforms have developed at the level of saturation (Edwards 1958, Bird and Dent 1966, Healy 1968a, Russell 1971). Two assumptions are fundamental to the Old Hat hypothesis:

1. that there exists a well defined level of saturation within coastal cliffs; and
2. that this saturation level abruptly separates a weak, weathered cliff in the oxidation zone above from an essentially unweathered, resistant zone below.

Until recently, geomorphologists have been mislead by the assumptions that acidity is essential for weathering, and that effective weathering is restricted to the oxidation (aeration, vadose) zone above the water table (Cotton 1942, Reiche 1950, p. 84). Reiche, for example, considered that significant alteration extends only down to the water table, although he recognized that groundwater circulation permits some weathering at greater depths in very permeable rocks. This reasoning was applied by Ruxton and Berry (1957) and by Linton (1955) to explain the development of tors. Below the water table, in the zone of saturation (the phreatic or reduction zone, where the pore spaces in the rock are filled with water), anaerobic bacteria, reduction, hydrolysis and ionic substitution may cause alteration of some minerals (Ollier 1969). The flow of water in the discharge belt towards the top of the water table also permits removal of solutes. It is probably sluggish, however, so that chemical reactions would proceed slowly and tend to attain equilibrium conditions. The rate of weathering in the upper vadose zone is therefore thought to be much greater than that in the zone of saturation. Birot (1968) for example, considered that the weathering rate above the water table may be 20 to 30 times greater than below. Nevertheless, several early workers believed that alteration takes place to depths which are well below the water table (Campbell 1917). More recently, Watson (1964) found that Rhodesian granites are altered well below the water table, and Nye (1955) noted that in western Nigeria, weathering has taken place more than 28 ft (8.5 m) into the saturation zone. Lelong and Millet (1966) distinguished between a *zone supérieure d'alteration* formed where there is rapid movement of ground water, and a *zone d'alteration inférieure*, where the percolation of water is slow. The relationship between the depth of weathering and water movement and availability may be complex. De Swardt and Casey (1963, p. 9) found little relationship between these factors in sandstones in eastern Nigeria, and Thomas (1974) commented that the same is true of many crystalline rocks in the tropics.

Even where there is a sharp transition between weathered and unweathered rock, the boundary is generally irregular, rather than flat and horizontal (Ollier 1969). Lelong (1966) noted that etch plains are undulating, with hummocks and hollows, ridges and grooves. In very porous rocks, or in those which contain a large amount of water because of their fissility, there may be a fairly well-defined water table. In rocks which lack intergranular porosity, so that water can only penetrate through cracks of various sorts, the water table is extremely difficult to define. In such cases, some cracks are dry while others are full of water. The groundwater is discontinuous and the boundary between weathered and unweathered rock is very irregular, with projections of impermeable rock standing above weathered rock nearby.

Not only is the contact between weathered and unweathered rock often irregular but it is also often gradual, and consequently poorly defined. The transition can be sharp in basic rocks in the tropics (Chorley 1969), but as in the case of granites, it usually takes place over several metres. Sharp transitions depend more on the position of the base of the groundwater zone than on its surface (Ollier 1969).

The height of the water table varies according to such factors as the structure and lithology of the rock, climatic and seasonal changes, and in the coastal zone, tides and wave conditions. Most of the cliff face occupies the intermediate belt (Ollier 1969) in which the rock is occasionally wet, but the presence of air facilitates many kinds of chemical reactions.

A capillary fringe or belt of fluctuation occurs within the zone which experiences tidal fluctuations, above the level of permanent water saturation. Oxidation and reduction can alternate in this zone in response to the ebb and flow of the tides. Recorders in wells fairly close to the coast show that the water level in coastal aquifers fluctuates according to tidal oscillations (Carr and Van der Kamp 1969, Thomson 1979). These wells, however, are too far from the cliff edge to determine the magnitude of the fluctuations in the cliff face. Wells placed in beaches (Dominick *et al*. 1973, Hanor 1978) show that intertidal water tables oscillate with tidal ebb and flow, but it is difficult to apply these observations to rock coastlines. The phreatic surface in the Chalk of the Isle of Thanet in southeastern England also fluctuates with the tide, but it is approximately at mean sea level close to the coast (Hutchinson 1972).

Trenhaile and Mercan (1984) monitored saturation levels over a tidal cycle in intertidal rocks in eastern Canada, and also in the laboratory, using a pressurized vessel to simulate tidal oscillations. Few rocks had become completely saturated with water when they were first exposed by the ebbing tide. The rocks progressively desorbed during their period of exposure, although less than half the contained water was usually desorbed before the rocks were again immersed by the rising tide. The rate of desorption was partly determined by ambient temperature and humidity, although it would probably be greater in coarser rock types, and in warmer, drier climates. Probably the outer portion of the rock dries out fairly completely in the intertidal zone, while at greater depths the water is largely retained. The saturation data suggest that chemical processes could operate on the undersaturated surface layer, down to the low tidal level, particularly in warm, dry environments. It appears, therefore, that there must be a gradual transition in the intensity of chemical weathering from the high to the low tidal levels, rather than an abrupt change from weak, weathered rock above the saturation level, to more resistant, unweathered rock below. This study provided no support for the contention that chemical weathering and water layer levelling in warm regions can only operate effectively down to a well-defined intertidal level of permanent seawater saturation.

Present evidence suggests that intertidal rocks at or near the platform surface can only be permanently saturated with sea water below the low tidal level, where they are constantly submerged. The intertidal zone experiences alternate wetting and drying, and a variety of associated physical and chemical weathering processes. Furthermore, as has been noted, weathering fronts are generally irregular and the transition between weathered and unweathered rock is usually gradual. Chemical weathering undoubtedly plays an important role in the erosion of marine cliffs in warm, wet climates (see, for example, Tricart 1972, Consentius 1975), but further work will be necessary to substantiate the assumptions that shore platforms can be produced by weak waves washing away weathered debris, or by stronger waves excavating the weathering front from beneath coastal cliffs.

3

The Solution of Limestones

Mechanical wave attack is a potent erosional agent on many limestone coasts. Some old limestones are quite resistant to mechanical breakdown, even in vigorous storm wave environments. Younger limestones, which are usually physically much weaker, are common in the tropics, where wave action is usually less vigorous. The nature and efficacy of the destructional mechanisms operating on limestone coasts, where mechanical wave action is judged to be ineffective, have long been contentious issues. The intertidal and splash–spray zones of many limestone coasts have features similar to those etched by corrosional processes associated with meteoric water. The field evidence therefore testifies rather convincingly to the efficacy of corrosional processes on some coasts consisting of rocks which are carbonate rich, or which have a carbonate cement.

Numerous processes and modes of operation have been invoked to account for the occurrence of corrosional features on limestone coasts. Surface sea water is normally saturated or supersaturated with calcium carbonate, particularly in the tropics and on the western sides of continents. It has been suggested that these features are the result of solution by fresh rain or groundwater, operating down to the low tidal level, and possibly assisted by the release of humic acids by decaying vegetation (Wentworth 1939, Kuenen 1950). The amount of fresh water in the intertidal and splash–spray zones, however, is most unlikely to exceed the amount of salt water, even in particularly wet climates. Corrosional features also occur at the foot of isolated blocks of rock in the intertidal zone. These blocks cannot provide a sufficient catchment area to sustain runoff or groundwater flow. Perhaps the greatest problem with the hypothesis that coastal corrosion features are largely the result of freshwater solution is the fact that well-developed forms occur in the Red Sea and in similar areas where the role of fresh water and decaying vegetation can be discounted (Macfadyen 1930, Guilcher 1953). It has also been proposed that water movement generated by wave action in the surf zone, facilitates solutional processes during occasional periods of seawater undersaturation. Thinning of the diffusion layer at the rock surface by fluid motion brings fresh solvent closer to the rock, and aids the movement of solvent and solute ions towards and away from it (Kaye 1957, 1959).

A drop in the air and water temperature could cause sea water to become undersaturated with respect to calcium carbonate, although Guilcher (1953, 1958a) doubted its significance. Undersaturation may be induced by nocturnal cooling, which raises the carbon dioxide content of enclosed water (Revelle and Emery 1957). In Western Australia, for example, Fairbridge (see Revelle and Fairbridge 1957) found that the pH of open sea water remained at about 8.2, while in pools it fluctuated between 9.4 and 7.5, as temperatures dropped 5 to 6°C from day to night.

Biochemical processes play an important role in modifying the chemistry of sea water, to the point at which it may become undersaturated with calcium carbonate. Ranson (1955a, 1959) considered that the solution of limestones by sea water is impossible without the intervention of biological influences. Much emphasis has been placed on the ability of marine fauna and microflora to control the chemical environment of pool waters. During the day, the extraction of carbon dioxide by algae from pools is greater than its production by faunal respiration (De Virville 1934, 1935). Bicarbonates are decomposed and transformed to carbonates, and the pH and the saturation level increase. Precipitation of calcium carbonate can only occur if the saturation is of sufficient magnitude and duration. At night, essentially the opposite takes place (Emery 1946). Continued production of carbon dioxide by faunal respiration, unbalanced by algal uptake during the hours of darkness, causes a reduction in pH and the transformation of calcium carbonate into more soluble bicarbonate. Reduction in the carbonate ion concentration and in the level of calcium carbonate saturation, the lower pH, and the increase in the solubility of calcium carbonate may facilitate solution at this time.

Carbon dioxide may react in water in several ways as shown in Figure 3.1 (Schneider 1976); Schmalz and Swanson (1969) found that diurnal pH, dissolved CO_2 and carbonate saturation cycles in sea water are dependent upon alternations of light and darkness in the presence of green vegetation. The phenomenon therefore appears to be the result of the extraction of carbon dioxide through photosynthesis, and variations in the rate of metabolic production. The amplitude of this diurnal variation depends upon the amount of water and the local biomass. Consequently, the effect

$$CO_2 + H_2O \rightleftharpoons H_2CO_3$$

$$H_2CO_3 \rightleftharpoons H^+ + HCO_3^-$$

$$HCO_3^- \rightleftharpoons CO_3^{2-} + H^+$$

$$CO_2 + OH^- \rightleftharpoons HCO_3^- \text{ (at pH 8 or greater)}$$

Figure 3.1. The reaction of carbon dioxide with water (Schneider 1976).

of biological activity on the carbon dioxide system is much less in the open sea than in rock pools, where the buffer capacity of the water may be insufficient to stabilize the pH condition at extreme respiration and assimilation rates.

Several workers have found that diurnal variations in the pH of tropical sea water are quite small (Verstappen 1960). In the Bay of Djakarta for example, a maximum pH of 8.4 was attained in the afternoon in the shallow moat of a coral island, but this was almost matched by a value of 8.2 in the early morning (Zaneveld and Verstappen 1952). Schmalz and Swanson (1969) found that the pH in open water, which is subject to mixing, varied by only 0.15. Newell and Imbrie (1955) and Newell (1956) also found consistently high pH values and insignificant fluctuations in the carbonate concentration of tidal pools and open sea water. Much greater variations have been recorded from the French coast (De Virville 1934, 1935), and from Onotoa Atoll, where the daytime pH was 8.6 in open shoal water and 9.1 in tidal pools, with corresponding night-time pH values of 7.3 and 8.0 (Cloud 1952). Even in rock pools, diurnal variations may be too small for effective solutional activity.

In an essentially non-tidal area, as in the western Mediterranean, where water in rock pools is renewed infrequently, evaporation and the growth of vegetation produce conditions which are not conducive to solution (Debrat 1974). Schneider (1976), for example, found that solution does not take place in rock pools in the spray and splash zone in the microtidal northern Adriatic, even when the water is undersaturated with respect to calcium carbonate.

Ginsburg (1953a) considered that the biochemical explanation is only appropriate in special cases, because vertical surfaces are just as intensely eroded as horizontal ones, despite the fact that the latter retain far more water. The erosional forms produced on vertical surfaces, however, are generally quite different from those associated with rock pools. Any attempt to explain the origin of features on vertical surfaces must consider not only the chemical and biochemical effects of spray and splash, but also the activity of boring invertebrate fauna and microflora (see Chapter 4).

Whether solution actually occurs depends not only upon the presence of an undersaturated solution, but also upon the availability of an active limestone surface, and sufficient time for the various chemical reactions to take place. Biochemical processes, which may play an important, if not crucial, role in facilitating solution, can also inhibit or prevent it from operating effectively.

A great variety of dissolved organic substances affect the solubility of calcium carbonate (Smith *et al.* 1968, Kitano *et al.* 1969, Chave and Suess 1970, Suess and Futterer 1972, Pytkowicz 1973, Williams 1975). These organic substances build complexes with calcium ions which are of unknown abundance and solubility. As much as half the carbon dioxide assimilated by algae during photosynthesis is quickly released in the form of dissolved organic substances. This coats the rock surface, hindering or preventing solution even if the water is undersaturated (Suess 1970). The solution of calcium carbonate is also determined by ionic pairing and complexing by other inorganic ions on the Mg^{2+}, Ca^{2+}, and CO_3^{2-} ions (Greenwald 1941, Revelle and Emery 1957, Revelle and Fairbridge 1957, Cloud 1965, Stumm and Brauner 1975).

Ion-pairs behave differently from their constituent parts. Ionic interactions affect the charge distribution of solutes and the solubility of minerals. Calcium carbonate may be as much as 10 times more soluble than it would be in the absence of carbonate ion-pairs (Garrels *et al.* 1961, Pytkowicz and Hawley 1974, Pytkowicz 1975). The presence of Mg^{2+} in solution, for example, reduces the saturation level with respect to calcium carbonate by forming ion-pairs which hinder the activity of the carbonate ions (Pytkowicz 1965, 1969, 1973, Berner 1967). The association of CO_3^{2-} and HCO_3^- anions with Na^+ and Mg^{2+} cations partly explains the high apparent supersaturation of sea water with calcium carbonate. It has been estimated that as much as 75% of the CO_3^{2-} ions in sea water are complexes in the form of $MgCO_3$, and 15% in the form of $NaCO_3^-$. Only about 10% of the carbonate ions, therefore, are free and susceptible to the reactions in which we are interested (Garrels *et al.* 1961, Debrat 1974).

Precipitates of different solubility are formed on calcite when it reacts with supersaturated sea water (Weyl 1967). It has been suggested that near the ocean surface, a magnesium carbonate of about 4 mole per cent $MgCO_3$ can form a film on calcite which is about 30% less soluble than pure calcite (Morse *et al.* 1979, 1980). Some surface films are more soluble than calcite or aragonite. Calcites which contains magnesium are unstable with respect to pure calcite, and above about 8 mole per cent Mg they are unstable with respect to aragonite (Wollast *et al.* 1980). Magnesian calcite of lower solubility than pure calcite forms if there is an increase in the magnesium-to-calcium concentration in a solution (Chave *et al.* 1962, Plummer and Mackenzie 1974, Berner 1975, Thorstenson and Plummer 1977, Wollast *et al* 1980, Mucci and Morse 1984). It has been found that magnesian calcite with a high magnesium content will precipitate or dissolve in solutions with a mean pH as alkaline as 8.15; this may be in response to small changes in the pH—as a result, for example, of photosynthesis (Koch and Disteche 1984).

The size and shape of the calcium carbonate particles are significant in determining their solubility. Sea water comes into contact with calcium carbonate in many forms. Mineral particles such as calcite, aragonite, and various magnesian calcites are the product of biological, physical, and chemical processes, and they occur as crystals or crystal aggregates of variable size and shape. Each of these factors plays an important role in determining how carbonate crystals interact with sea water (Schmalz and Chave 1963, Chave and Schmalz 1966).

Recent work (Cooke 1977) has suggested that chemical solution is possible in sea water, although some field investigations have found that it is of fairly minor importance, even in stagnant tidal pools with a large biomass relative to the amount of water. Undersaturation of tidal pools and inshore waters with respect to calcite occurs at night on Aldabra Atoll, and may occur at any time with respect to aragonite and high magnesian calcite. Trudgill (1976b) calculated that about 10% of the total erosion of this coast can be attributed to solution, but Torunski (1979) considered that even this small amount is exaggerated. The absence of any solution of inorganic limestones in rock pools in the northern Adriatic may be because of oversaturation of the pool waters and the presence of organic coatings or dense microfloral covers which inhibit contact between the water and the calcite substrate (Schneider 1976).

Solution produces minute etch patterns on rock grains, which may be distinguished under an electron microscope from those resulting from other processes. These solutional traces have been identified on grains from the shallow waters of the North Sea, but they appear to be absent from those obtained from the warmer waters of the Mediterranean and Caribbean seas (Alexandersson 1969, 1972, 1975, 1976). These results suggest that limestone solution is more effective in cool, temperate latitudes than in tropical and other warm waters. In view of the complex and still poorly understood biochemical and chemical environments of rock pools and open ocean waters, this conclusion is premature. Lower temperatures and lower levels of calcium carbonate saturation facilitate solution in cool waters, but the generally weaker wave action in tropical waters, and the prevalence of aragonite and magnesium-rich coral, which are more soluble than pure calcite (Revelle and Fairbridge 1957, Chave *et al.* 1962), may favour limestone solution in these latitudes.

It is now possible to quantify the role of solution on limestone coasts (Trudgill 1976b, Schneider 1976), but there is at present insufficient data with which to compare the efficacy of corrosional processes in different morphogenic environments. We are still largely dependent therefore, on morphological evidence, which tends to be ambiguous; although it often suggests that chemical solution is a significant erosional process, on many coasts it may instead reflect the bioerosional activities of a variety of marine organisms.

4

Bioerosion and Other Biological Influences

Bioerosion has been defined as the removal of the lithic substrate by direct organic activities (Neumann 1966). It is probably of greatest importance in the destruction of coastal rocks in the tropics, where an enormously varied marine biota live on calcareous substrates which are particularly susceptible to biochemical and biomechanical processes. Their activities can produce 'biokarst', a surface composed of features produced by direct biological erosion and/or deposition of calcium carbonate (Schneider 1976, Viles 1984). Outside the tropics, bioerosion is usually less effective because of generally less suitable rocks, vigorous wave action in temperate regions, and an impoverished bioeroding community in cold, high latitudes (Bromley and Hanken 1981).

A number of factors determine the nature, significance, and rate of bioerosional activity. The distribution of marine organisms in the vertical and horizontal planes is of crucial importance (Doty 1957). The availability of moisture is the main factor which controls their distribution; tidal characteristics largely determine how long the substrate is inundated or exposed and subject to desiccation, which partly depends upon whether the tide is diurnal, semi-diurnal, or mixed (Johnson and Sparrow 1961). A vigorous wave environment also increases the availability of moisture, and the width of each organic zone increases with increasing exposure to waves. In some humid regions, for example, bioerosion is more intense where there is frequent wetting of the substrate by waves and rain than in drier, sheltered areas (Schneider 1976). In some cases, however, bioerosion is greatest in sheltered sites because wave shock and mobile sedimentary particles driven by waves and currents can prevent colonization of exposed areas by some organisms (Golubic *et al.* 1975). Other important site considerations include the salinity of the water, the temperature, and the degree of exposure to sunlight. These and other factors account for differences in the species composition and consequently in the intensity and type of bioerosional activity in exposed and sheltered locations (Doty 1957).

The efficacy and zonation of erosive organisms also vary according to the nature of the substrate. Soft, fine-grained calcareous rocks are especially vulnerable to boring organisms, and limestones can be riddled with holes down to depths of at least 40 m below the water surface (Revelle and Fairbridge 1957, Warme 1970). Even organisms which penetrate the substrate by purely chemical processes can bore into rocks which consist of less than 5% carbonate material, and with grain sizes varying from that of claystones to medium-grained sandstones (Warme 1970). The widespread occurrence of carbonate rocks in the tropics favours chemical borers, but mechanical borers are active on a variety of substrates in temperate and boreal waters, although hard conglomerates and igneous rocks are generally avoided (Evans 1970, Warme 1975). Other lithological and structural factors affect the efficiency of bioerosional processes. Porosity influences the moisture availability, and bedding planes, joints, and other structural weaknesses provide access to the rock interior for some organisms (Warme and Marshall 1969, Hodgkin 1970). Much remains to be determined about the degree to which bioerosional activity is determined by the chemical and physical properties of the substrates which are occupied. Elucidation of the nature of this relationship depends upon greater understanding of the way in which organisms penetrate substrates.

Some invertebrates break down rocks by purely chemical or mechanical processes. Microflora (including algae, fungi, lichens, and bacteria) and fauna which lack hard parts may erode by purely chemical means, but many fauna use a combination of mechanical and chemical processes. This generally involves initial weakening of the rock by secreted fluids, followed by abrasion by teeth, bristles, hooks, shields, valvular edges, and other hard parts (Carriker and Smith 1969).

Microflora

Algae are probably the most important erosive organisms in the intertidal and supratidal spray zones (Duerden 1902, Nadson 1927, Ginsburg 1953b, Ranson 1955a,b, 1959, Hodgkin 1970, Dalongeville 1977). A great variety of algae are involved, including species of Chlorophyta (green algae) in the intertidal zone; Rhodophyta (red algae), primarily in the lower intertidal and subtidal zones; and especially the minute Cyanophyta (blue–green algae), which often form a slimy covering over the rocks in the upper intertidal and supratidal zones (Boney 1966). Genera such as *Calothrix, Phormidium, Gloecapsa, Nodularia, Rivularia, Plectonema, Oscillatoria*, and *Lyngbya* can be associated with *Verrucaria* and *Lichena* lichen in a zone of blackened rock (Lewis 1964). This zone is very common on rocky coasts (Stephenson and Stephenson 1949), although it can be poorly developed or absent on granites, on tropical coasts where desiccation is severe, and on Antarctic shores where abrasion by sea ice prevents its development (Fogg 1973). Species belonging to other genera, including *Entophysalis*, *Scytonema*, *Hormathonema*, *Hyella*, *Mastigocoleus*, *Kyrtuthrix*, *Solentia*, *Pleurocapsa*, and *Schizothrix* are also important components of floral communities in the upper portions of rock coasts (Prescott 1968,

Lukas 1969, Chapman and Chapman 1973, Round 1973, Debrat 1974, Golubic *et al*. 1975, Schneider 1976). Mollusca and other grazing organism limit the occurrence of Cyanophyta in the intertidal or eulittoral zone, although many species are abundant at these lower elevations. Intertidal *Rivularia* colonies are common in summer in temperate regions, but Cyanophyta, such as *Entophysalis deusta, Mastigocoleus testarum, Coccochloris stagnina*, and *Gardnerula corymbosa*, are generally more frequent in the eulittoral of warmer waters (Fogg 1973). Eulittoral endolithic Cyanophtes are particularly common on limestone coasts.

Epilithic algae inhibit the rock surface, but endolithic algae penetrate the rock and are connected to the surface by a network of fine filaments. Endolithic cyanophyta can bore into rock to a depth of about 500 to 900 microns, in populations ranging from 150,000 to 1,000,000 individuals per cm^2. As much as one-third of a rock can be perforated (see, for example, Schneider 1976, Le Campion-Alsumard 1979, Schneider and Torunski 1983). The reason why some algae are rock borers is unclear, although it may be to escape intense sunlight, in which they fail to photosynthesize, and to prevent excessive drying out (Schneider 1976). Endolithic algae are distributed from the upper sublittorial zone into the spray zone (Golubic 1969), but the shallow, predominantly horizontal borings in the supratidal zone are more destructive to the substrate than borings perpendicular to the rock surface in the lower intertidal zone (Frémy 1945, Purdy and Kornicker 1958). Epilithic, as well as endolithic, algae contribute to rock destruction. The products of metabolism and organic waste cause considerable variation in the chemistry of water in intimate contact with the rock under algal mats (Newell and Imbrie 1955; Newell 1956; Revelle and Fairbridge 1957; Neumann 1966, 1968). Extreme conditions can cause these algal colonies to become slowly etched into the substrate.

Other microflora such as fungi and lichen, are effective rock borers; fungi can bore much deeper than algae, but as they need the algae for nutrients their depth of penetration is effectively limited by the occurrence of the algae (Kohlmeyer 1969). Much of the erosion attributed to algae may in fact be the result of fungi, particularly on the bottom of pools and in other wet sites, where they are most common.

Endolithic lichen are found up to several millimetres beneath the rock surface. Lichen are important members of the biome in the zone extending from the upper intertidal zone into the area of spray and splash. Among the most common genera are *Verrucaria, Lichina, Caloplaca, Xanthorina*, and *Arthopyrenia* (Johnson and Sparrow 1961). One of the most common lichens is *Arthopyrenia sublittoralis*, which forms pits which are generally similar to those of algae although usually larger (Bromley 1970, Sarjeant 1975). Schneider (1976), making a detailed examination of lichens on the northern Adriatic coast of Yugoslavia, recognized three distinct zones. A lichen zone characterized by endolithic Cyanophytes and *Verrucaria* and *Arthopyrenia* species extends from the upper intertidal zone into the area of wave splash. In the upper splash zone is found the Halophyte zone, and the first appearance of species of, for example, the genera *Caloplaca, Amphoridium, Lecania*, and *Collema*. Above the Halophyte zone, which marks the beginning of a soil cover, is the lichen zone proper: a zone moistened by spray, but seldom by splash. The area is blanketed by lichens, particularly by *Amphoridium calcisedum* and other endolithic forms, which have assumed the role of the endolithic Cyanophytes in the upper intertidal zone.

It has been suggested that bacteria indirectly contribute to the breakdown of carbonates in the intertidal zone, and they are certainly important as the initial colonizers of substrates (Ginsburg 1953b, Schneider 1976). Bacteria are also known to directly attack the organic matrix of coral skeletons (DiSalvo 1969).

Grazing

Epi- and endolithic microflora, which are usually the pioneer colonizers of the inter- and supratidal zones, permit subsequent occupation by a host of grazing organisms. Algae, lichen, and fungi are sources of food for these grazers; and they prevent the development of anaerobic conditions in rock pools thereby making them suitable for colonization. A great variety of organisms graze on the microflora of the intertidal and supratidal zones, although the number may be even greater below the low tidal level. The grazing activity of chitons and gastropods, including limpets and periwinkles, has usually been emphasized in the literature. Other organisms are also significant, such as echinoids (Fig. 4.1*i*), crabs, and on coral reefs, starfish (*Acanthaster planci*). Fish of the family Scaridae (Parrot-fish), and to a lesser extent the families Acanthuridae (Surgeon-fish) and Chaetodontidae (Butterfly-fish), can also accomplish considerable erosion in the subtidal zone (Bardach 1961, Bromley 1970, Gygi 1975, Neudecker 1977, Reese 1977, Frydl and Stearn 1978) (Table 4.1). Grazing on epilithic microflora and on the ends of the endolithic forms can cause mechanical rasping of rock surfaces which have been weakened by the biochemical processes associated with the microflora, or by the penetration of their filaments into the rock. Organisms which graze on epilithic microflora account for the least amount of rock destruction, but considerable erosion can be caused by the consumption of epi-endolithic or shallow endolithic forms. Large herbivorous gastropods, for example, can excavate grooves 0.5 mm deep and 1 mm wide in a single traverse (Newell and Imbrie 1955).

The microflora and the associated grazing fauna can coexist in a homeostatically regulated system. Each grazing organism may harvest microflora

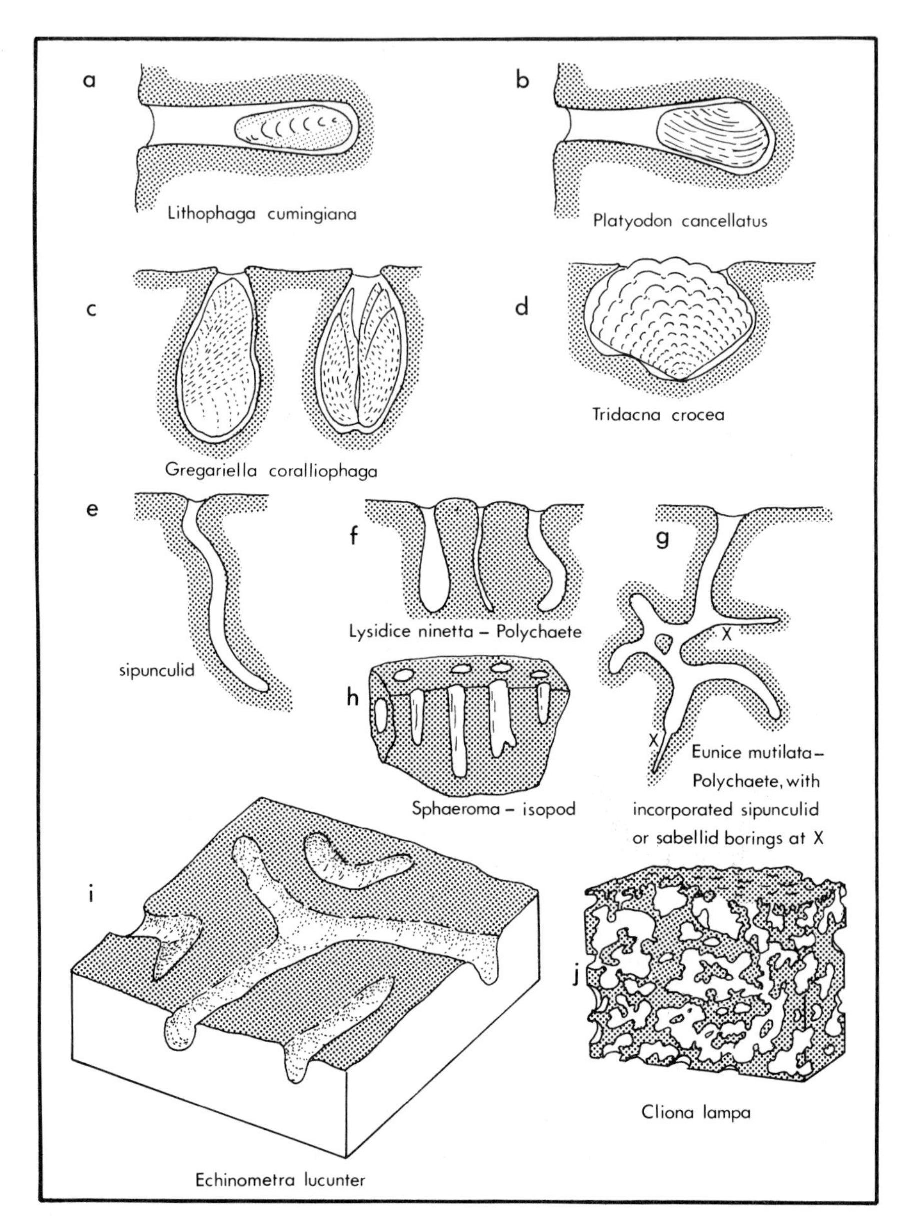

Figure 4.1. Examples of the bioerosional effects of boring and grazing marine organisms (after: (*a*) Otter 1937; (*b*) and (*d*) Yonge 1963, 1936 respectively; (*h*) Bromley 1970; and the remainder Bromley 1978).

Table 4.1. Grazing Organisms

Phylum	Class	Genus	References
Mollusca	Amphineura (chitons)	*Acanthochithon* *Acanthopleura* *Acanthozostera* *Middendorfia* *Mopalia* *Syphrochiton*	Bertram 1936, Otter 1937, Ginsburg 1953b, Molinier and Picard 1957, Healy 1968b, Debrat 1974, Warme 1975, Schneider 1976, Taylor and Way 1976, Trudgill 1976b, 1983
	Gastropoda (snails)	*Acmaea* *Cellana* *Hipponix* *Littorina* *Lunella* *Melagraphia* *Melarhapha* *Monodonta* *Nerita* *Nodilittorina* *Patella* *Siphonaria* *Tectarius* *Turbo*	Welch 1929, Otter 1937, Ginsburg 1953b, Newell 1956, Molinier and Picard 1957, Southward 1964, Yonge 1963, McLean 1967a, Schneider 1976, Trudgill 1976b, Torunski 1979
Arthopoda	Malacostraca (crabs, lobsters)	*Grapsus* *Ochthebius*	Kaye 1959, Trudgill 1976b Schneider 1976
Echinodermata	Stelleroidea (starfish)	*Acanthaster*	Chesher 1969, Dart 1972, Polunin 1974, Endean 1977
Chordata (superclass Pisces)	Scaridae* (parrotfish)	*Astraeapora* *Montipora* *Porites* *Pseudoscarus* *Scarus* *Sparisoma*	Russell and Yonge 1949, Otter 1937, Cousteau 1952, Nesteroff 1955, Newell 1956, Cloud 1959, Hiatt and Strasburg 1960, Stephenson 1961, Bardach 1961, Emery 1962, Bromley 1970, Gygi 1969, 1975, Randall 1974, Neudecker 1977, Reese 1977, Frydl and Stearn 1978
	Balisidae*	*Balistipus*	
	Chaetodontidae*	*Chaetodon*	
	Labridae*	*Labroides* sp.	
	Canthigasteridae*		
	Monacanthidae*		
	Tetradontidae*		

*Family.

at a particular depth, thereby ensuring minimum competition between the various species. On Aldabra Atoll, gastropods tend to consume epilithic and shallow endolithic algae, echinoids and crabs the epilithic and shallow algae, and chitons the deeper endolithic forms (Trudgill 1976b) (Fig. 4.2). The rate of rock destruction is also determined by the mutually adjusted rates of grazing and microfloral penetration. If the grazing rate is greater than the rate at which microflora penetrate the substratum, the food resource gradually becomes exhausted and there will be a corresponding decline in the number and activity of the grazers (Golubic *et al*. 1975). The

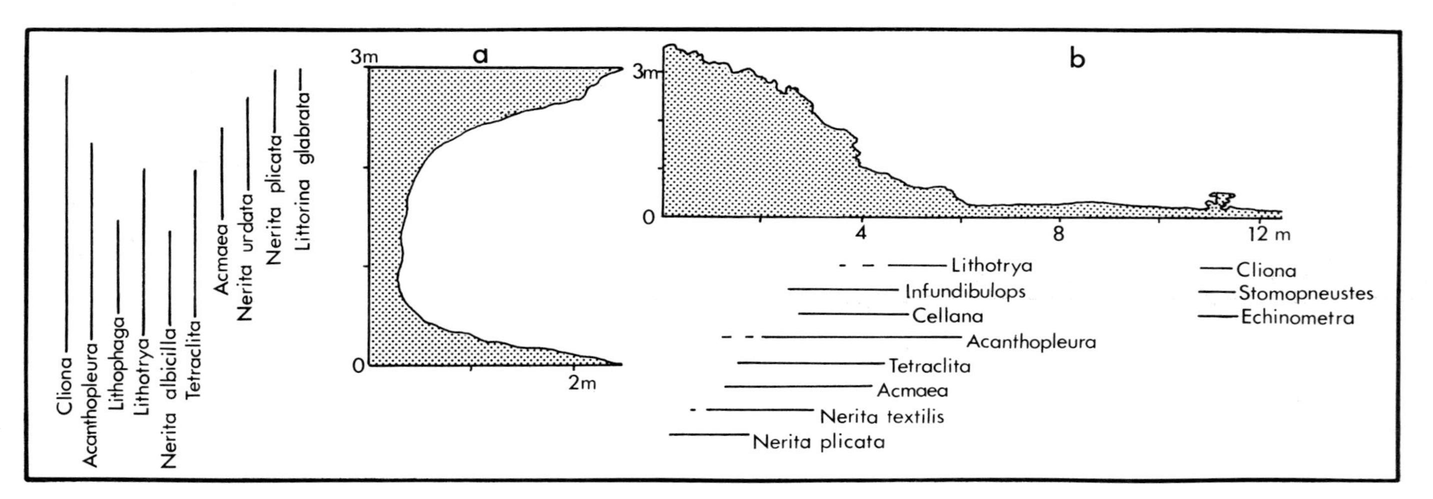

Figure 4.2. Distribution of the erosive organisms on Aldabra Atoll on (*a*) the northwestern coast; and (*b*) exposed coasts (Trudgill 1976b).

drier the site, the fewer the number of grazers and algal species; and in response to the limited grazing activity, the more numerous the epilithic algae. Alternatively, because microflora are able to bore deeper into rock in wet sites, the grazing population is denser, and the epilithic forms fewer (Schneider 1976) (Fig. 4.3).

The direct erosional role of grazing organisms is of particular significance on tropical and warm temperate limestone coasts, where wave attack may be fairly weak. On Aldabra Atoll, for example, Trudgill (1976b) has calculated that grazing organisms account for about one-third of the surface erosion in the mid-tidal zone when sand is available for abrasion, and as much as two-thirds of the erosion if sand is absent (see Chapter 10). Grazers also play an important indirect role in facilitating microfloral penetration of the substratum. Grazing prevents the attainment of climax endolithic populations, which, if ever attained, would produce a quasi-static situation in which the rate of rock destruction would be determined only by slow microfloral corrosion (Schneider 1976).

Faunal Boring and Burrowing

The terms 'boring' and 'burrowing' lack a universally acceptable definition, and many workers have used them synonymously. However, a number of workers (Bromley 1970, Warme 1975, Golubic *et al*. 1975) have used the term 'boring' for excavation in hard materials, and 'burrowing' for soft materials and this distinction will be followed here. An alternative definition proposed by Carriker and Smith (1969) has achieved little acceptance, but is worthy nevertheless of discussion because of its emphasis on criteria of geomorphic significance. They defined a burrower as an organism which excavates a space for the purpose of inhabitation, anchorage, nutrition, or sexual maturation, whereas a borer is one which excavates primarily for food. Burrowers often make excavations larger than themselves, but the rate of penetration is usually slow and intermittent, their activity being determined by biological and environmental factors. Each burrowing individual generally excavates only one or a single series of burrows during its lifetime. Borers excavate many boreholes during the life of each individual, but although the boring rate is quite fast and continuous, the excavation is usually much smaller than the organism which cuts it. Implicit in this classification is the general association of borers with soft substrates, since there is little food value in most rock materials.

The ability to excavate or penetrate through the substratum is widely distributed throughout the major invertebrate divisions, and at least 12 faunal phyla contain members which are endowed with this facility (Warme 1975). Particularly frequent mention has been made of the penetrating habits of *Lithotrya* and other boring barnacles, sipunculoid,

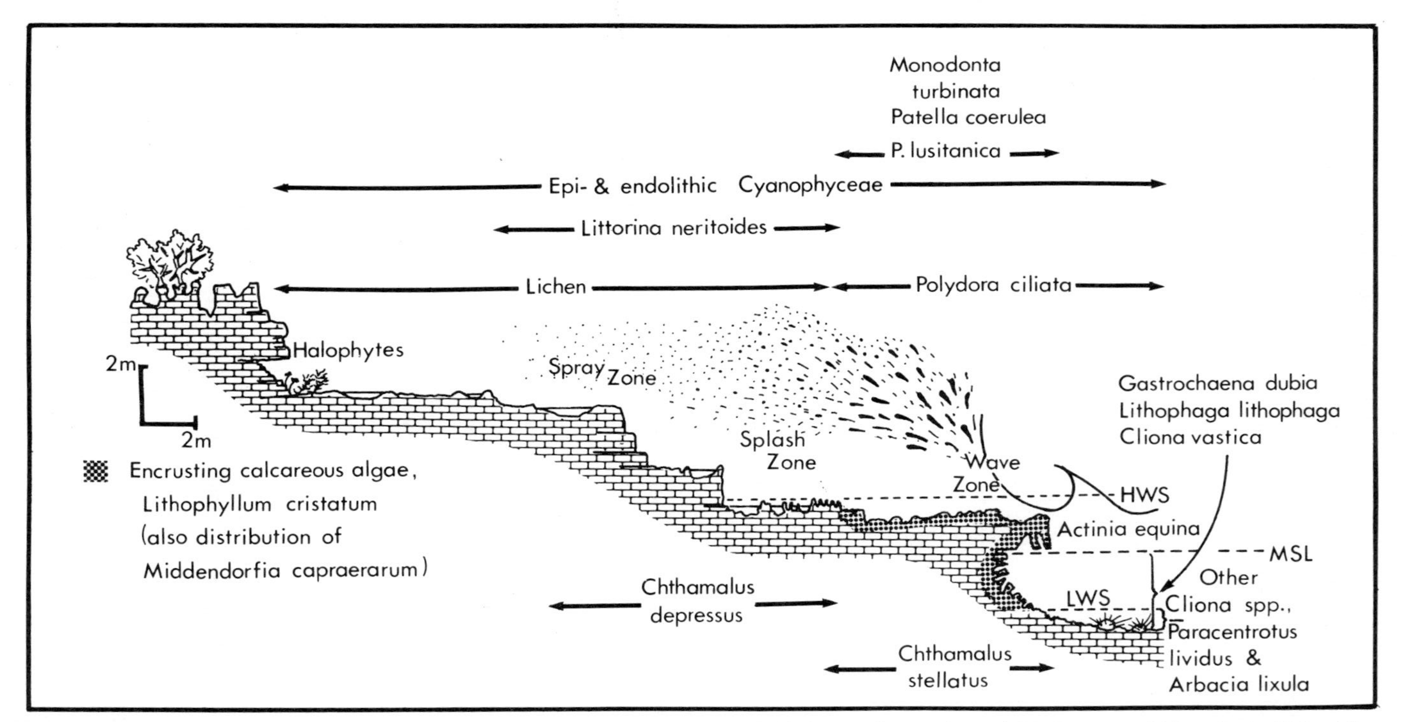

Figure 4.3. Distribution of the most important littoral organisms on Garzotto Point on the northern Adriatic coast of Yugoslavia (Schneider 1976).

and polychaete worms, gastropods, echinoids (sea urchins), bivalve molluscs and Clionid sponges (Table 4.2). Although the efficacy of these penetrating organisms varies greatly according to local geographical and geological factors, the bivalves or pelecypod molluscs and the Clionid sponges are the most important excavators of rock in most areas (Yonge 1951). The majority of the approximately one hundred species of boring sponges belong to about 14 genera in the family Clionidae. The Clionids bore microscopic to macroscopic excavations in limestones; other sponges, such as *Siphonodictyon*, produce much larger excavations (Warme 1975), although, as with other organisms, the size and shape varies according to such factors as the age and size of the individual. Clionid sponges play a particularly important bioerosional role, partly because of their abundance and widespread distribution in warm and cold seas. The boring habit is supremely developed in three pelecypod families: the Pholadidae, which are mainly mechanical borers; the Gastrochaenidae; and the Mytilidae, of which the *Lithophaga* genus is particularly prominent. Mytilids are largely chemical borers that require a calcareous substrate (Yonge 1955). Other bivalve borers and nestlers, which occupy and modify previously formed hollows, are found among families which do not consist primarily of boring species; these include the Petricolidae, Tridacnidae, and Hiatellidae (Warme 1975). Several phyla of worms are also active borers in calcareous and non-calcareous substrates. The size of their excavations varies with the size of the individual, ranging from less than 1 mm in diameter for small forms such as *Polydora* to 5 mm or more in diameter, and several decimetres in length, for larger genera. Sipunculids are particularly well adapted for boring hard materials, and are often found in substrates which are too resistant for excavation by other organisms. Some other fauna may alternate between boring and grazing activities. Gastropods (including *Patella* and *Acmaea* limpets), chitons, and echinoids excavate depressions which, although generally rather shallow, may, as in the case of some chitons, be up to five times deeper than the height of the animal (Warme 1975). These depressions are formed when grazing organisms regularly return to their particular spots or 'homes' after foraging at night for algae (Otter 1932, Healy 1968b, Bromley 1970, Warme 1975, Trudgill 1983). Useful tabulations of marine borers have been provided by Neumann (1968), Carriker and Smith (1969), and Schneider (1976), and a very extensive bibliography by Clapp and Kenk (1963).

Rock borers play a direct and an indirect role in the disintegration of rock substrates, particularly in the lower portions of the intertidal zone (Fig. 4.4). Boring directly removes rock material, rendering the residual mass much more susceptible to breakdown by wave action and other destructive mechanisms. Borers, however, also enhance a rock environment for algal colonization, and tremendously increase the area of rock

Table 4.2. Boring Organisms

Phylum	Class	Genus	References
Protozoa (Foraminifera)		*Cymballopora*	Warme 1975
		Rosolina	
Porifera (Sponges)		*Acanthacarnus*	Jehu 1918, Otter 1937, Yonge 1951, Ginsburg 1953b, Ranson 1955a, Neumann 1966, 1968, Healy 1968b, Cobb 1969, Hodgkin 1970, Rutzler 1974, 1975, Hein and Risk 1975, Warme 1975, Schneider 1976, Groot 1977, Moore and Shedd 1977, MacGeachy 1977, Pomponi 1977, Bromley 1978.
		Acarnus	
		Adocia	
		Alectona	
		Anthosigmella	
		Cliona	
		Oceanopia	
		Siphonodictyon	
		Spheciospongia	
		Spirastrella	
Coelenterata	Anthozoa (sea anemone)	*Palythoa*	Bromley 1978
Ectoprocta		*Immergentia*	Silen 1946, Soule and Soule 1969, Boekschoten 1970, Warme 1975
		Penetrantia	
Brachiopoda		*Chlidonophora*	Rudwick 1965, Bromley 1970
		Terebratulina	
Mollusca	Gastropoda	*Acmaea*	Otter 1937, Yonge 1951, 1955, 1958, 1963, Ginsburg 1953b, Ranson 1955a, Revelle and Fairbridge 1957, Neumann 1966, 1968, Carriker 1969, Evans 1968, 1970, Healy 1968b, Ansell and Nair 1969, Carriker and Smith 1969, Soliman 1969, Warme and Marshall 1969, Hodgkin 1970, Warme 1970, 1975, Vita–Finzi and Cornelius 1973, Hein and Risk 1975, Hamner and Jones 1976, Schneider 1976, Bromley 1978
		Capulus	
		Leptoconchus	
		Magilopsis	
		Okadaia	
		Patella	
		Purpura	
		Siphonaria	
		Urosalpinx	
	Pelecypoda		
	Mytilidae	*Adula*	
		Botula	
		Lithophaga	
		Modiolus	
		Mytilus	
	Pholadidae	*Barnea*	
		Penitella	
		Pholas	
		Nettastromella	
		Zirphaea	
	Myidae	*Platyodon*	
	Gastrochaenidae	*Gastrochaena*	
	Tridacnidae	*Tridacna*	
	Petricolidae	*Petricola*	
	Hiatellidae	*Hiatella*	
Annelida	Arcidae	*Arca*	
	Polychaeta	*Boccardia*	Jehu 1918, Calman 1936, Otter 1937, Yonge 1951, 1963, Ginsburg 1953b, Newell and Imbrie 1955, Ranson 1955a, 1959, Newell 1956, Revelle and Fairbridge 1957, Woelke 1957, Healy 1968b, Neumann 1968, Blake 1969, Haigler 1969, Jones 1969, Rice 1969, Bromley 1970,
		Caobangia	
		Dasybranchus	
		Dodecaceria	
		Eunice	
		Marphysa	
		Morpha	
		Polydora	
		Stylarioides	

Phylum	Class	Genus	References
Platyhelminthes	Turbellaria	*Pseudostylochus*	1978, Hodgkin 1970, Warme
Nemathelminthes	Nematodo		1970, 1975, Blake and Evans
Phoronida		*Phoronis*	1973, Hein and Risk 1975,
		Talpina	Schneider 1976, Van der Pers
Sipunculoidea		*Aspidosiphon*	1978
		Cloeosiphon	
		Dendrostonum	
		Lithacrosiphon	
		Paraspidosiphon	
		Phascoloma	
		Phascolosoma	
Echiuroidea		*Thalassema*	
Arthopoda	Cirripedia	*Concholepas*	Yonge 1951, 1963, Ginsburg
	(barnacles)	*Cryptophialus*	1953b, Tomlinson 1953, 1969,
		Fissurella	Newell and Imbrie 1955, Ranson
		Kochlorine	1955a, 1959, Newell 1956,
		Lithotrya	Revelle and Fairbridge 1957,
		Trypetesa	Neumann 1968, Bromley 1970,
		Weltneria	1978, Ahr and Stanton 1973
	Malacostraca	*Sphaeroma*	Calman 1936, Higgins 1956, Warme 1975
Echinodermata	Echinoidea	*Arbacia*	Fewkes 1890, Jehu 1918, Otter
(grazer-borers)	(sea urchins)	*Diadema*	1932, 1937, Yonge 1951,
		Echinometra	Ginsburg 1953b, Ranson 1955a,
		Echinostrephus	Revelle and Fairbridge 1957,
		Eucidaris	Kaye 1959, McLean 1964, 1967b,
		Heterocentrotus	Neumann 1966, 1968, Healy
		Lytechinus	1968b, Hunt 1969, Dart 1972,
		Paracentrotus	Krumbein and Pers 1974, Warme
		Stomopneustes	1975, Schneider 1976, Trudgill
		Strongylocentrotus	1976b, Benayahu and Loya 1977, Stearn and Scoffin 1977, Bromley 1978, Focke 1978, Glynn *et al*. 1979, Torunski 1979.

surface exposed to other physical and chemical processes (Ginsburg 1953b, McLean 1974). Although it is difficult to substantiate, the indirect role of rock borers may be of greater significance to the destruction of coastal rocks than is the direct removal of material.

Rates of Biological Erosion

The efficacy of biological processes in the destruction of coastal rock and their contribution to the development of coastal landforms are dependent upon the rate at which they operate. Not only is this information generally lacking at present, but our even greater ignorance of the efficacy of the other processes which operate on rock coasts makes it difficult to assess the

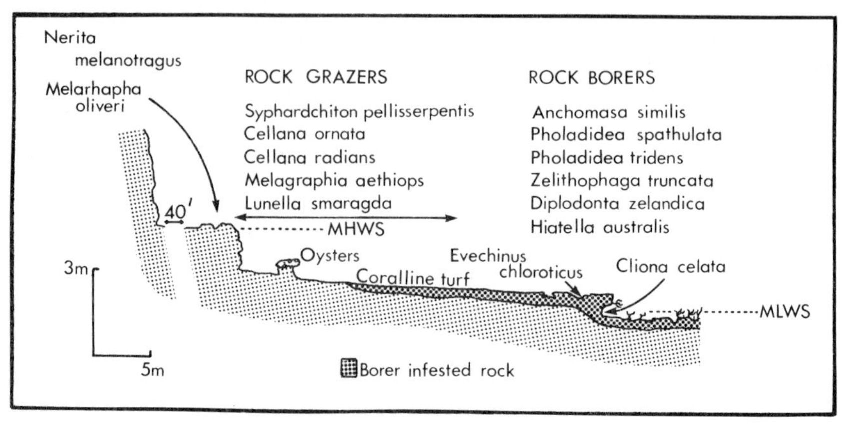

Figure 4.4. Distribution of bioerosional organisms on the shore platforms of the Whangaparaoa Peninsula, north of Auckland, New Zealand (Healy 1968b).

relative role of biological processes on different substrates in a variety of morphogenic environments.

Some pertinent information is available in the published literature (Table 4.3), but its reliability varies considerably. Some measures of bioerosional activity are simply estimates, which are often based upon debatable assumptions. Data which refer to the erosion rates associated with single organisms cannot be compared with rates which refer to the total biological contribution, or to the total rate of destruction accomplished by all erosional agencies. Further, the lifestyle of the organism must be considered: grazers tend to lower the whole rock surface, while the effect of borers is concentrated into very small areas. Other organisms, including some limpets, chitons, and sea urchins, erode fairly rapidly at their homesite, but they also graze over a much larger area (McLean 1974, Torunski 1979). It is also important to distinguish between the initially rapid rate of penetration, and the slower rates which may follow at a later stage. The time-dependency of boring rates has, for example, been documented for sponges (Rutzler 1975) and algae (Golubic *et al*. 1975). It is therefore difficult to derive conclusions from the diverse information which is presently available.

Population density, the environment, the characteristics of the substrate, and the effect of grazers make it particularly difficult to determine the boring rate of marine flora (Kobluk and Risk 1977). A number of factors also affect the erosional efficiency of single faunal elements. Animal size, the depth of algal penetration, rock hardness, and probably other factors such as porosity, grain size, and mineralogy, for example, affect the efficiency of gastropods (McLean 1967a). Assessment of the rate and role

Table 4.3. Rates of Bioerosion

Organism	Rate	Rock	Site	Elevation	Author	Comments
Algae (*Hyella*)	30–50 microns in 3–4 weeks	Carbonates			Golubic *et al*. 1975	Initial boring rate
Algae (*Hyella caespitosa*)	50 microns in 4 weeks	Icelandic spar			Le Campion–Alsumard 1975	
Algae (*Ostreobium*)	Several 100 microns in 253 days	Icelandic spar-calcite	Jamaica		Kobluk and Risk 1977	
Patella vulgata (limpet) and other gastropods	Up to 0.5 mm per yr.	Limestone	Plymouth UK	Intertidal	Southward 1964	
Patella	1.5 mm per yr.	Chalk	Dover UK		Hawkshaw 1878	
Patella coerulea	0.51–0.76 mm per yr.	Limestone	N. Adriatic	High tide	Torunski 1979	
Acmaea	1.51 mm per yr.	Beachrock	Barbados	Intertidal	McLean 1967a, & Trudgill 1976b	
Notoacmea onychitis and *Patelloida alticostata*	1 mm per yr.	Dune rock	W. Australia	Base of notch at mean sea level	Hodgkin 1964	Limpets and other agents
Limpets	up to 6 mm per yr.				Neumann 1968	Estimate
Siphonaria exulum and *Melanerita melantragus*	0.6–1 mm per yr.	Beachrock	Norfolk Isl. S. Pacific	Upper intertidal	Hodgkin 1964	Total erosion
Littorina scutulata and *Littorina planaxis*	0.6 mm per yr.	Sandstones	S. California	Pools at and above high tide	North 1954	0.25 mm yr^{-1} for *L. planaxis* alone
Littorina planaxis	0.07 mm per yr.	Calcareous sandstones	S. California	Shallow intertidal pools	Neumann 1966 Emery 1946	Predominant species
Mainly *Littorina neritoides*	0.25–1 mm per yr.	Limestone	N. Adriatic	Supra- and intertidal	Schneider 1976	

Table 4.3. Continued

Organism	Rate	Rock	Site	Elevation	Author	Comments
Mainly *Littorina neritoides*	0.07–0.13 mm per yr.	Limestone	N. Adriatic	Supratidal	Torunski 1979	
Nerita tesselata	0.1 mm per yr.	Beachrock	Barbados	Intertidal	McLean 1974	
Browsing snails	1 mm per yr.	Calcareous aeolianite	Puerto Rico		Kaye 1959	Maximum
Gastropod/ Chiton grazers	1–2 mm per yr.	Beachrock	Barbados	Intertidal	McLean 1974	
Amphineura	*c*. 6 mm per yr.				Neumann 1968	Estimate
Acanthopleura gemmata (chiton)	0.02–0.7 mm per yr.	Coral pinnacles	One Tree Isl. Great Barrier Reef	Upper notch	Trudgill 1983	Sedentary oyster pedestals
Lithophaga	9.1 mm per yr.	Limestones	Aldabra atoll	Intertidal notch	Trudgill 1976b	
Lithophaga lithophaga	0.01 mm per yr.	Dolomite/ oolitic limestone	N. Adriatic	Subtidal horizontal surface	Kleemann 1973	
Lithophaga lithophaga	0.15–0.40 mm per yr.	Dolomite/ oolitic limestone	N. Adriatic	Subtidal vertical surface	Kleemann 1973	
Lithophaga cumingiana and *Lithophaga obesa*	9 mm per yr.	Dolomite/ limestone	Oman	Intertidal notch	Vita–Finzi & Cornelius 1973	Rate for cutting boreholes
Lithophaga cumingiana and *Lithophaga obesa*	0.15–0.4 mm per yr.	Dolomite/ limestone	Oman	Intertidal notch	Vita–Finzi & Cornelius 1973	Rate for total notch retreat
Lithophaga cumingiana	15 mm per yr.	Beach sandstone	Great Barrier Reef	High tidal	Otter 1937	
Penitella penita	12 mm per yr.	Greywacke sandstone	Oregon	Intertidal	Evans 1968	
Pholas	13 mm per yr.	Cherty chalk	Norfolk UK	Subtidal	Reid 1907	

Pholas and *Barnea*	4 mm per yr.	Flysch	Sochi USSR	Subtidal	Nikitin 1951, see Zenkovitch 1967	
Petricola and *Barnea*	120 mm per yr.	Limestone	Sevastopol USSR		Zernov 1913, see Zenkovitch 1967	Blocks 'thrown into sea'
Cliona lampa	10–14 mm per yr.	Calcarenite	Bermuda	Subtidal notch	Neumann 1966, 1968	Maximum rate
Cliona lampa	0.1–1 mm per yr.	Icelandic spar-calcite	Bermuda	Notch and sublittoral	Rutzler 1975	
Lithotrya	8.4 mm per yr.	Limestone	Aldabra atoll	Intertidal notch	Trudgill 1976b	
Echinometra lucunter	49.3 mm per yr.	Beachrock	Barbados	Intertidal	McLean 1967b & Trudgill 1976	
Paracentrotus lividus (grazing)	1.1 mm per yr.	Limestone	N. Adriatic		Torunski 1979	
Paracentrotus lividus (boring)	1,000 mm per yr.	Limestone	France		Cailliaud 1865, see Torunski 1979	
Strongylocent-rotus lividus	10 or more mm per yr.	Limestone	Atlantic France		Cailliaud 1865, see Otter 1932	
Diadema antillarum	5.3 mm per yr.*	Coral	Barbados	Subtidal	Stearn and Scoffin 1977	
Echinoids	*c.* 6 mm per yr.				Neumann 1968	Estimate
Macroborers	0.8 mm per yr.*	Coral	Barbados	Subtidal	Stearn and Scoffin 1977	
Mainly fish	0.5 mm per yr.	Beachrock	Great Barrier Reef	Near low neap tide	Stephenson 1961	
Fish	1.3 mm per yr.	Calcarenite	Bermuda	3–4 m below low tide	Bromley 1978	
Fish	0.01 mm per yr.	Calcarenite	Bermuda	Subtidal	Gygi 1975	
Sparisoma viride	2.37 × 10E-2 to 9.9 × 10E-2 mm per yr.	Coral	Barbados	Subtidal	Frydl and Stearn 1978	
Total bioerosion	10 mm per yr.	Siltstone and sandstone	Northern New Zealand	Upper mid-littoral and superlittoral	Healy 1968b	
Total bioerosion	6–7 mm per yr.	Coral	Florida	Subtidal	Hudson 1977	Mainly sponges, also fish and sea urchins

*Calculated from author's data, using a value of 1.69 g cm^{-3} for rock density (density of Bermuda calcarenite provided by Neumann 1966).

of boring activities is rendered even more difficult by the fact that for many organisms, such as *Cliona* sponges, echinoids, *Polydora* annelid worms, *Hiatella*, and some *Petricola* bivalves, boring is faculative rather than obligatory (Otter 1932, Yonge 1963): for example, individuals of a particular species may be active borers where they require protection from desiccation or wave attack, but not where they occupy sheltered or tideless environments.

Several workers have found that bivalve molluscs rapidly penetrate calcareous substrates, and there is some evidence to suggest that their efficiency is matched by echinoids and *Lithotrya* barnacles. The boring rate of the important annelid and sipunculoid worms has not been determined, and that of the boring sponges may have been exaggerated. Neumann (1966) obtained a high figure for the boring rate of the *Cliona lampa* sponge, which has often been quoted by others. Neumann recognized that his figure reflected the maximum possible rate, which is attained during initial penetration; the much lower rates found by Rutzler (1975) may therefore be more representative of the true activity of the species (Schneider 1976, Moore and Shedd 1977). Although boring organisms have a direct erosional role in replacing solid rock with empty spaces, their contribution to the retreat of entire rock surfaces is dependent upon the removal of the weakened honeycombed material by wave action and other processes.

Boring organisms do not inevitably facilitate rock destruction. The rock is certainly weakened by the unlined borings of algae, porifera, and polychaete and sipunculoid worms, but borings which are lined with dense forms of calcium carbonate by bivalves such as *Lithophaga* and by some polychaetes bind the rock together, thereby increasing its resistance to disintegration (Otter 1937). Grazing organisms, in contrast, always contribute directly to the erosion of rock surfaces. Therefore, although the rates of erosion attributed to grazing molluscs are generally much lower than the corresponding figures for borers, the data are not comparable. This is because rates for the grazers usually refer to the lowering of entire rock surfaces by numerous individuals of a particular species, rather than to the rate of penetration of a single boring individual at one spot on the rock surface. The rate of destruction of a substrate also depends upon the number of organisms boring or grazing on it, as well as the activity of each individual. Thus, although the magnetite-capped teeth of chitons make them more efficient as individuals than most other organisms, they are not sufficiently numerous in most areas to make the contribution to erosion that has often been ascribed to them (Taylor and Way 1976).

A number of studies have found that intertidal limestones in several areas are eroding at the rate of about 1 mm yr^{-1}. Hodgkin (1964) has claimed that this value is characteristic of hard limestones in the tropics and subtropics, on both vertical and horizontal surfaces, whether exposed to the sun or in shaded sites, and irrespective of the degree of exposure to wave action. Later studies have recorded erosion rates for limestones in sheltered environments where biological activity is dominant which are broadly similar to Hodgkin's (Schneider 1976, Trudgill 1976b, Torunski 1979), but much higher rates have been recorded where wave action is more vigorous. It has been proposed that the erosion rate is determined by the maximum boring rate of endolithic microflora, which is about the same on all carbonate coasts despite variations in the nature of the substrate and in the grazing populations (Schneider and Torunski 1983).

Protective Marine Organisms

Marine flora and fauna also assume a protective role on rock coasts. A number of organisms contribute to the formation of organic crusts which protect the underlying rock from wave and physico-chemical attack in the lower intertidal and upper subtidal zones.

The Rhodophytic (red algae) Corallinaceae family contains most of the prominent carbonate-encrusting algae (Bosence 1983). Crustose coralline algae have often been indiscriminately classified as *Lithothanium*, although this genus is most suited to the subtidal and deep water zones of cool or cold Arctic waters. *Tenaria* colonizes shallow subtidal warm waters, such as the Mediterranean, and *Lithophyllum, Porolithon* and *Neogoniolithon* the intertidal and shallow subtidal zones of tropical waters, although the latter genus may extend down to somewhat greater depths. Genera such as *Melobesia* and *Jania* are able to live in temperate and tropical areas (Adey and MacIntyre 1973). The Chlorophytic Codiaceae family, of which the best known genus is *Halimeda*, are important in sheltered tropical lagoonal environments (Johnson 1961, Wray 1977). *Halimeda*, for example, is particularly abundant in Florida, Bermuda, in the eastern Bahamas, off Yucatan, and elsewhere in Atlantic algal reefs and reef complexes (see, for example Ginsburg *et al*. 1972).

Vermetid gastropods, including *Vermetus triquetrus* and *Dendropoma petraeum*, Serpulidae and Sabellariidae polychaetes, cirrepeds such as *Tetraclita squamosa*, pelecypods such as *Mytilus*, *Brachyodontes*, *Lasaea* and *Saccostrea*, zooanthid, hydroid and ascidian/tunicate Coelenterates and other fauna, can be associated with algal crusts (Safriel 1974, Trudgill 1976b, Pérès 1968), which form corniches projecting outwards from steep cliff faces, or trottoirs or ledges resulting from differential erosion of the cliff above the level protected by the encrustations (see Chapter 10).

Focke (1978) found that there is little variation in the distribution of encrusting organisms on the sheltered and exposed coasts of Curaçao in the Netherlands Antilles. On the sheltered coasts, however, the crusts are quite thin, and the substrate is intensely bored. On the more exposed trade wind coasts, the much thicker crusts have been altered by marine diagenesis. The prominence of crustose coralline ledge coatings on the exposed portions of outer coasts, which has been noted elsewhere (Adey and MacIntyre 1973), may be because of increased nutrient supply in turbulent water, the pumping of sea water through the accretions, or both. Lithification of this crust provides effective protection to the substrate against wave erosion and biodegredation.

Any dense organic cover or mat can prevent the drying out of a rock surface at the low tidal level, thereby providing protection to a surface which would otherwise be susceptible to water layer levelling (Hills 1949, 1971) (see Chapter 2). Individual organisms can also protect the immediately underlying material. Encrusting barnacles, such as *Tetraclita squamosa*, can occupy residual rock pinnacles which they have apparently protected from the erosional processes reducing the surrounding surfaces (MacFadyen 1930, Hodgkin 1964). Large sublittoral brown algae can function as a baffle to incoming waves at the low tidal level. The common *Fucus* genus, however, is probably fairly unimportant in this regard, since species such as *Fucus serratus* do not flourish in the lower intertidal zone where wave action is vigorous, or where there are large populations of grazing limpets. Large species of genera such as *Himanthalia*, *Laminaria*, and *Saccorhiza* may be important (Guilcher 1958a). It has been shown that as much as 20% of a wave's energy is required to bend artificial sea grass (Wayne 1974, Kirk 1977), but strong wave action can dislodge algae, together with portions of the rock to which they are attached (Bertram 1936, Healy 1968b). The strength of the attachment between seaweeds and the rock stratum (Barnes and Topinka 1969) is such that rocks as heavy as 9 kg have been thrown onto shore platforms during storms, and the very buoyancy of the floating giant kelp *Macrocystis* can float boulders free of the sea bed (Yonge 1936, Edwards 1941, 1951, Emery 1963).

5

Frost and Related Mechanisms

The assumption is often made in geomorphological texts that rocks in cold climates are split and shattered by the alternate freezing and thawing of water (White 1976). There is no doubt that the approximately 9% expansion which accompanies the change in phase from water to ice is capable of generating considerable pressures against confining walls (Bridgman 1912, 1914). Under ideal laboratory conditions, at −5°C ice can exert a maximum pressure of 62 MPa. This increases to 115 MPa at −10°C, and 214 MPa at −22°C. Any further drop in temperature provides only a very slight increase in pressure, to a maximum of 215 MPa at −40°C (Winkler 1973). Rocks are much weaker in tension than in compression, and although the tensile strength of saturated rocks increases as the temperature decreases (Mellor 1971), no rock would be able to withstand the pressures which could be generated under ideal conditions. Stresses which are less than one-tenth the theoretical maximum exceed the tensile strength of even the most resistant rocks (Hardy and Jayaraman 1970, Rzhevsky and Novik 1971). These maximum values, however, probably do not obtain in the field. Grawe (1936) objected to the uncritical acceptance of the view that rock weathering in cold climates is primarily the result of alternate freezing and thawing of contained water. He noted that the attainment of maximum ice pressures requires a closed system containing only water, temperatures as low as −22°C, and rock which is strong enough to resist the pressure. It has been suggested that a closed system can be produced in a rock crevice by rapid freezing of the surface water layers, thereby forming an ice plug (Battle 1960, Pissart 1970). Even if this closes the system, high pressures can only be generated if the frost penetrates deeply enough to freeze all the contained water (Tricart 1970).

It is evident that many types of rock are severely damaged by their exposure to air at temperatures which fluctuate about the freezing point, but the mechanisms which accomplish this deterioration are still the subject of considerable debate (McGreevy 1981). A distinction should be made between the damage induced by ice-filled capillaries and the shattering caused by the freezing of water contained in joints, bedding planes, and other rock crevices (Tricart 1956, Wilson 1968). Tricart argued that although it is more difficult to freeze the water in a rock crevice, because of its volume, the potential for rock damage is greater than that which result from the freezing of water in rock pores. Most of the literature on frost action is concerned with rocks in cool, freshwater environments, or with concrete and other building materials in fresh and saline water (McGreevy and Whalley 1984). There has been little work on the role of frost action in cool coastal areas, but some inferences can be made by reference to the periglacial and engineering literature. It is therefore necessary to first consider the literature which deals with the mechanisms and efficacy of frost action.

The Mechanisms of Frost Action

Collins (1944) suggested that frost damage to concrete is the result of the growth of ice crystals and the segregation and concentration of the ice into layers parallel to the cooling surface. This mechanism is similar to that proposed as an explanation for frost heave in soils (Taber 1930), and for the deterioration of porous brick (Gill and Thomas 1939). The growth of ice lenses is probably possible only in rocks with a very low tensile strength. Macroscopic ice segregation, for example, only occurs in concretes which are of very low quality, or which are in the first stages of hydration (Nerenst 1960). Ice lenses, however, have been found to develop in saturated mortar which is subjected to prolonged periods of freezing interrupted by rapid thawing (Cordon 1966). Ice segregations have also been reported in shales and other rocks which have numerous planes of weakness and low tensile strength (see for example, Potts, 1970). Conditions which facilitate ice segregation include rocks which have low tensile strength and high porosity, with a proportion of small voids; access to unfrozen water on the side away from the freezing front; rock temperatures several degrees below freezing point; and a period of freezing long enough to permit the slow growth of the ice lenses. Collins (1944) considered that waterline exposures are particularly suitable for ice segregation.

Although Powers (1945) doubted that ice lenses could develop at the macroscopic scale in concretes because of its tensile strength, he did suggest that segregation could occur at the submicroscopic scale. Water which is contained in very small capillaries is virtually unfreezable. In theory, ice cannot form if the vapour pressure over a water meniscus is less than that over ice at the same temperature. The degree to which water can be supercooled in a capillary is determined by the size of the capillary. If it is assumed that the capillary is round, the relative pressure (p/p_0) is given by Lord Kelvin's equation (Scheideggar 1957):

$$\ln (p/p_0) = 2\gamma M/\rho rRT$$

where p is the pressure over a concave water meniscus; p_0 is the saturated vapour pressure; T is the temperature (°K) and γ the surface tension of the water at that temperature; M is the molecular weight of water and ρ is its density; R is the gas constant; and r is the radius of the capillary. The equation demonstrates that the relative vapour pressure increases as the size of the capillary decreases; this causes the freezing point to be depressed in small capillaries. As an example, in a capillary with a radius of 0.01 microns, water has a relative vapour pressure of 0.90. As this is equivalent to the relative vapour pressure of ice and supercooled water at $-11°C$, water will remain unfrozen in all smaller capillaries at this temperature (Hudec 1973).

Because of supercooling, the growth of ice crystals in the larger capillaries can be sustained by the diffusion of unfrozen water from the smaller voids (Powers and Brownyard 1947, Powers and Helmuth 1953). This mechanism may account for the commonly observed postfreezing dilation of cement pastes, and the flow of water towards the freezing sites. In rock, Powers (1955, 1975) has argued that because the capillaries are generally much larger than those in cement paste, most water freezes within a narrow range of temperatures near the normal melting point. He believed there would be very little unfrozen water available to sustain crystal growth after the initial freeze. An external source of unfrozen water may permit continued dilation (Taber 1950), although Thomas's (1938) experimental results suggest that it is not essential in all cases. Powers considered that the conditions necessary for crystal growth are rarely satisfied in the field, although he suggested that it can occur in some very fine-grained rocks during slow freezing (Powers 1975).

Powers (1945, 1949, 1965) proposed that the destruction of concretes and rock aggregates is accomplished by hydraulic pressures. They are generated by the flow of water away from an advancing ice front, where it is displaced by the growth of ice crystals. The magnitude of these pressures is determined by the porosity and permeability of the material; the rate of freezing, which controls the rate of flow; the amount of water which must be expelled from the freezing zone; and the distance the water must travel to reach a joint, bedding plane, or other escape boundary. The degree of saturation of the rock is particularly important: if an interconnected void system is less than 91.7% saturated, assuming a uniform distribution of water, crystal growth could be accommodated without expulsion of water, and without the generation of destructive hydraulic pressures. Fagerlund (1975) pointed out that although there is a well-defined critical saturation level for all porous materials, the value is different for different materials, and is considerably less than 91.7% in some cases. Nevertheless, the greater the degree of saturation above the critical level, the greater the amount of water which must be expelled, and the greater the pressures which will be exerted against confining capillary walls (Powers 1955). High hydraulic pressures may be generated if the exterior of the rock is frozen, thereby inhibiting expulsion of the excess water. Verbeck and Landgren (1960), however, have made the important point that complete saturation is generally very difficult to attain under natural circumstances. If the rock is undersaturated, it may survive the freezing season without ever having been subject to damaging hydraulic pressures. This is likely to be a very common situation in the field. Mellor (1973) provided further support for the hydraulic pressure mechanism, although he noted that water will freeze if it flows into large capillaries or crevices in the rock. Ice contained in the capillaries of cold rock will undergo phase transformation to polymorphs of higher density when subjected to high hydrostatic pressures.

Litvan (1972a,b, 1973a, 1976) argued that the adsorbed water, or water contained in the fine capillaries of porous solids, cannot freeze without redistribution. When the temperature falls below the bulk freezing point, ice crystals form on the exterior of a rock, while liquid water is supercooled inside. The vapour pressure of water which is cooled below the bulk freezing point is greater than that of the ice on the exterior surfaces. Because the equilibrium state cannot be restored by solidification of the adsorbed, supercooled water, water must be expelled to the larger capillaries or to the external boundaries of the solid to reduce the vapour pressure to that of the ice on the exterior. The difference between the vapour pressure of the water and ice increases with decreasing temperature, so the amount of water which must be expelled from a rock increases as the temperature continues to fall below the bulk freezing point. Rocks could be damaged by the freezing of water expelled into crevices, or, if the water cannot be expelled fast enough, by the generation of high hydraulic pressures. If the distance to the nearest escape boundary makes it impossible to expel water at the required rate, hydraulic pressures may crack the rock, providing new avenues of escape. Water then migrates to these cracks, desiccating the rock mass, and, as it freezes, exerting pressures which serve to propagate these new fractures. The mechanism is facilitated by rapid cooling, high levels of saturation, and low permeability. Very low porosity reduces the level of saturation. A rock with a combination of very large and very small capillaries is probably more durable than one which has the same overall porosity, but largely contains capillaries of medium size. This is because large capillaries are seldom full of water, and the water contained in small capillaries can only be frozen at very low temperatures. The minimum temperature determines the proportion of the capillaries which can contribute to the freezing process. The freezing rate is also crucial. If a rock freezes slowly, desorption can take place without generating high hydraulic pressures. Alternatively, a drop in temperature of only a few degrees, if accomplished very quickly, can cause the rock to be damaged.

Litvan's thesis has not gained general acceptance. Everett (1961) has shown that two phases can exist simultaneously, with their corresponding vapour pressures. This is because the pressure of ice in a capillary which also contains water is raised by the formation of a curved meniscus or interface. It has been argued, therefore, that frost damage can be explained by hydraulic pressures generated by water migrating from the smaller capillaries to ice crystals nucleated in the larger voids, and by the pressures generated by the growth of these crystals (Beaudoin and MacInnes 1974).

Temperature-dependent Wetting and Drying

The formation of ice, whether on the internal or external surfaces of rocks, may not be necessary to account for the damage which is normally ascribed to frost action. It has been suggested that damage can result from temperature-dependent wetting and drying, and the pressures associated with the adsorption of water molecules onto the surfaces of small rock capillaries (Dunn and Hudec 1965, 1966).

Mineral surfaces possess a charge which is related to unequal molecular attraction and broken or unsatisfied bonds (Hudec 1977). These charged surfaces exert an electrostatic attraction on other dipolar or ionic molecules. Water is a highly polar fluid which is attracted to mineral surfaces by Van der Waals forces. Water molecules have two positive hydrogen ions at one end, and a single negatively charged oxygen ion at the other (Fig. 5.1). When a surface is wetted, existing non-ionic bonds may be weakened or replaced by the insertion of water molecules (Vos and Moddle 1976). This weakens the structural cohesion of the substance, and produces an ordered stacking of the water molecules. Ordered water (also referred to as adsorbed, sorbed, or interfacial water, or as the electric double layer) is the result of the attraction between the positively charged ends of the water molecules and the negatively charged surfaces (Anderson

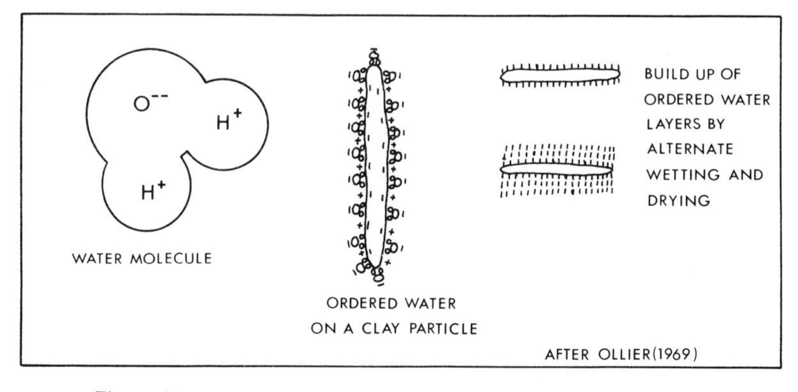

Figure 5.1. The adsorption of water molecules by clay particles.

and Hoekstra 1965, Scott 1969, Anderson and Morgenstern 1973, Anderson *et al.* 1974). This attraction causes the water molecules to become oriented against the mineral surfaces like numerous small bar magnets (Anon 1966). Although the electrostatic attraction declines exponentially with the distance from the capillary walls (Ruiz 1962), alternate wetting and drying can produce an increasingly ordered, rigid layer. These layers can be several molecules thick, according to the polarity of the molecules, the strength of the Van der Waals forces, the temperature, and the relative humidity (Hudec 1973, 1977). According to the Brunauer–Emmett Teller equation, the ordered water tendency increases with decreasing temperature (Brunauer 1943, Cady 1969).

Adsorbed water has a lower vapour pressure than normal water, and is virtually unfreezable down to temperatures of −40°C or more (Dunn and Hudec 1965, 1966, Litvan 1972a,b). However, the growth of unfrozen but rigid, ordered, water layers exerts damaging pressures against confining walls. This may be enhanced by the generation of repulsive forces associated with like charges at the free boundaries of the adsorbed layers (Fahey 1983). A study of the durability of carbonate rocks found that the greatest deterioration under freeze–thaw conditions occurred in the specimens containing the largest amounts of unfreezable, adsorbed water (Dunn and Hudec 1972). The internal surface area, pore size, and the pore size distribution have an important bearing on the amount and type of water contained in a rock. The greater the internal surface area, the greater the amount of water which can be contained in the adsorbed state. A given porosity can be the result of either a small number of large pores, or a large number of small pores. In the former case the internal surface area is small, whereas in the latter it is large. The susceptibility of a rock to weathering has been found to be directly related to the internal surface area and the average pore size (Hudec 1978a). If the pore is small, the rigid ordered layers may fill it and exert pressure against the confining walls.

Clay minerals provide very high internal areas in shales, siltstones, some dolomites, and other fine-grained argillaceous rocks. Clay particles have a high surface-to-volume ratio, and because of imperfections and atomic substitution in the clay lattice, they have positive edge charges and negative face charges (Fig. 5.1). As the ratio of the edge to the plane is small, clays possess a net negative charge (Wayman 1967) and water is adsorbed by clay surfaces by hydrogen bonding. The expansion of carbonate rocks when wetted is determined by their clay and chert content (Hudec 1977). In some clay-rich rocks, isothermal expansion is equivalent to a thermal expansion of 230°C (Hudec 1980a).

Helmuth's (1960) conclusion that adsorbed water contracts on cooling led Powers (1975) to suggest that Dunn and Hudec (1965, 1966) had inadvertently measured the effects of osmotic pressure and clay particle swelling rather than the dilation induced by adsorbed water. Anderson and Hoekstra (1965) and Hudec (1973, 1980a) found that the number of ordered water layers increases with rising temperatures. Hudec (1977) determined that the maximum amount of adsorbed water occurs at temperatures of between 20 and 25°C. Hudec's view that fine-grained argillaceous rocks adsorb and expand as temperatures rise, and desorb and contract as temperatures fall, is in accordance with the known behaviour of adsorbed water molecules. Temperature-dependent wetting and drying ultimately causes rocks to fail as a result of fatigue. Freezing can therefore be considered as a drying process, so that freezing and thawing is simply a special case of wetting and drying, which occurs about the freezing point (Hudec 1980a). When the temperature falls, the number of adsorbed water layers declines in saturated sorption-sensitive rocks. Water is expelled into the larger voids or cracks; being no longer adsorbed, it freezes, thereby contributing to the breakdown of the rock by normal frost action. Alternate freezing and thawing and wetting and drying may therefore operate together (Hudec 1973).

Larsen and Cady (1969) and Cady (1969) have compared Power's hydraulic pressure model with Dunn and Hudec's adsorbed water model. They suggested that the expansion of concrete specimens as heat is released by freezing provides support for the hydraulic pressure model. They also found, however, that dilation continues after freezing, a phenomenon which is difficult to reconcile with the hydraulic pressure mechanism. To account for postfreezing dilation, they proposed that bulk water which remains unfrozen becomes ordered. The increase in volume associated with this ordered water generates hydraulic pressures. Dunn and Hudec (1972) acknowledged that Larsen and Cady's modification of their model may be valid, but they questioned whether frozen or ordered water is sufficiently mobile to be able to generate hydraulic pressures.

Larsen and Cady (1969) proposed that the pore and sorption characteristics of a porous solid determine whether the freezing–hydraulic pressure mechanism or the ordered water–hydraulic pressure mechanism is dominant. They noted that Dunn and Hudec used a very narrow range of carbonate rocks, and they pointed out that clay-rich dolomitic limestones are unusually susceptible to the ordering of water molecules. Because the Van der Waals forces can only operate over very small distances, the ordered water hypothesis is only applicable to rocks with very fine capillary systems. Mellor (1970), for example, found that 5–7% of the water taken up by sandstone specimens was adsorbed, compared with 45% by granites and 90% by shales.

Dunn and Hudec (1972) and Hudec (1973) classified rocks as being sound, frost-sensitive, or sorption-sensitive, according to their degree of saturation after 24 hours' immersion in water; the amount of water adsorbed at 100% relative humidity; and the additional amount of water taken on upon immersion. Sorption-sensitive carbonates were found to be those which critically saturate with mainly adsorbed, unfreezable water. True frost-sensitive carbonates also critically saturate, but with mainly bulk, freezable water. Larsen and Cady (1969), however, found that the parameters used by Dunn and Hudec did not describe the performance of other carbonate and non-carbonate rocks. In an experimental study, it was found that hydration accounted for about one-quarter of the production of fines from schist aggregates produced across a freezing cycle (Fahey 1983).

Dunn and Hudec (1972) recognized the special characteristics of those rocks which deteriorate as a result of the ordered water mechanism. They acknowledged that clay rejection during dolomitization is a special situation, which produces rock textures unlike those of other rock types. Nevertheless, Hudec (1973) and Hudec and Sitar (1975) considered that true frost action is a minor destructive mechanism in fine-grained sedimentary rocks, which constitute the majority of the continental rocks exposed to weathering. True frost action, which requires that ice forms on the internal or external surfaces of a rock, may be restricted to coarse-grained lithologies with large capillaries which can fill with bulk, freezable water. Hudec and Sitar (1975) have argued that not only is true frost action limited to fewer rock types than is temperature-dependent wetting and drying, but it is also restricted in the field by the need for a sufficient supply of water to attain critical saturation of the rock. Sorption-sensitive rocks can saturate under conditions of high humidity. The sorption–desorption mechanism could explain the apparent efficacy of 'frost' action in areas which rarely experience freezing temperatures (Hudec 1978b).

Osmotic Pressures

Rocks do not usually saturate completely when they are immersed in water because of air bubbles trapped in the capillaries. Osmosis takes place when vapour is transported across the air bubble from an area of high vapour pressure in a large capillary to a smaller capillary where the pressure is lower. The osmotic pressure generated within the smaller capillary is proportional to the difference in the vapour pressures between the menisci of the two capillaries. The osmotic pressure, therefore, is sufficient to equalize the pressures in the capillaries. Adsorbed water is drawn between rock grains by ice nucleation, or by the osmotic action of the ordered water layers (White 1976), which, being much denser and about 20 times more viscous than ordinary water, can function as an ionic screen or membrane. The adsorption of exchangeable ions to clay crystals (Fig. 5.2) can also create osmotic pressures in shales and argillaceous carbonates (Winkler 1973). Isothermal expansion of wetted carbonates and shales with a large

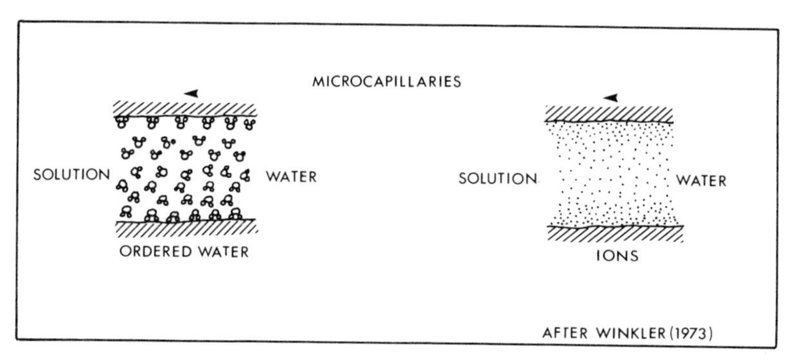

Figure 5.2. Osmosis in rock capillaries and the role of ordered water and adsorbed, exchangeable ions.

proportion of small capillaries may be partly a result of osmotic pressures (Hudec 1977, 1980b, 1982).

Osmotic pressures are also generated when saline solutions are frozen. When water in the larger capillaries of a rock begins to freeze, the salinity of the unfrozen solution increases. The unfrozen water which is in these large capillaries therefore has a higher salt concentration than the unfrozen solution in the smaller capillaries. This produces a pressure differential in the opposite direction to that of the flowing water, which permits further growth of the ice crystals. It is generally thought that osmotic pressures supplement the hydraulic pressures induced by freezing (Powers 1955, Cordon 1966), although in cement pastes, and possibly in very fine-grained rocks, frost damage can result from localized osmotic pressures generated at the sites of ice formation (Helmuth 1960). Winkler and Singer (1972) and Winkler (1973) have calculated the maximum osmotic pressures which can be exerted by different solute concentrations for a variety of solutions, and temperatures. The osmotic pressures exerted by NaCl solutions increase particularly rapidly with increases in concentration, and decrease with temperature; although the effect of temperature variations is very slight within the range of concentrations to be expected of sea water. In coastal environments, osmotic pressures related to the concentration of salts near the rock surface may cause scaling, a phenomenon which is common on concretes exposed to de-icing salts.

The Effect of Salts

The presence of salts affects the freezing point of solutions as well as the osmotic pressures generated during freezing. Salts lower the freezing point of water and probably reduce the number of freeze–thaw cycles which can occur in a given period (Dunn and Hudec 1965, 1972, Larsen and Cady

1967, Litvan 1975), but the effect of salts on the efficacy of frost action remains to be determined.

It has been found that the occurrence of salts in solution inhibits frost action, if the salt supply is limited and the amount of salt is kept essentially constant (McGreevy 1982). This may partly explain why saline solutions facilitate frost weathering in some cases but not in others (eg. Goudie 1974, Williams and Robinson 1981, Fahey 1985). Although any reduction in ice formation would cause a corresponding reduction in the hydraulic pressures generated, the lower freezing point also provides longer periods of thaw, during which water from an outside source could be absorbed. This would raise the saturation of a rock to a higher level than could be attained in a freshwater environment (Powers 1945). Several workers have found that the saturation level can also be raised directly by the presence of salts (Litvan 1976, MacInnes and Whiting 1979, Hudec 1980b), although McGreevy (1982) found that the saturation level was about 10% greater with water than with a NaCl solution. Drying or freezing a rock will concentrate a solution into progressively smaller capillaries, increasing the capillary tension (Litvan 1975, Hudec 1977). Concrete retains more adsorbed water at all relative humidities after being saturated by a solution of NaCl than when it is saturated in distilled water (Whiting 1974). The ability of carbonates to absorb water upon immersion is about 13% greater in a 3% NaCl solution than in fresh water. Increases in absorbed and adsorbed water in the presence of salts can be attributed to the hydration of salt ions and to the ion exchange phenomenon on the internal surfaces of the rock (Hudec and Rigbey 1976).

The presence of salts in coastal environments may accelerate the deterioration of rocks (Tricart 1970, Williams and Robinson 1981), although it has been suggested that it reduces the efficacy of frost action by making the ice cellular and rather soft (Guilcher 1958a). Several studies have indicated that the greatest deterioration is accomplished by solutions which contain 2 to 6% of their weight in salt (Arnfeld 1943, Verbeck and Klieger 1957, Litvan 1975, 1976, Browne and Cady 1975, MacInnes and Whiting 1979, Trenhaile and Rudakas 1981, McGreevy 1982). This may indicate the presence of a threshold for NaCl solutions, and possibly for other salts (McGreevy 1982), related in part to the increased saturation and broadening of the temperature range of freezing (Litvan 1973b). These effects oppose each other, providing a complex relationship between the level of salinity and the rate of rock deterioration.

Experimental Work

Experimental studies have generally failed to distinguish between deterioration of specimens as a result of ice formation, and deterioration caused by wetting and drying about the freezing point. Engineers and geologists have attempted to determine the durability of concretes, bricks, tiles, and rock aggregates, often when saturated by de-icing salt solutions. With the exception of Hogbom's (1899) work, however, it was not until the 1950s that experiments were specifically designed to consider the geomorphological effects of frost action (Tricart 1956, Masseport 1959). So many research designs have been employed that it is extremely difficult to compare the results of these studies. Even when standardized tests are used, such as those of the American Society for Testing Materials (ASTM), the severity of the experimental conditions is usually so great that it is impossible to relate the results to natural environments. Even those geomorphological studies which have attempted to simulate natural freeze–thaw cycles are concerned with freshwater environments rather than with the saline conditions of coastal regions. Furthermore, the use of cores, cubes, and other unconfined samples in most experimental work facilitates expansion of the specimens, and expulsion of the excess water during freezing. In the field, this may be impeded by the presence of surrounding material, which could cause destructive pressures to be generated in the rock.

The use of air temperature fluctuations to represent freeze–thaw cycles provides a further source of error. This can involve large discrepancies between the frequency and intensity of freeze–thaw cycles at and beneath a rock surface, and in the air above (Thorn 1982, McGreevy 1985b). Unfortunately, only a few measurements have been made of bedrock temperatures (Gardner 1969, Thorn 1979, Hall 1980). Extreme examples of the difference in air and rock temperatures probably occur in climates in which there are many low-amplitude fluctuations about the freezing temperature (Washburn 1973). Insulation on dark rock surfaces can induce thawing when air temperatures are below zero. Other factors—such as the damping of temperature fluctuations with depth within the rock, and the insulating effect of snow and vegetation on the surface—must also be considered.

Laboratory experiments have been largely concerned with the effects of frost in intact rock or concrete cores, rather than with frost riving associated with joints and macrocracks (McGreevy and Whalley 1982). These studies have therefore emphasized the role of capillary structure, particularly the size distribution and continuity, in determining the water absorption and retention properties of a rock. Capillary structure influences the degree of saturation which can be attained under given conditions, and consequently a rock's susceptibility to frost attack (Sweet 1948, Guillien 1949, Lewis *et al*. 1953). Experiments have confirmed that the susceptibility of a rock varies according to the amount of water which is available to it (Guillien and Lautridou 1970; Potts 1970; Douglas 1972 cited by Embleton and King 1975, p. 9; Trenhaile and Rudakas 1981). In general, the greater the degree of immersion of a rock sample in water, the greater is its deterioration. Many experiments have grossly exaggerated the efficacy of frost action as rock specimens were saturated before being exposed to frost action, although as has been previously noted, complete saturation of a rock is very difficult to achieve under natural conditions. In the Colorado Front Range for example, moisture-rich areas associated with seasonal snow patches lack adequate freezing intensity, while adequately frozen snow-free vertical sites lack sufficient moisture (Thorn 1979).

Experimental studies have also considered the effect of various freeze–thaw cycles on rock breakdown. The results, however, are contradictory. Thomas (1938), for example, found that two rock types may reverse their relative susceptibilities to frost action under different cycles. Mellor (1970) found that the bulk freezing strain in a rock increases as the freezing rate increases, and most workers have found that rapid freezing is more deleterious to rock samples than slow freezing (Thomas 1938, Battle 1960, Wiman 1963, Arni 1966, Potts 1970, Lautridou 1978). Several workers, however, have found that the greatest damage occurs with slow rates of freezing, despite the fact that there can be far fewer slow cycles in a given time period than fast cycles (Taber 1950; Tricart 1956; Williams 1969 cited by Sparks 1971, p. 374; Williams and Robinson 1981).

The use of standardized methods has not eliminated contradictions between the results of different workers. For example, several attempts have been made to compare the severity of four freeze–thaw tests on concrete specimens (ASTM 1952, 1953). Arni *et al.* (1956) found that the slow freezing and thawing in water test (C 292–54T) was the most severe in terms of the breakdown per unit time. Flack (1957) found that the slow freeze in air–thaw in water cycle (C 310–53T) was the most severe test on a per cycle basis, whereas Arni *et al.* concluded that it was the least destructive. Flack found that the second most severe test was the rapid freeze in air–thaw in water cycle (C 310–53T), followed in turn by the slow freeze and thaw in water cycle, and the rapid freeze and thaw in water cycle. Flack's data appear to support Arni *et al.*'s conclusion that the rapid cycles are the most severe in terms of the deterioration accomplished in a specific time period. Flack noted that on a cycle-by-cycle basis, the length of the cycle and the amplitude of the temperature fluctuations had little effect on the amount of deterioration accomplished in each test. Of far greater significance was whether air or water was used as the freezing medium, the former being more severe. Lautridou (1971) also found that at temperatures below −5°C, the length of the freezing period has no effect on the breakdown of saturated rocks. At temperatures between 0 and −5°C, significant damage only occurs if the freezing period is at least 10 hours. Anon (1959) found that freezing and thawing in water is more severe than freezing in air and thawing in water, contrary to Arni *et al.*'s conclusion. They found there was little difference between the deterioration per cycle of the fast and slow water freeze methods, although the rapid air freeze method was more severe than the slow air freeze method. The results also support the view that rapid cycles are the most severe per unit time.

However, even the 'standardized' ASTM tests permit a wide range of experimental conditions, particularly with regard to the rate of freezing: for example, there is no restriction on the maximum rate of cooling in the rapid air freeze–water thaw test, and insufficient control of the freezing rate in the rapid freeze and thaw in water test. Cooling rates may therefore differ quite markedly between different laboratories (Powers 1955). Partly because of these difficulties, these tests have been superceded by C 666–77, which includes both the rapid freezing in air–thawing in water and the rapid freezing and thawing in water tests (ASTM 1980).

Because of differences in experimental design, and in the collection and preparation of rock samples, experimental data can only provide a very rough basis for comparison of the susceptibility of various rock types to frost action. Rock durability is particularly sensitive to the degree of saturation and to the characteristics of the freeze–thaw cycle which is used. Furthermore, considerable variation in texture, mineralogy, in the size and continuity of the capillaries, and in the number and prominence of planes of weakness produces considerable variation in the durability of apparently similar rock specimens.

The susceptibility of many rocks depends upon whether they were previously chemically weathered (Martini 1967). In the field, rock dip and the characteristics of joint systems and other structural factors also affect the water retention properties of a rock outcrop, and consequently its durability. These factors make it extremely difficult to classify the susceptibility of rock types to frost action in the field. Nevertheless, the results of experimental studies are generally consistent with field observations (Alexandre 1958, Waters 1964).

Particularly susceptible rocks have usually been found to include well-foliated, fine-grained metamorphic rocks such as slates, phyllites and argillites; fine-grained quartzites; fine-grained low-rank greywackes; fissile, fine-grained shales, siltstones, claystones and friable chalks; and other fissile sedimentary rocks containing mica or illite clays, and soft fibrous serpentinites. Unfoliated metamorphics (such as medium and coarse-grained quartzites), very fine-grained, low porosity lithographic and other crystalline limestones, and igneous rocks such as basalts, coarse-grained granites, peridotites and pyroxenites have generally been listed among the most durable rock types. Moderately coarse-grained schists and crudely foliated, coarse-grained gneiss, fine to medium-grained granites, porphorytic granites, mudstones, greywackes, and other sandstones usually occupy intermediate positions in the scale of frost susceptibilities (Tricart 1956, Wiman 1963, Godard and Houel-Gangloff 1965, Rognon *et al.* 1967, Martini 1967, 1973, Stewart 1970, Potts 1970, Keeble 1971, Lautridou and Coutard 1971, Goudie 1974, Aubry and Lautridou 1974, Charre and Lautridou 1975, Trenhaile and Rudakas 1981, Swantesson 1985). It appears that, in a general sense, if the same amount of water is available to all rock types, the rate of deterioration will increase with decreasing grain size and tensile strength, and with the number and prominence of lines of weakness in the rock.

Frost Action in Coastal Environments

Although there is little agreement about the mechanisms which cause the breakdown of rocks in cool climates, there is general acceptance of the factors which facilitate this deterioration. Porous rocks which can be critically saturated with weakly saline solutions are particularly susceptible. The least durable rocks are generally fine-grained, often with numerous planes of weakness and low tensile strength. Low permeability and rapid freezing are conducive to the generation of high hydraulic pressures, but slow freezing is necessary for the disruptive growth of ice crystals or lenses. Processes which are directly or indirectly dependent upon the formation of ice in the rock may be restricted to coarse-grained lithologies with large capillaries. The deterioration of fine-grained argillaceous rocks could be the result of alternate wetting and drying, although the process is enhanced by temperature fluctuations within the freezing and thawing range.

Coastal environments may provide nearly optimum conditions for these weathering processes. Coastal engineers recognize the severity of conditions in the intertidal zone (Cook 1952, Kennedy and Mather 1953, Gjorv 1965, ACI 1980), which can be as severe as any to which concretes are exposed (Mather 1969). The saturation level in coastal rocks can be high because of a ready supply of water. It has been reported that frost action is particularly intense on salt- and freshwater coasts where rocks are frequently wetted (Taber 1950, Mackay 1963). Taber found that the presence of a mud matrix permits high levels of saturation to be maintained in beach gravels during freezing, and Martini (1975) thought that an underlayer of sand performs a similar function.

In cool regions, rocks in the intertidal zone can freeze in air during low tidal periods, and thaw in water during high tides. In coastal Maine, for example, concrete specimens placed at the mid-tidal level were alternatively immersed in water at about 2.8°C and exposed to air at between −2 and −23°C (Cook 1952). Rocks in the intertidal zone are frozen much more frequently than those which only experience fluctuations in air temperatures. In coastal Maine, more than 200 freeze–thaw alternations can occur in one year in the intertidal zone (average 133), compared with only 30 to 40 inland (Kennedy and Mather 1953). If the tidal regime is semi-diurnal, thawing of intertidal rocks in water can provide two freeze–thaw cycles every 24 hours. Not only are freeze–thaw alternations more frequent in the intertidal zone than inland, but the rapid change in temperature induced by sudden emergence or submergence provides particularly severe conditions which have few terrestrial counterparts. Furthermore, the salinity of sea water is within the range which has been found to be most deleterious to rocks. These factors suggest that frost action must assume an important role in the development of rock coasts in cool environments.

Frost action is unlikely to operate with uniform intensity throughout the coastal zone. The characteristics of freeze–thaw cycles, and the amount of time available to a rock to absorb water, are determined by the tidal regime and by the rock's position within the intertidal zone. The number of freeze–thaw cycles is probably greatest near the high tidal level. This is because the upper portion of the intertidal zone is exposed to the air for the greatest amount of time. This area therefore experiences most of the cycles related to air temperature fluctuations, as well as those which are induced by tidal ebb and flow. The minimum and maximum temperatures of the latter cycles are determined by the air and water temperatures, and are therefore essentially the same for all sites between the high and low tidal levels. Near the high tidal level, the rocks are immersed, and therefore thawed, for only a very small portion of the tidal cycle. Just above the low tidal level, rocks are submerged and therefore thawed for nearly all the cycle, whereas at the mid-tidal level, the periods of exposure and submergence are the same. It has been found that saturated rocks must be frozen for at least 10 hours to effect rock breakdown at temperatures between 0 and −5°C (Lautridou 1971). The greater number of frost cycles and the time available for freezing suggest that frost action is most severe in the spray and the upper portions of the intertidal zones, assuming that the rocks are able to attain critical levels of saturation.

The tide also influences the salinity of rock pools. Over most of the intertidal zone, pools are periodically flushed and replenished with sea water, with a normal salinity of about 35‰. Much higher salinities can be attained in the splash and the upper portion of the intertidal zones because of the evaporation of pool waters, although this may not be significant at the time of year when frost is most common. Probably of greater significance is the dilution of pool waters and the washing away of salts from rock surfaces in the splash and upper intertidal zones by meltwater emanating from the ice foot or running down or out of the cliff face.

True frost action, or temperature-dependent wetting and drying, can only be significant erosive mechanisms if the rocks attain critical levels of saturation. Frost sorption-sensitive carbonates have been found to be those that are at least 80% saturated after 24-hours immersion in water (Hudec 1973). Not only is the length of the absorption period greatest at the low tidal level, but because of the pressures exerted by the tidal head, the rate at which rocks take up water is also probably greatest at this level. In a semi-diurnal tidal regime, the period of immersion varies from almost 12 hours near the low tidal level to almost zero at the high tidal level. The lower the absorption efficiency of a rock, therefore, the closer it must be to the low tidal level to become saturated in the time available. Effective frost action also requires that rocks remain saturated during the freezing process. Although rocks with large capillaries absorb water most readily, they also lose it most quickly (Fagerlund 1975). The ability of porous materials to attain and retain critical levels of saturation may be enhanced in the presence of salts (Litvan 1976, MacInnes and Whiting 1979). True frost action in the intertidal zone is dependent upon the attainment of critical saturation levels, and rapid freezing upon exposure to air temperatures.

Trenhaile and Mercan (1984) measured saturation levels in the intertidal rocks of mesotidal Gaspé, Québec, and in the macrotidal Bay of Fundy, Nova Scotia (Fig. 5.3). Laboratory experiments were also conducted on cylindrical rock cores by simulating tidal oscillations in a pressurized vessel. None of the limestone cores attained high levels of saturation under pressures up to the equivalent of 3 m of water. The saturation of the gneiss and schist samples was probably high enough for effective frost action at the mid-tidal level and below, although rocks of low porosity are not particularly susceptible to frost action (Lautridou and Ozouf 1982). Shales, silty shales, greywackes, siltstones, argillite, and several other rock types were more than 90% saturated when first exposed by the falling tide on the mid-tidal platforms of Gaspé. In Fundy, initial saturation levels were between 75 and 80%, irrespective of whether the rocks were situated 4.7 or 1.7 m below the high tidal level. Effective frost action in the basalts and conglomeratic sandstones of Fundy, therefore, depends upon whether water is approximately equally distributed within these rocks. There was little difference in the saturation levels of pool and platform surface rocks in Gaspé, although the former were immersed for a much greater period. In the laboratory, there was also little difference in the saturation of rocks immersed for either one or two hours. This suggests that rocks absorb water quite quickly, and that the amount of water taken up is determined by the pressure of the tidal head rather than by the length of the immersion period, although this was not confirmed by the Fundy data.

Sorption-sensitive rocks can become critically saturated during periods of high humidity. Cool coastal exposures facilitate adsorption and the attainment of high levels of saturation. In Gaspé, shales in the cliff were

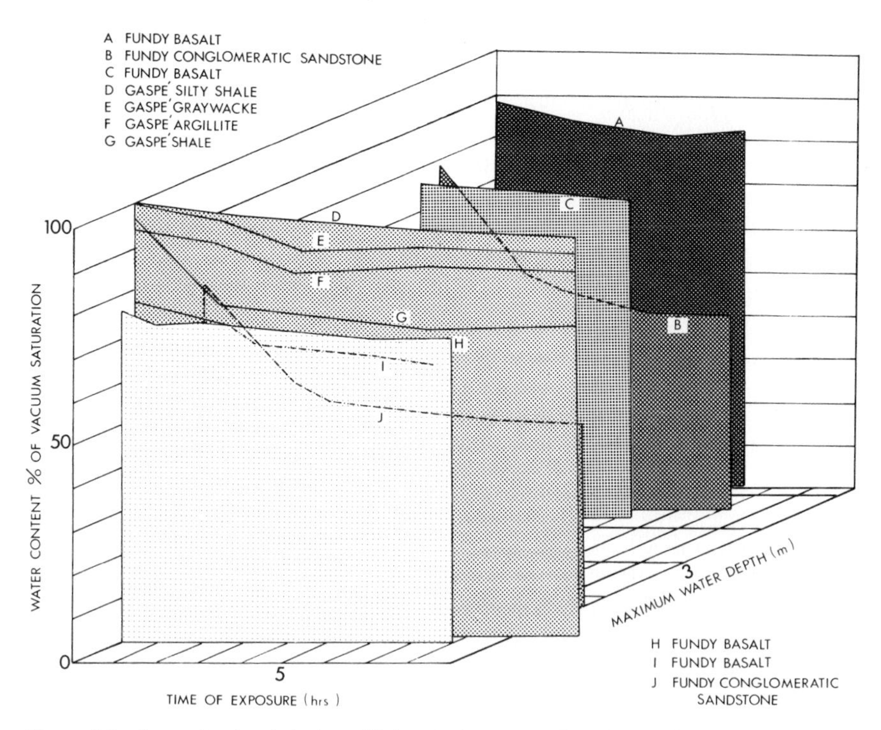

Figure 5.3. Saturation levels in intertidal rocks in eastern Canada as a function of the depth of water at high tide at each measurement site, and the time that the rocks had been exposed to the atmosphere at the time of measurement.

saturated (100%) when the relative humidity was 100%, but slates and greywackes were only 69 and 18% saturated, respectively. When the relative humidity was 88%, the shales and greywackes were only 79 and 7% saturated, respectively (Trenhaile and Mercan 1984). Although sorption-sensitive rocks are easily saturated in all portions of the intertidal zone, the efficacy of the mechanism may be dependent upon the desiccation of the rocks by freezing.

A distinction must be drawn between the upper and lower portions of high marine cliffs. On the upper cliff, fresh water can be supplied by snowmelt, rain, or groundwater seepage. At lower elevations, saline spray increasingly supplements those freshwater sources, particularly when strong onshore winds occur during periods of high water. It is presumably more difficult to saturate vertical than horizontal rock surfaces. Nevertheless, shale beds and other suitable strata in the cliff face can be saturated by groundwater seepage, or possibly by adsorption. Whether destruction is accomplished by true frost action, osmotic pressures or wetting and drying, weathering processes are clearly responsible for the breakup of shales, slates, and other fine-grained argillaceous rocks in the cliffs of cool coastal regions (Trenhaile 1978).

The role of true frost action and temperature-dependent wetting and drying in coastal environments depends not only on the lithology, but also on sufficient wave action to remove the weathered debris (Davies 1972). Unless wave action is vigorous, material will accumulate and progressively bury the cliff under a debris apron (Bird 1967, Howarth and Bones 1972). In Spitsbergen for example, cliff retreat is rapid because the intense weathering in spring is followed by the removal of debris by waves in the summer. Frost action, therefore, is the main process responsible for the retreat of coastal cliffs in this area (Jahn 1960a,b, Moign 1974a, Guilcher 1974a). The data on rates of slope retreat in cold coastal regions tend to be contradictory and inconclusive on the efficiency of frost action. The rate of retreat of a limestone sea cliff in Spitsbergen was found to be between 2.5 and 5 cm yr^{-1} (Jahn 1961), compared with only 4 to 6 mm yr^{-1} on the cliffs of Gotland Sweden (Rudberg 1967), 6 to 8 mm yr^{-1} in the South Shetland Islands (Hansom 1983), and only 1 to 2 m in the entire postglacial period on the sides of valleys in northern Scandinavia (Rapp 1960a). Cliffs of limestone, sandstone, and chert lying above the present level of wave action along a fiord in western Spitsbergen may have retreated by about 0.34 to 0.5 mm yr^{-1} in the last 10 ka. In the same area, but about 5 km from the coast, the rate of retreat is about 0.05 to 0.5 mm yr^{-1} (Rapp 1960b). In northern Sweden, rates of retreat in a variety of rock types is between 0 and 0.9 mm yr^{-1} (Rapp and Rudberg 1964). In northern Finland, mean Holocene recession rates may have been about 0.04 to 0.94 mm yr^{-1} (Soderman 1980). Intense frost action can produce overhanging cliff profiles at the base of cliffs, where groundwater and spray and splash provide an abundant supply of water, particularly at the periphery of the ice foot and other snow patches (Gardner 1969). It has been suggested that convex profiles develop in rocks which are moderately resistant to weathering, and vertical profiles in those which are particularly susceptible to frost action (Flores Silva 1952, Tricart 1969b).

Several workers have suggested that frost action can produce shore platforms (see Chapter 9). A number of early investigators emphasized the role of frost to account for the formation of platforms in polar areas where the wave fetch is limited. Nansen (1922) accorded a major role to frost action in the development of the Norwegian strandflat, and Zenkovitch (1967) suggested that benches result from frost action undercutting the cliff in the frequently wetted spray zone above the level of the high spring tides.

Cliff recession in Spitsbergen and extensive limestone platforms on the southern coast of Anticosti Island in the St Lawrence Estuary of eastern Canada have been ascribed to a combination of solution and frost action at the cliff base, with the removal of the debris by the ice foot (Corbel 1954, 1958). In western Scotland, it has been suggested that frost shattering during the Loch Lomond stadial (Younger Dryas), produced intertidal shore platforms in hard rocks in sheltered environments (Sissons 1974, Gray 1978, Dawson 1980). Andersen (1968) and Sollid *et al.* (1973) have arrived at similar conclusions for the Norwegian Main Line platform. Even in the fairly mild climate of southern England, wave-cut platforms in the Chalk were damaged by frost during a particularly severe winter (Williams and Robinson 1981). A preliminary attempt has been made to assess experimentally the role of temperature fluctuations about the freezing point in the development of a rock coast in Gaspé, eastern Québec, Canada (Trenhaile and Rudakas 1981). Although physical weathering, whether by frost or water adsorption, may play an important role in cliff and platform erosion, no evidence was found to suggest that these mechanisms are able to planate intertidal areas consisting of a variety of rock types to form smooth, subhorizontal platform surfaces.

The 0°C isotherms for air and surface sea water are in fairly close proximity wherever they occur (Fig. 5.4). Freezing of intertidal rocks in air

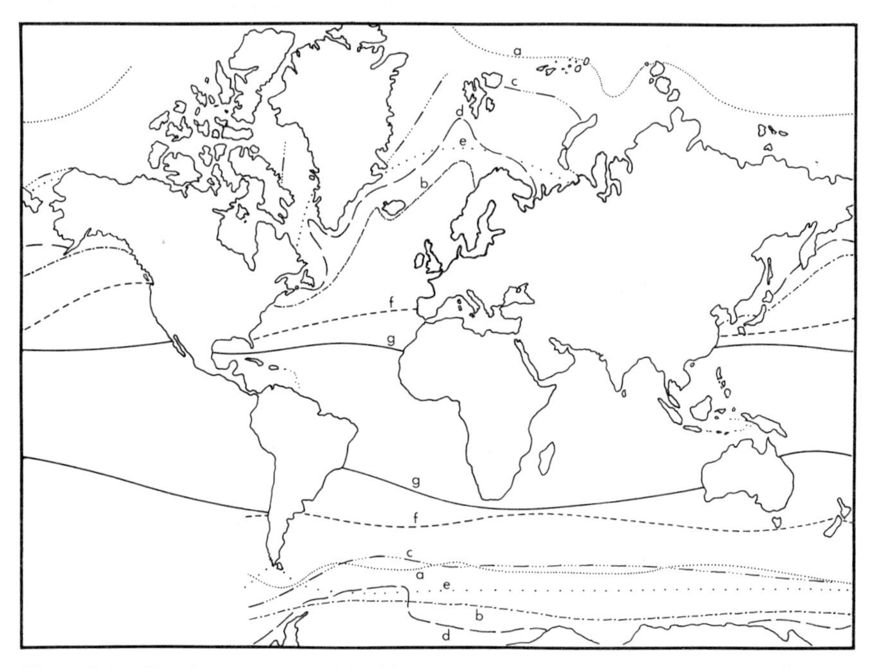

Figure 5.4. Climatic parameters of significance to coastal frost action: *a* and *b* are 0°C mean monthly air isotherms for July and January, respectively; *c* and *d* are the 0°C mean monthly water isotherms for July and January, respectively; *e* is the polar limit of 60 frost-free days per year; *f* is the equatorial limit where frost does not occur every year; and *g* is the equatorial limit of frost-free coasts (partly based upon Davies 1972).

and thawing in water is inhibited polewards of these isotherms, because of low water temperatures. Towards the equator, the frequency of freeze–thaw cycles progressively diminishes, up to the point at which coasts are completely frost free. Freeze–thaw induced by air freeze and water thaw in the intertidal zone is probably most active, therefore, in the area immediately equatorwards of the 0°C water isotherm. At different times of the year, this area encompasses the coasts of Alaska, northern British Columbia, northern New England and eastern Canada, southern Greenland, Iceland, northern Fennoscandia, Spitsbergen, Japan, Korea, the Yellow Sea coast of China, eastern Siberia, the southern tip of South America, and in winter, parts of the coast of Antarctica (Trenhaile 1983a).

Frost action in the intertidal and supratidal zones is also induced by fluctuations in air temperatures. Several workers have considered the occurrence and characteristics of atmospheric freeze–thaw cycles in North America (Russell 1943, Visher 1945, Fraser 1959, Williams 1964). Peltier's (1950) temperature and precipitation parameters suggest that maximum freeze–thaw activity occurs in northern Canada (Fraser 1959), but it has been found that the frequency of freeze–thaw cycles actually tends to increase southwards in Canada and northwards in the United States. On the Atlantic coast, the maximum number of cycles per year occurs in the northeastern United States, eastern Québec, and the Maritime provinces of Canada. The general pattern is similar on the Pacific coast, although maximum frequencies occur further north than on the Atlantic, around the Alaskan Panhandle (Williams 1964). Williams's data also show that although the frequency of freeze–thaw cycles is generally greatest near the central portions of the Atlantic and Pacific coasts, the length of the freezing period and the minimum temperatures which are attained increase northwards. Although I am not aware of similar data for Eurasia, the general pattern is likely to be similar to that in North America.

Much of the area in which freeze–thaw cycles associated with fluctuations in air temperatures are most frequent overlaps with the area in which cycles due to water thawing are also frequent. The zone of frequent atmospheric cycles, however, extends polewards of the 0°C water isotherm. The efficacy of frost action probably gradually diminishes polewards in this zone, but because some mechanisms, such as ice crystal or lense growth, are enhanced by long periods of freezing, some rock breakdown may be attributed to these processes in high latitudes.

Some of the areas in which frost action is apparently most active are also storm wave environments, where wave action is particularly vigorous. Furthermore, those rocks which are vulnerable to wave action are also generally those which are susceptible to frost. Vigorous wave action on vulnerable rocks can therefore obscure or inhibit the work of frost action in exposed areas. In particularly sheltered sites, intertidal or supratidal weathering may assume a more dominant role, but without sufficient wave action to remove the weathered debris the processes cannot operate efficiently.

The role of frost action and temperature-dependent wetting and drying in the coastal zone remains to be determined. It has been suggested that in the absence of vigorous wave action, the development of shore platforms and ledges in sheltered environments must be the result of frost shattering and planation. It has not yet been adequately demonstrated, however, that rocks can be critically saturated in the intertidal zone. Physical weathering probably plays an important role above the high tidal level, wherever suitable lithologies are saturated by groundwater seepage. Cliff recession would produce a rugged intertidal ramp or platform, but it is difficult to attribute the formation of fairly smooth, horizontal surfaces, cut across a variety of rock types, to frost or temperature-dependent wetting and drying acting alone. Very high latitudes do not provide optimum conditions for frost action. Nevertheless, it is possible that because of weaker wave action, frost plays a proportionally greater role in those areas than in the mid-latitudes, where wave action is more vigorous.

6

Mass Movement

Mass movement is the transport of material downslope in response to gravitational stress. It has been defined as unit movement of a portion of the land surface as in creep, landslide, or slip (AGI 1957, p. 179). Mass movement occurs as slopes try to attain stable, equilibrium forms. Marine cliffs may possess short rather than long-term stability because of undercutting, oversteepening, and the removal of basal debris by wave action. Mass movements, ranging according to local circumstances from the quasi-continuous fall of small debris to infrequent but extensive landsliding, play an important role in the development of rock coasts.

Numerous attempts have been made to classify the various forms of mass movement (Sharpe 1938, Hutchinson 1968a, Savage 1968, Carson and Kirkby 1972, Varnes 1975). On rock slopes, the main distinction to be made is between the free fall of material from steep cliffs, and deeper-seated movements which include a variety of slips.

Free Falls and Topples

Fresh rock faces and the presence of talus at the base of cliffs attest to the importance of rock falls on many coasts. Their occurrence, however, has not been documented to the same degree as landslips. Rockfalls often occur on remote stretches of rugged coasts, where they do not pose an immediate threat to human activity. Consequently, the majority of rock falls, and particularly the many small falls which take place, are not observed or recorded. It is difficult, therefore, to assess the relative significance of falls and slips on the development of coasts. Falls tend to be more widespread than deep-seated slips, and their frequency may compensate for their smaller size. Although they are underrepresented in the literature, falls are probably the dominant form of mass movement on most coasts. More than 50 falls have been reported in Kent since 1800, but this is undoubtedly a very small proportion of the total. Some falls are quite large; one in St Margaret's Bay in southeastern England involved about 254 Gg of chalk, which was thrust up to 360 m into the sea (Hutchinson 1972).

Rock falls involve the detachment and fall of surficial material from steep rock faces. Falls occur in well-fractured rocks (Middlemiss 1983), particularly where notches have been produced by wave action in lithologically or structurally weaker rocks at the cliff base. In the tropics, where wave action is often quite weak, cliff recession in coral limestones is frequently a result of the collapse of large blocks of rock over a deep solution notch. Large and small rock falls are common in the Liassic limestones and shales of the Vale of Glamorgan, south Wales. Deep undercutting of the cliff along thick shale beds in the *bucklandi* produces large rock falls, particularly where there are faults (Williams and Davies 1980). Most of the larger falls have occurred in shallow bays, where the *angulata* shales underlie the more resistant *bucklandi* limestones. Large rock falls are most frequent in the western portion of this coast, probably because of the alternation of resistant limestones with weak shales, although wave action is also most vigorous in this area. In the eastern portion, which is sheltered from the dominant westerly waves, there is a steady fall of marl fragments, whereas to the west the falls are less frequent but larger (Trenhaile 1969).

Granular disintegration occurs in rocks which lack a small-scale fracture pattern. In these rocks, the destruction of the rock cement by weathering results in the slow release of the grains. Coastal subsidence can also occur where deep cave systems have penetrated into the cliff zone. The term 'sagging' has been used to describe the downward settling of sandstones caused by failure in the underlying shales in the cliffs of north Devon (De Freitas 1972). Mobile flow slides in the upper and middle Chalk of southern England can be generated by falls from cliffs more than 40 or 50 m high. This causes debris to spread out as a thin sheet over the foreshore, possibly as a result of the generation of excess fluid pressure by crushing under the impact of these weak, nearly saturated rocks (Hutchinson 1980). Carson (1976) distinguished between the release and fall of rock rubble, and the failure and fall of an initially intact slab of rock. Slab falls require massive cohesive rocks, and deep tension cracks parallel to the cliff face. They are common in the chalk cliffs of southern England where undercutting has occurred, and in the basalts of the Bay of Fundy, where they comprised almost half the identified forms of slope failure (Carson and Kirkby 1972, Thomson 1979). On Surtsey, vertical fractures have developed parallel to the lava cliff faces, and vigorous wave action results in the formation and collapse of the overhanging portions (Norman 1980).

Slabs of rock also topple or overturn by forward tilting. Toppling involves the overturning of interacting rock units when the centre of gravity of each unit overhangs a pivot point contained within it (De Freitas and Watters 1973). Toppling is characteristic of rock masses which consist of columns defined by joints, cleavage or bedding planes. It occurs where the discontinuities dip into the rock mass, and on slopes which can be lower than those associated with landslips. Erosion at the toe of slopes facilitates toppling by reducing the resistance to overturning provided by more stable elements. Coastal toppling has been described on the castellated granite cliffs of Land's End, and on the cliffs of north Devon, England (De Freitas 1972, De Freitas and Watters 1973, Goodman and Bray 1977).

Several types of toppling failure have been identified (Goodman and Bray 1977, Evans 1981) (Fig. 6.1), including:

1. Flexural toppling on slopes consisting of a semi-continuous series of columns which break as they bend forwards. It usually occurs on slates, phyllites, and schists.

2. Block toppling in thickly- bedded sedimentary rocks, and in columnarly jointed volcanics broken by widely spaced joints.

3. Block flexure toppling in interbedded sandstones and shales, interbedded chert and shales, and thinly bedded limestones. Pseudo-continuous flexure of the columns takes place along the cross-joints. Sliding occurs along the joints in the toe of the slope, and it accompanies overturning in the rest of the rock mass. Block flexure toppling in the Carboniferous sandstones and shales of north Devon, has been described by De Freitas and Watters (1973).

Secondary toppling can be induced by a number of independent factors. These include block sliding on a bedding plane; the slump or creep of material above steeply dipping, layered rock; and by sliding, which transmits a load to the toe of a slope. The development of tension cracks parallel to the cliff edge also causes blocks to topple.

Rock and slab falls, sags, and topples are essentially surficial failures induced by frost action and other types of weathering, basal erosion, unloading, and hydrostatic pressures exerted by water in rock clefts. Many falls are the result of the reduction in confining pressures caused by slope erosion and retreat, and the formation of tension cracks parallel to the erosion surface. The development of these cracks reduces the strength of the rock mass and facilitates slab falls and topples. A small failure in the chalk cliff at Joss Bay on the Isle of Thanet, Kent, investigated by Hutchinson (1968b, 1972), was found to be due to the occurrence of a tension crack parallel to the cliff edge, and probably also because of the development of a shallow wave-cut notch. Shear took place along a fairly plane slip surface, inclined at an angle of 60–65° to the horizontal, when the crack had extended down to about half the height of the cliff. Hutchinson has recognized this pattern in other falls (Fig. 6.2).

Most of the rock falls in cool regions occur when frost is thought to be most active. This relationship has been noted in the mountains of Norway (Bjerrum and Jorstaad 1968), Sweden (Rapp 1960a) western Canada (Luckman 1976), coastal Finland (Soderman 1980), Northern Ireland (Douglas 1980), Gottland, Sweden (Rudberg 1967) and the Bay of Fundy, Canada (Thomson 1979). In Kent, in southeastern England, there is a

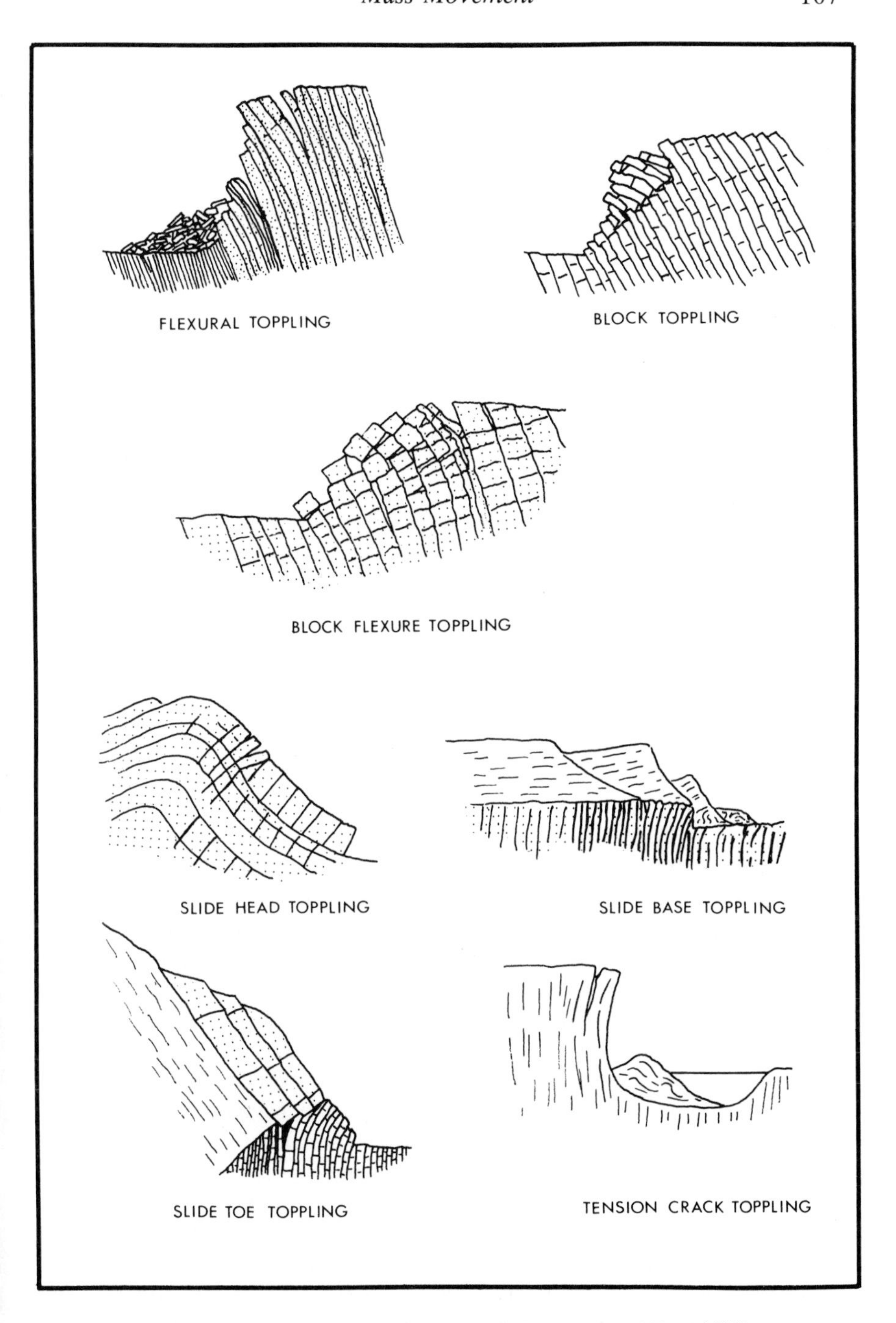

Figure 6.1. Types of toppling failure (after Goodman and Bray 1977).

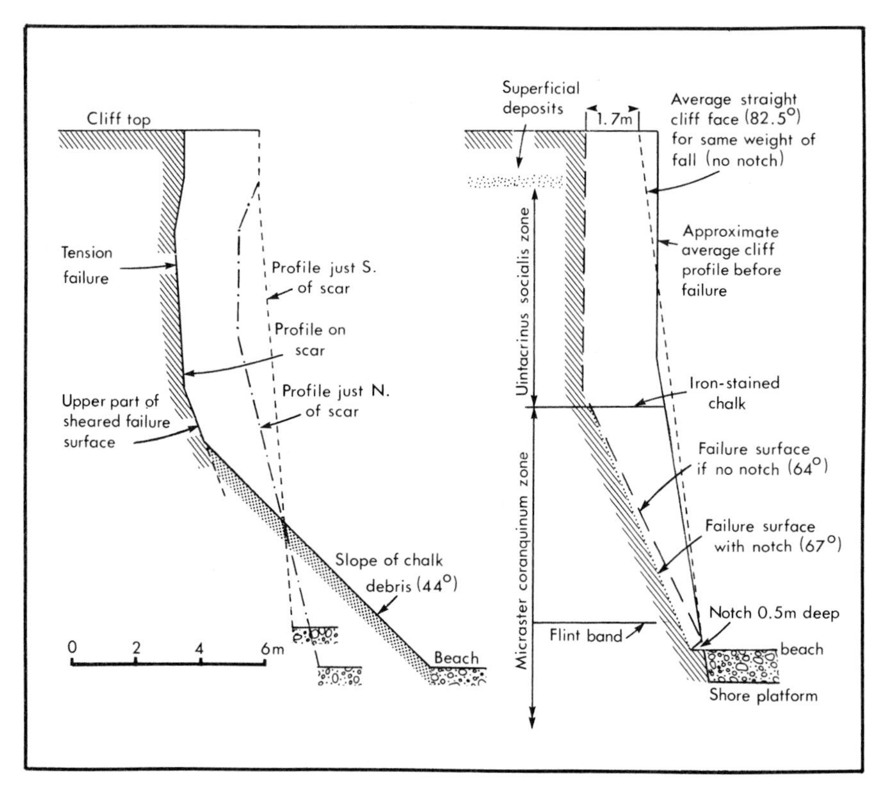

Figure 6.2. Cross-sections of a chalk fall in Joss Bay, Kent, England (Hutchinson1972).

close correspondence between the occurrence of chalk falls and periods of frost and rainfall. Most falls occur between October and April, largely as a result of rainfall in autumn and frost in spring, although storm waves are also most frequent at that time (Hutchinson 1972, Middlemiss 1983). In Sussex, on average, about four times as much land is lost to cliff erosion in winter than in summer (May 1971).

Varnes (1958) suggested that water in rock clefts exerts sufficient pressure to cause slabs of rock to be ejected from the cliff face: in an area to the south of the Loire Estuary in western France, for example, fractures parallel to the cliff in gneiss and micaschist are filled with water from October to December and from March to April. This causes falls through rock decay and the pressures exerted by the water in the cliff face (Gautier 1971). Terzaghi (1962) also believed that the reason that most rock falls in Norway occur in spring and autumn is not because of frost action, but because snow melt and heavy precipitation cause high water pressures in clefts. The presence of open tension joints permits water to penetrate most easily into the rock mass close to the cliff edge. The water table is therefore raised particularly high in this area by rainfall, and it can be raised further during the period of snowmelt by ice plugging the joint exits in the cliff face. Slope failure begins at the cliff foot, where the cleft water pressure is greatest; the upper portion of the cliff collapses as it is deprived of support (Terzaghi 1962).

The tide also induces water pressure fluctuations in clefts in coastal cliffs. In the Bay of Fundy, the frequency and size of the rock and slab falls increase eastwards, as the tidal range increases (Thomson 1979). This increase does not appear to reflect variations in rock type or wave activity, but it remains to be determined whether it is necessarily a response to tidally induced increases in cleft water pressure.

Landslips

Deep-seated mass movements are common in some coastal regions, where geological conditions are suitable. Landslips occur when the compressive strength of a rock is exceeded by the load it must bear. Failure begins when the stress at some point within the rock mass exceeds the strength of the rock, so that cohesion at that point becomes zero. Failure spreads as the additional load is transferred to the adjacent portions of the rock, increasing the shearing forces (A. Young 1972). Easily sheared rocks with low bearing strength are particularly susceptible to sliding. They include clays, and sedimentary rocks with a high proportion of clay minerals, such as shales, mudstones, and marls; porous volcanic rocks, such as tuffs and breccias; and plate-like or foliated rocks such as talc, schist, phyllite, and serpentine (Savage 1968, Simonett 1968). Rocks which have a fine joint system, those which are deeply fissured, and brecciated rocks are also potentially unstable. Slope failures are often associated with alternations of permeable and impermeable strata; the presence of massive rocks overlying incompetent materials; and the occurrence of sedimentary rocks dipping seawards. Muir-Wood (1971) has provided a useful summary of the factors which determine the stability of coastal slopes.

The strength or cohesion of a rock is progressively weakened by alternate wetting and drying and the swelling of clay minerals, and by deep chemical weathering (see, for example, Kenney 1975). Durgin (1977) found that the type of mass movement which occurs on granite outcrops is determined by the degree of decomposition. Rockfalls, rockslides, and blockglides occur on fresh granite; rockfalls and rolling rocks characterize the corestone phase; debris flows, avalanches, and slides occur on decomposed granites; and slumping on saprolite. Slope failures tend to occur in the advanced stages of decomposition, when the regolith has attained a critical depth. Landslips are therefore particularly common in the humid tropics, where chemical weathering is most intense. Slips on vegetated coastal cliffs frequently occur in granites, gneisses, and lower micaschists in these areas, usually following a period of heavy rainfall (Tricart 1972).

A distinction is usually made between translational slips and rotational slumps. Translational slips are structurally controlled, and involve failure along a straight, predetermined slip surface. This is often a bedding, cleavage, or joint plane, particularly if it is occupied by a weak, argillaceous filling (Hutchinson 1968a). Translational slips tend to occur where there is a considerable difference between the strength of surficial and underlying materials. The terms 'rock slide' or 'rock glide' are used to refer to the movement of many units down a slip surface, and 'block slide' or 'block glide' to the movement of a small number of fairly undeformed units. Deep-seated slumps generally occur in thick and fairly homogeneous deposits of clay or shale. Slumps in rock are usually more irregular than those in unconsolidated materials. This is because of the influence of discontinuities on the slide surface, and the presence of beds with varying degrees of resistance to shear. Rock slumps commonly combine some of the characteristics of slumps with those of translational slips. The slumping of a rock mass removes lateral support to the material behind, inducing further slumps inland. Multiple rotational landslides on the Isle of Wight and at Folkestone Warren in southern England, for example, have occurred in thick deposits of stiff, fissured clay underlain by hard rock and capped by a jointed rock stratum (Hutchinson 1980).

Major landslips have taken place in several areas on the southern coast of England, but particularly between Folkestone and Dover in the southeast, and around Lyme Regis in the southwest. These slides have usually occurred on cliffs consisting of chalk underlain by the upper Greensand, Gault Clay, and the lower Greensand. This sequence is seen in four areas of the breached Weald–Boulonnais anticline, three on the Isle of Wight, and several more in southwestern England and further south in France (Hutchinson *et al.* 1980). The very large landslip area at Folkestone Warren occupies about 3.2 km of the coast, where the North Downs meets the sea. Instability in this area may be partly the result of the very low angle of dip of the beds (1°), and partly the local absence of the upper Greensand and the correspondingly greater thickness of the Gault Clay (Osman 1917, Ward 1945, Muir-Wood 1955, Viner-Brady 1955, Toms 1946, 1953). Borings have confirmed that the slips are deep-seated rotational failures, in which the chalk has slipped through the Gault Clay, the slipping plane penetrating to the base of the Gault (Ward 1945, Toms 1953, Muir-Wood 1955). Failures are of three main types: deep-seated renewal of movement throughout almost the whole of the rotational landslip units which constitute the undercliff; renewal of deep-seated rotational slipping in the seaward part of the undercliff; and falls of chalk from the rear scarp (Hutchinson 1969).

Cretaceous, arenaceous, and calcareous sedimentary rocks, including the Chalk, lie unconformably on early Mesozoic shales, marls, mudstones, and limestones in the landslip area near the border of Devon and Dorset (Arber 1940, 1941, 1973, Denness *et al.* 1975). Although Ward (1945) considered that the slips were rotational, Arber (1940, 1962) originally disagreed with this interpretation. She thought that unlike Folkestone Warren, where the Gault Clay underlying the Chalk facilitates rotational failure, in the Lyme Regis area the seaward-dipping Lias and upper Greensand limestone beds would resist rotational shear. Steers (1962a,b) supported her view that sliding in this region took place along the unconformity between the Gault and the lower Lias. Arber (1971), however, later retracted her objections to Ward's hypothesis, accepting that slipping has occurred along a surface deep within the Liassic or even the Rhaetic rocks. No major landslips have occurred in this area for almost 150 years, but minor changes still take place (Arber 1940). Brunsden (1974) monitored the area for five years, and found that mudslides, rotational slips, and wave erosion are still active. Pitts (1979, 1983) made a morphological map of the area. He found that each slip unit initially takes the form of a rear scarp slip extending to the cliff top. Subsequent activity involves numerous complex events which break up the slipped masses. Whole slope events, involving failure of the rearmost cliffs, were found to have a maximum frequency of once in 120 to 150 years in this area. Brunsden and Jones (1980) proposed that process and form are well integrated on a large slide in west Devon which seems to be in a state of dynamic equilibrium; the size and position of the landform elements change through time in response to rotational slippage, mudslides, gulleying, toppling, sagging, rock falls, and debris accumulation, although the gross forms appear to persist over a 100-year time period (Fig. 6.3).

A number of other major landslips have occurred on the southern coast of England. On the Lizard in the southwest, landslips are associated with the penetration of serpentine by dykes (Steers 1964). De Freitas (1972) has also described a slump in Devonian shales and sandstones on the western side of the Lizard. At Brixham in Devon, a slip which occurred in mechanically weak micaceous slates and shales was monitored for eight years (Derbyshire *et al.* 1975, 1979), and rotational shear sliding, slab failure, and surficial sliding were recorded. Wave undercutting periodically causes rotation of the tilt blocks and shallow sliding, which serve to maintain an essentially constant overall gradient. On Portland Island, large blocks of limestone have squeezed out the underlying Portland Sand, and possibly the Kimmeridge Clay, causing them to slide seawards (Steers

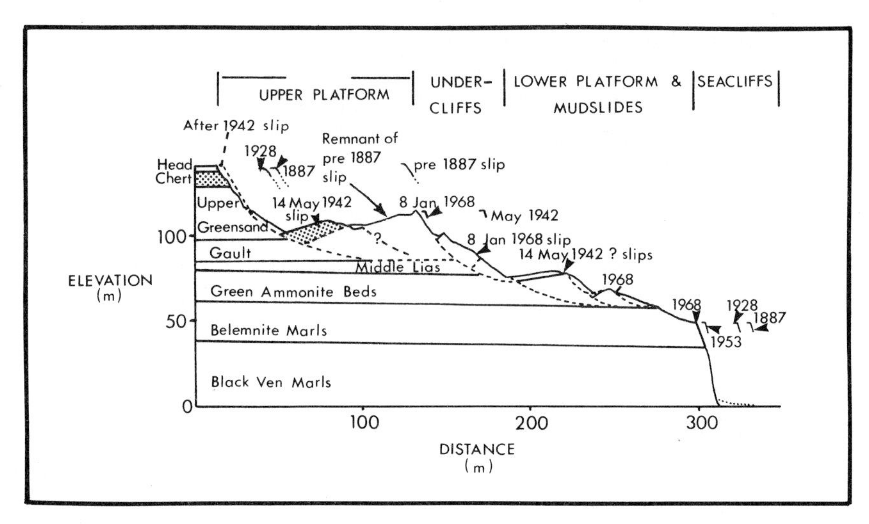

Figure 6.3. Fairy Dell, Dorset, England, showing the retreat of the landslip scar and sea cliff (Brunsden and Jones 1980).

1964). Further east, on the western side of St Alban's Head, thick limestones have slipped down over the impermeable Kimmeridge Clay (Trueman 1949).

Conditions on the south coast of the Isle of Wight are similar to those at Folkestone, and several large slips have occurred (Trueman 1949, Hutchinson *et al.* 1980). Between Shanklin and Niton in the south, upper Greensand in the cliff has slid forward over the underlying Gault Clay (Steers 1964, Wright 1969). The multiple rotational landslips near Ventnor are reputed to be the largest in southern England (Hutchinson 1980). Other slips have occurred at Boulder in the northwestern part of the island, between Yarmouth and the Needles in the west, and at Thorness Bay near Cowes in the north. Wave erosion of the clay marls has also produced deep-seated rotational slips near Seaview in the northeast (Ward 1962).

At Newhaven in Sussex, movements have taken place in the Woolwich and Reading Beds, and in the Thanet Sands, but earlier losses were attributable to erosion of the underlying chalk cliff, which is now protected by a shingle beach (Ward 1948). Two large slips occurred in 1827 and 1893 in the lower Greensand at Sandgate, Kent. Topley (1893) thought that the slips were caused in part by the washing out of the fine sand from the Sandgate Beds, thereby undermining the Folkestone Beds above. Blake (1893), on the other hand, proposed that the later slip was a simple slide associated with wet, slippery clay bands within the Sandgate Beds. More recently, it has been suggested that the slips were more deeply seated in the Atherfield Clay, at the base of the Greensand (Hutchinson 1968b).

Coastal landslips have been reported and investigated less frequently around the rest of the coast of Britain. This may be in part because they are less common in the resistant rocks on the western and northern coasts. Possibly of greater significance is the remoteness of many of these areas, and the fact that unlike some of their counterparts on the densely populated coast of southern England, they rarely threaten residential areas or transportation routes. Arber (1911) mentioned the existence of several landslip areas on the coast of north Devon, the most extensive occurring in alternating upper Carboniferous sandstones and shales in the Boscastle district. Slips have also been reported in the well-jointed, seaward-dipping Silurian Aberystwyth Grits in the Allt Wen cliffs of Cardiganshire (Steers 1964). At St Bees near Whitehaven in northwestern England, the dip in the Triassic Bunter Sandstone allows joint blocks to slide seawards (Steers 1964). Several slips have been reported in western Scotland. They occur in limestones, green schists, and intrusives at Aignish on the Kintyre Peninsula; in lavas overlying Mesozoic sediments in Mull; and in columnar pitchstone overlying lavas on Eigg (Steers 1973). Steers considered that the slips on the eastern side of Skye are the finest and most extensive in Britain. In this area the lavas, which have vertical joint planes, rest on Mesozoic rocks which fail under the weight of the flows above. In northeastern Skye, doleritic columns have foundered on incompetent sedimentary rocks, but the slips are much smaller here (Richards 1969). Between Melvich and Berwick in northern Scotland, several slips have taken place in breccia, where the vertical joints allow water to penetrate to the underlying mudstones. The junction between the two rock types in this area is unconformable. Slips in Runswick Bay, northern Yorkshire, have occurred in upper Lias shales, overlain by lower Deltaic Series sandstones (Agar 1960, Steers 1964). Wave erosion at the cliff base has induced slumping and flow in the cliff, and the development of vertical tension cracks in the sandstones (Rozier and Reeves 1979). Further south, slips have been reported south of Robin Hood's Bay and Filey Bay (Steers 1964).

Coastal landslips have been documented in several areas on the Pacific coast of the United States (Shepard and Wanless 1971). Many slips have occurred in seaward-dipping igneous and sedimentary rocks on the coast of northern Oregon (Byrne 1963, 1964). A number of slips, some of which are still active, have taken place in Ecola State Park, in Tertiary volcanics and thinly bedded sandstones and shales overlain by massive siltstones. A large slip occurred on Humbug Mountain on the Oregon coast (Dicken 1961, Shepard and Wanless 1971). On the coastal cliffs of southern California, landslips and sheetwash may be more important erosive mechanisms than direct wave action (Emery 1960). On the Pacific Palisades for example, 20 landslips have been mapped along an 8 km stretch of coast. Many of the slips in California have occurred along seaward-plunging synclines. One of the largest and best documented is at Portuguese Bend on the southwestern edge of the Los Angeles Basin (Merriam 1960, Easton 1973). This slip has taken place in moderately indurated sandy and tuffaceous shales and siltstones. There are also several bentonite beds, and at least one provides an incompetent horizon along which slipping has occurred. This slip was initiated by wave erosion, which caused rockfalls, rockslides, and block glides (Merriam 1960). The larger block glides eventually became slumps which extended further inland. Earthflows further modified the debris and the clayey and brecciated material produced by earlier slips. A major failure has also occurred at Point Firmin, about 8 km to the east of Portuguese Bend (Miller 1931, Varnes 1958). The failure is essentially a large block glide in Tertiary shales and sandstones. The area is on the flank of a plunging anticline, and the rocks dip seawards at between 10 and 22°. The major slip surface or surfaces lie below the base of the cliff, and are apparently coincident with the bedding planes of the shales. A large slip has been reported on Santa Catalina Island, off the coast of southern California. The Parson's Landing slip is in shattered talcose gneisses, with irregular masses of harder, less talcose rocks and quartzite. Present movements are restricted to shallow slips induced by wave erosion (Slosson and Cilweck 1966). Many landslips and other slope failures in Alaska have been triggered by earthquakes. In Lituya Bay, near Sitka, an earthquake in 1958 caused a rock slide which extended about 900 m up the side of the inlet, and involved 81.7 Tg of rock (Miller 1960). Many more slides were associated with the 1964 earthquake and other tremors (Shepard and Wanless 1971).

One of the few slips reported from the Atlantic coast is in the Bay of Fundy in New Brunswick, Canada, where large sandstone blocks glide seawards over impermeable beds which have been undercut by the sea (Ruitenberg *et al*. 1976).

Little reference has been made to coastal landslips elsewhere in the world. Despite the large Australasian coastal literature, only a few workers have mentioned the occurrence of landslips, although they are a common element of coastal scenery in many areas. A large, late Holocene slump occurred in Pleistocene sediments overlying Miocene limestones south of Adelaide. Saturation of the sediments may have been assisted by the presence of a fairly impermeable Pleistocene limestone at the base of the slip mass. A smaller, modern slump has also occurred in weathered basalt overlying Tertiary limestone at Portland, Victoria (Bourman and May 1984). A major slip has been reported in the well jointed siltstones on the Otway coast of Victoria, Australia, where the steeply seaward-dipping strata (40°) are particularly susceptible to wave erosion (Gill and Clarke 1979). Strong weathering along the bedding planes and joints help to make the Cretaceous arkoses and mudstones unstable on the Victorian coast at Windy Point (Joyce and Evans 1976). Slumps and debris slides have occurred in the mudstone and sandstone cliffs of the Gisborne coast of New Zealand (McLean and Davidson 1968). Debris slides are caused by the movement of weathered material over the unweathered material below, although unaltered bedrock may be incorporated in a matrix of fine debris. North of Christchurch, New Zealand, a 4-ha block of sandstone, 10 m in thickness, slid over an impermeable clayey siltstone dipping seawards at 22° (Smale *et al.* 1982). Elsewhere, Zenkovitch (1967) noted the occurrence of slips in thick Neogene limestones overlying clays on the Crimean Peninsula, in flysch on the Caucasian Black Sea coast, and on the Caspian coast. Semi-elliptical bays have been produced by fault-induced slides in coral limestone over marls on the Maltese Islands (Paskoff and Sanlaville 1978). In western Provence, France, slides have been attributed to the effects of rainwater and undercutting of the cliffs during storms. Extensive sliding to the west of Toulon is partly the result of strong chemical alteration of the phyllite rocks (Froget 1963). Other coastal slips have been discussed in western Provence (Blanc 1975), Israel (Arkin and Michaeli 1985), on coral islands (Verstappen 1960), and in Japan (Horikawa and Sunamura 1967).

Cohesive masses will not fail until a critical combination of slope height and angle in relation to the strength and bulk density of the mass has been attained. The critical height (H_c) for a vertical cliff in unweathered, mechanically intact rock, is given by:

$$H_c = q_u/w,$$

where q_u is the unconfined compressive strength of the rock; and w is its unit weight. For the weakest rock under the worst conditions, where q may be 352 kgf cm^{-2} (5,000 lb in^{-2}), and w is 2,723 kgf m^{-3} (170 lb ft^{-3}), then:

$$H_c = 1{,}300 \text{ m}.$$

The critical height of a cliff in granite and other hard rocks would be much greater; but vertical slopes are unable to attain such heights, and even gentler slopes which are much lower than the critical height have failed (Terzaghi 1962). This is because the presence of joints, bedding planes, faults, and other discontinuities in the rock is more important than the strength of the rock *per se*. A cross-section through a rock consists of joints separated by areas of intact rock. As the cohesion along the joints is zero, the effective cohesion, C_1, is given by:

$$C_1 = C \cdot A_g/A,$$

where C is the cohesion of the intact rock, A_g is the area of the intact rock in the gaps between the joints in a section, and A is the total area of the section. Unfortunately, the effective cohesion cannot be determined in the field.

The critical angle (ϕ_c) of a cliff is the steepest gradient which can be maintained in long-term stability. Terzaghi (1962) analysed the effect of the joint pattern on critical slope angles. Joints can be essentially random in some massive rocks such as granites, and in some limestones and dolomites; or there can be several well-defined joint sets. When the joints are random, he suggested that the critical slope angle for hard massive rocks is about 70°, assuming that seepage pressures are not acting upon the joint walls. On stratified sedimentary rocks, the critical slope angle is determined by the angle of friction along the joint walls (ϕ_f), the dip angle (α) and its direction, and the relative spacing and offset of the bedding and cross joints.

Possible cases provided by Terzaghi's analysis, as summarized by A. Young (1972) include those presented in Fig. 6.4 and Table 6.1. The critical slope angle increases with increasing values of C/D, so that when rocks are thinly bedded and the cross-joints are widely spaced, the cliff will be nearly vertical, irrespective of the dip of the rocks.

Terzaghi also considered the case in which slippage along bedding planes is resisted by an effective cohesion (C_1), as well as by friction. In such cases, a composite cliff profile will develop, in which the lowest portion of the slope will be vertical up to a height H, whereas above, it will slope at an angle α. It can be shown that the vertical portion of the cliff cannot be stable unless:

$$H \leqq C_1/[w \cos\alpha\,(\sin\alpha - \cos\alpha\tan\phi_f)].$$

Continued erosion and recession of the vertical face resulting in an increase in its height would cause slipping along the plane BB, through the base of

Table 6.1. Critical slope angles according to rock dip and joint pattern

	Dip angle α	Critical angle ϕ_c	Figure
Horizontal bedding	(0°)	90°	6.4*a*
Dip towards cliff	($<\phi_f$)	90°	6.4*b*
	($>\phi_f$)	α	6.4*c*
Dip into cliff	$(90-\alpha)<\phi_f$	90°	6.4*d*
	$(90-\alpha)>\phi_f$:		
1. Cross-joints not offset		$(90-\alpha)$	6.4*e*
2. Cross-joints offset $C/D \geqq 1$		90°	6.4*f*
3. Cross-joints offset $C/D < 1$		$(90-\alpha)-\tan^{-1}(C/D)$	6.4*g*

Note

C is the degree of cross-joint offset between adjacent strata measured parallel to the bedding planes, and D is the bed thickness.

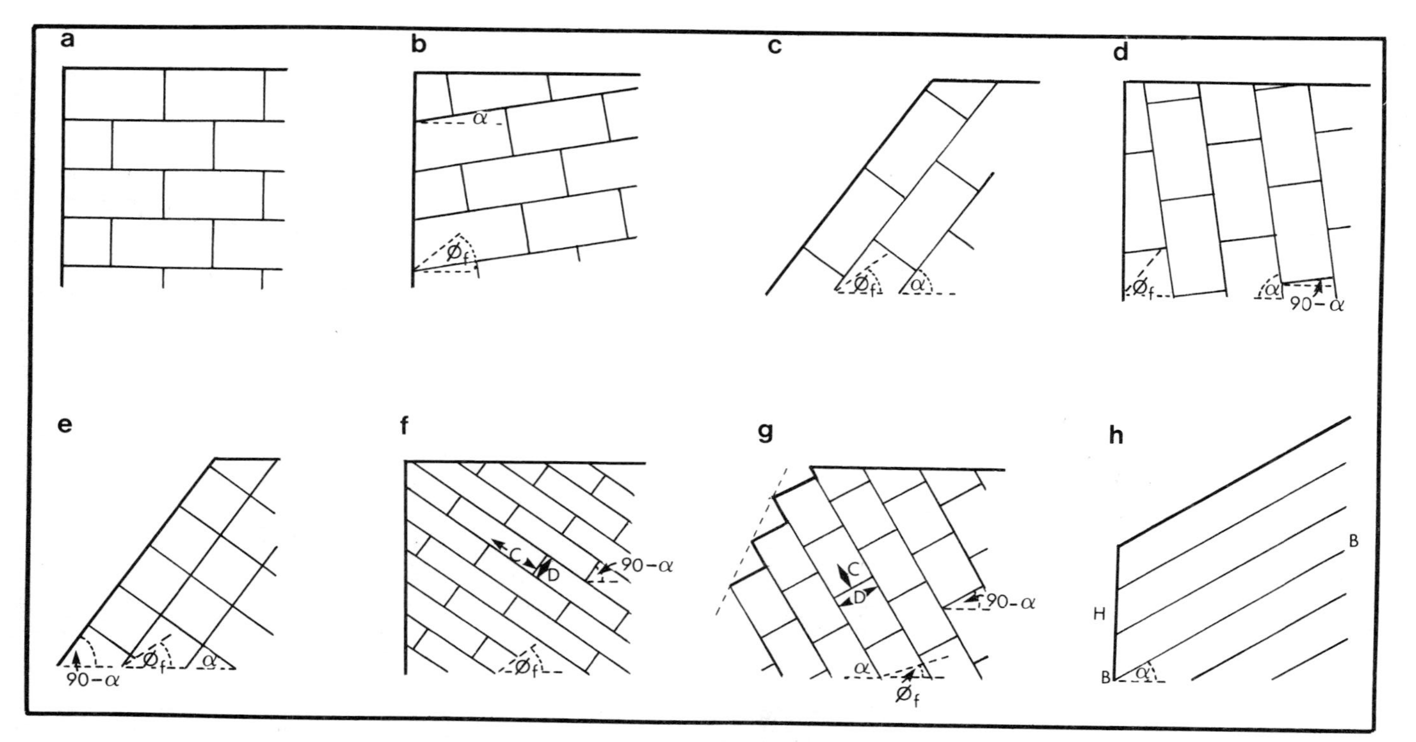

Figure 6.4. Critical angles of coastal cliffs in resistant bedded and jointed rocks (Terzaghi 1962, A. Young 1972).

the slope (Fig. 6.4*h*). In cases where the rocks dip away from the slope, the critical slope angle depends upon the joint pattern. Terzaghi's pioneering work showed that slope stability in jointed rocks largely depends upon the joint pattern and orientation, and to a lesser extent upon the effective cohesion.

Sliding along faults depends upon the nature of the rock within the fault zone. The crushed rock which often occupies fault zones varies greatly from one part of the fault to another. It also varies through time, as a result of chemical weathering, or hydrothermal alteration. These factors make it extremely difficult to predict the stability of faulted slopes.

Cruden (1975) has extended Terzaghi's analysis. Theoretically, large translational slips should involve highly cohesive masses along discontinuities dipping at low angles which are only slightly greater than the peak angle of friction along the discontinuities. He noted that the critical height of the lower, vertical portion of a slope becomes very large as the dip either approaches the vertical or becomes equal to the angle of friction. When $\alpha < \phi_f$, or when the discontinuity dips away from the slope, failure cannot take place by slipping.

Most landslipping events appear to be associated with the buildup of groundwater (Varnes 1950, Záruba and Mencl 1969, Watson 1976, Smale *et al*. 1982), and the instability caused by undercutting at the base (Ward 1962, Froget 1963, Derbyshire *et al*. 1979). Numerous workers have found that slope failures occur during or following periods of high precipitation or snowmelt. All the deep-seated movements at Folkestone Warren in the last two hundred years, for example, took place during periods of seasonally high groundwater levels (Hutchinson 1969). The Great Dowlands slide near Lyme Regis occurred on Christmas Day in 1839, following a period of heavy rainfall which had begun in June (Arber 1940). The 1893 Sandgate slide occurred in March, after a very wet February (Topley 1893, Blake 1893).

Water entering and flowing through potential slip masses can undermine cliffs by washing out underlying beds of sand. This happened at Newhaven in Sussex, and in the Downlands slip, where the washing out of the Foxmould Sands allowed the overlying upper Greensand and Chalk to slip seawards (Ward 1948). Groundwater also softens colluvial materials, facilitating their flow (Merriam 1960, Rozier and Reeves 1979); induces swelling and the generation of pressures in argillaceous rocks (Muir-Wood 1971, Pitty 1971); and produces high pore and cleft water pressures.

It has been suggested that water lubricates the surfaces of some materials, inducing slipping where they are overlain by massive formations (Holmes 1965). As Terzaghi (1950) has noted, however, water actually acts as an antilubricant in contact with many common minerals, including quartz. Furthermore, only a very thin film of moisture would be necessary to provide the full lubricating effect, and as this is always present in humid environments, it cannot explain the correlation between landslipping events and periods of heavy rainfall. The increase in the weight of potential slip masses as a result of the absorption of water, is of fairly minor importance (Carson 1976). Water is most likely to initiate slope failure when permeable rock lies on top of impermeable materials (Denness *et al.* 1975). The weight of massive, heavy rocks on wet, impermeable clay or shales can cause slipping (Holmes 1965, Muir-Wood 1971), which can likewise occur when drainage is concentrated along a fault zone (Kidson 1962).

Groundwater reservoirs provide a supply of water which is fairly independent of seasonal rainfall (Denness *et al.* 1975). Ponding of water in surface depressions created by earlier landslip events can promote further instability (Slosson and Cilweck 1966). Slope failure may have been aggravated by the addition of water from septic tanks and cesspools associated with residential development in the Portuguese Bend slip area of southern California (Merriam 1960), and by excessive watering of cliff-edge gardens in Israel (Arkin and Michaeli 1985). Irrigation water entering the shale zones may have increased the instability of the Point Firmin area in California (Miller 1931).

The effect of water on the instability of sedimentary materials is partly determined by their texture, mineral content, and other physical and physio-chemical characteristics. Landslips can occur when gypsum is removed by circulating water, or by its recrystallization in minute fractures in otherwise impermeable clays and shales (Varnes 1950). Landslips around Algiers Bay have been attributed to water percolating down through limestones, liberating the potassium and fixing calcium ions in the underlying glauconite marls. Deflocculation and the hydrolyzation of the alumino-silicates makes the marls fluid, so that they are unable to support the limestones (Proix-Noé 1946, Drouhin *et al.* 1948). Other slides have occurred along bedding planes covered by flakes of mica (Varnes 1950).

Coastal slopes are potentially unstable because of oversteepening by wave action (Sharpe 1938, Viner-Brady 1955, Ward 1962). Wave erosion and coastal landslipping therefore have a close cause-and-effect relationship. On the coast of northern Oregon, nearly all slips occurred in winter and early spring, when wave height and precipitation are greatest (Byrne 1963). It is difficult to understand the logic of McLean and Davidson's (1968) argument that cliff retreat and platform development are the result of mass movement rather than wave action. Some form of mass movement must occur in all cases where a cliff is exposed to effective wave action. In the absence of wave attack which periodically trims back and oversteepens the cliff, the gradient of the slope would decline until it had attained a more stable form. Platform development could not continue under these conditions, other than at a very slow rate.

The occurrence of very large landslips can actually inhibit platform development, by protecting the cliff under a thick mantle of debris. Alternatively, the more frequent fall or slipping of smaller amounts of material provides abrasives which facilitate cliff erosion (Wright 1969). The degree of protection afforded to a cliff by beach accumulations can be critical in determining its stability. At Newhaven in Sussex, England, the construction of a long jetty resulted in the buildup of shingle on a wave-cut platform. This accumulation provided protection to the cliff behind, which has now been fairly stable for more than a century (Ward 1948). Alternatively, interruption of the littoral drift following the development of Folkestone harbour probably accounted for the increased activity of slips in this area in the latter part of the nineteenth and the early part of the twentieth centuries (Hutchinson *et al*. 1980). Similarly, extensive groyning interfered with the longshore movement of shingle at nearby Sandgate, exposing the cliff to wave action (Topley 1893).

Changes in sea level have at various times alternately exposed and protected coastal slopes from wave action. On Santa Catalina Island off the coast of southern California, wave erosion 12 to 18 m below the present level of the sea removed support from the base of what later became the Parson's Landing slip (Slosson and Cilweck 1966). The initial movements in this area probably occurred after 17 ka BP. Emery (1967) has also considered the relationship between landslipping activity and variations in sea level. He assumed that large landslips could only occur when waves were able to attack steep slopes. During the glacial stages of the Pleistocene when sea level was low, wave activity was restricted to the gently sloping continental shelves, and large landslips could not occur. During interglacial periods of high sea level, waves were able to attack steeper hill slopes, and landslips took place on the oversteepened marine cliffs. Following a long period of stability, the present period of landslipping began about 5 ka BP, when the sea had almost attained its present level. Coastal landslips must therefore have been periodic, in response to fluctuations in sea level.

Blanc (1975) emphasized the effect of sea level on the pressure release or unloading mechanism, rather than on the efficacy of wave action. He argued that pressure-release forces are greatest when the decompression zone in the cliff is extended during a marine regression. A transgression reduces the decompression zone, stabilizing the slope. Blanc proposed that landslips in Provence, France, were therefore most frequent during glacial periods of low sea level.

Changing sea and tidal levels also induce fluctuations of the water table within potential slip masses. On Santa Catalina Island, the postglacial rise in sea level after 17 ka BP caused a rise in the groundwater table, which may have saturated and weakened the rock further inland (Slosson and Cilweck 1966). Tidal ebb and flow induce changes in pore and cleft water pressures. The 1893 Sandgate slip took place during, or began at the time of, low spring tides on two consecutive days (Topley 1893). Muir-Wood (1971) considered that the worst conditions for stability correspond to high water tables and low tides, when pore water pressures are high. He suggested, however, that in some cases, slips occur at the time of high tides, because of the increased ratio between pore pressure and the total overburden at the base of the slip. It has been noted that a slip near Christchurch, New Zealand occurred during a high tidal period, possibly because of an increase in the bouyancy of the toe of the slope (Smale *et al*. 1982). Easton (1973) found that the Portuguese Bend slip in southern California is most unstable during high tidal periods, particularly when the duration of high water is greatest, and especially during spring tides. He considered that the effect of the tide is related to:

1. saturation of material near the base of the slip;
2. saturation of the basal shear plane of the slip;
3. reduction of the load on the submerged toe of the slip because of the buoyant effect of water; and
4. increased pore water pressures.

In reply to Easton, Durgin (1974) emphasized variations in pore water pressure associated with artesian conditions, which are responsive to tidal fluctuations and other external pressure influences. He suggested that the weight of the water on the artesian zone at the time of high tidal levels increases the pore pressures, so that failure could result from residual high pressures as the tide flows. He also questioned whether the weight of the water on the toe of a slip wouldn't more than compensate for any buoyant effect, although Bromhead (1972) thought that it would compensate for the increase in pore water pressures.

7

Changes in Relative Sea Level

The scenery of a slowly changing rock coast is the legacy of processes which have been at work for millions of years. The form of the coast has been fashioned by innumerable shifts in the focus of wave attack, induced by changing sea level and movements of the land. Marine deposits and platforms above the present level of the sea testify to the occurrence of relative sea levels up to several hundred metres above today's. Emerged coastal terraces are an important scenic element in many areas, and because of the associated reduction in local cliff height, their presence aids contemporary erosion at the present level of the sea. Where a former sea level was similar to today's, shore platforms and other erosional features may be largely inherited, with only minor contemporary modification. Rock coasts preserve the evidence of former relative sea levels selectively. The hardness of the rock essentially functions as a filter, which determines the amount of time that the sea must be at a particular elevation to produce significant erosional forms. There may be no evidence of a fairly brief sea level stand in an area consisting of resistant rocks, but it may be recorded in the cliff profiles of more easily eroded materials. However, erosional features cut into resistant rocks are more durable than those in weaker rocks.

The last two decades have witnessed an enormous increase in the literature concerned with high Pleistocene sea levels (see, for example, the annotated bibliographies of Richards and Fairbridge 1965, Richards 1970, 1974). This upsurge in interest has been stimulated by developments in geophysics, Quaternary chronology, climatic modelling and other related fields. Of particular significance has been the introduction and widespread application to coastal studies of radiocarbon dating in the 1960s, and uranium series dating, and to an increasing degree amino acid analysis, in the 1970s. These developments have made it possible to date events more reliably in the upper and late-middle Pleistocene. The oxygen isotopic analysis of deep sea cores, speleothems, and glacial ice has permitted independent corroboration of changes in sea level inferred from dated strandlines. These techniques have necessitated a fundamental reassessment of the traditional tenets of Pleistocene 'eustasism'.

The basis of the glacio-eustatic theory was propounded more than a century ago, when it was realized that the growth and decay of immense Pleistocene ice sheets must have induced considerable changes in sea level

(Maclaren 1842, Tylor 1869). Much of the early work on Pleistocene sea levels was based upon the following assumptions:

1. Some parts of the world are geologically so stable that changes in sea level recorded in these areas must be eustatic.
2. Pleistocene changes in sea level measured on 'stable' coasts were synchronous and of uniform magnitude around the world.
3. Following from the first two assumptions, it was frequently assumed that strandlines in stable areas can be correlated on the basis of their elevation.

This approach is epitomized by the early work on the emerged terraces and raised beaches of the Mediterranean, and particularly by attempts to correlate them with similar features around the world.

The Mediterranean

De Lamothe (1911) recognized eight shorelines in Algeria and Tunisia, ranging up to 325 m above present sea level. Similar work in the Italo-French Riviera by Depéret (1906), and elsewhere by others, suggested that de Lamothe's four lowest levels can be traced around the Mediterranean. Both workers considered that the shorelines were produced by eustatic changes in sea level rather than by tectonic movements of the land, although Depéret believed that the raised beaches correspond to Alpine fluvio-glacial terraces, and to glacial rather than to interglacial periods. The four levels became known by the names used by Depéret (1918) to refer to the sedimentary cycles:

1. Sicilian, 90–100 m asl;
2. Milazzian, 55–60 m asl;
3. Tyrrhenian, 28–32 m asl;
4. Monastirian, 18–20 m asl.

A fifth, unnamed level was recognized about 7 or 8 m above present sea level, and later a sixth, only a few metres above today's level. The Monastirian and the two unnamed levels below, became known as the Main, Late and Epi-Monastirian. Zeuner (1952a) also recognized an early Tyrrhenian stand at about 45 m asl in the Mediterranean, and in Britain. In the absence of adequate dating, many of the terraces were correlated on the basis of palaeontological evidence. The Tyrrhenian, Monastirian, and the shorelines at lower elevations contain warm water *Strombus* (*S. bubonius*), Senegal, or 'Tyrrhenian' fauna, whereas the pre-Tyrrhenian fauna contain some species which are now restricted to cooler, more northerly regions. The term 'Emilian' has been used by some workers to refer to the occurrence of more temperate fauna between the Calabrian and Sicilian stages. The Milazzian may include 'Sicilian' species, but the examples at Milazzo are neither typical of the Sicilian nor Tyrrhenian, although they are now regarded as Tyrrhenian (Ruggieri and Sprovieri 1977). Fairbridge (1972) proposed that the Sicilian and other early Pleistocene deposits contain cold water species because they were laid down during glacial periods, when, reflecting a general decline in sea level, glacial levels were higher than the interglacial level today. The absence at even higher elevations of warm water species from the early interglacial periods, and the fact that the Sicilian and Milazzian fauna are not really associated with any shorelines, suggest that these cold species do not represent marine transgressions (Ruggieri and Sprovieri 1977).

Opinions have varied on the age of the strandlines. Those between 180 and 200 m above present sea level in southern Italy, Sicily, and Egypt were termed Calabrian (Zeuner 1952a), and these and others which extend up to 300 m or more in north Africa and elsewhere have been attributed to the early Pleistocene or the late Tertiary. The Sicilian has generally been attributed to the Donau/Günz interglacial, but Guilcher (1969) assigned it to the later Günz/Mindel. The Milazzian was usually placed in the Günz/Mindel interglacial, and the Tyrrhenian in the Mindel/Riss (Zeuner 1952a,b, West 1968), but Guilcher (1969) and Fairbridge (1961a) placed them both in the latter period. Zeuner thought that the Main and the Late Monastirian shorelines were formed in the Riss/Würm interglacial, and the Epi-Monastirian in an interstadial of the Würm, or last glacial period; whereas other workers considered that all the Monastirian levels belong to the last interglacial period.

Butzer and Cuerda (1962) revised the Mediterranean nomenclature on the basis of evidence collected on Majorca. They divided the Sicilian into an early segment at about 110 m above present sea level, and a later section consisting of three strandlines at elevations between 50 and 72 m. All the other raised shorelines of Majorca, ranging in elevation from the oldest at 33 to 34 m asl to the youngest at 0.5 m asl, were placed within four divisions of the Tyrrhenian. Tyrrhenian I was assigned to the Mindel/Riss interglacial, and Tyrrhenian II and III (Zeuner's Late Monastirian) replaced the Monastirian in the Riss/Würm. This classification is similar to the palaeontological classifications of Zeuner (1952a) and others, although Zeuner preferred Depéret's altimetric classification for Pleistocene correlation. The traditional Mediterranean nomenclature was later completely discarded by Butzer (1975), as a result of the study of a complex series of strandlines in Majorca (Butzer and Cuerda 1962). Because of the lack of *Strombus* deposits older than the last interglacial period in the Mediterranean, it is becoming common to restrict the term Tyrrhenian (or Eu-Tyrrhenian) to those of Riss/Würm age. The Late Monastirian is then referred to as Neo-Tyrrhenian (Nilsson 1983).

Radiometric dating suggests that *Strombus* sediments were deposited in at least two warm stages. The Eu-Tyrrhenian (Main Monastirian) and the Neo-Tyrrenhian (Late Monastirian) have been dated at between 115 and 125 ka BP and 75 and 95 ka BP, respectively (Stearns and Thurber 1967, Lalou *et al*. 1971). In central Italy, the first *Strombus* beach has been dated at about 177 ka BP, a second at 127 ka BP, and a thira at 90 ka BP (Ambrosetti *et al*. 1972). North of Rome, sea level oscillations have been identified between 320 and 280 ka BP (Aurelian I and II), between 230 and 200, 200 and 185, and 180 and 170 ka BP (Eu-Tyrrhenian I), and between 140 and 100 ka BP (Eu-Tyrrhenian II) (Radtke *et al*. 1981). Marine sediments about 4 to 5 m above sea level on the Sorrento Peninsula have been U-series dated at 129 ka BP (Brancaccio *et al*. 1978). Three late Pleistocene marine levels have been identified near Taranto in southeastern Italy. The oldest is more than 350 ka old. The oldest *Strombus* level, which is about 156 ka old, is overlain by a second, more recent *Strombus* deposit (Dai Pra 1982).

Correlation of coastal terraces on the basis of their altitude (see, for example, Zeuner 1952a, 1959, Cooke 1971) is extremely hazardous, and in many cases quite meaningless. The presence of young fold mountains in this area, and in some cases vulcanism, suggest that crustal movements played an important role. The type localities of the Mediterranean strandlines, in the Gulf of Palermo and at Cape Milazzo in Sicily, on the Oran coast of Algeria, and at Monastir in Tunisia, are distant from each other. Extrapolations are therefore necessary, although it is difficult to determine the amount of tectonic disturbance since their formation (Gill 1968). The problems involved in correlating Pleistocene shorelines in the Mediterranean, with particular emphasis on the tectonic stability of the area, have been discussed by Fairbridge (1972), Hey (1978), and by others in a Quaternary shoreline edition of *Quaternaria* (1971, vol. 15). The Mediterranean basin experienced considerable tectonic activity during and following the Pleistocene. The elevations of the terraces often converge or interchange along coastal tracts because of tilting or warping. For example, uplift has carried the Sicilian terrace up to 300 or 400 m in some areas, and to 800 m on Mt Etna. Calabrian sediments are found at up to 1,400 m in southwestern Calabria (Bowen 1978), and the Milazzian is up to 200 m above sea level in some areas. Even the type areas of the Milazzian and the Monastirian are probably disturbed (Gill 1968); orogenic movement is still very active in the Mediterranean.

Until the tectonic history of the Mediterranean is elucidated, eustatic changes in sea level cannot be reliably determined from the elevation of the marine terraces and deposits. Nevertheless, it has been tacitly assumed that large portions of the Earth's surface have been sufficiently stable during the Pleistocene to allow identification of the local counterparts of the Mediterranean terraces, partly at least on the basis of their elevation

(Newman 1968). On the Atlantic coast of the United States, the Wicomico terrace has been correlated with the Mediterranean Tyrrhenian (Daly 1934); the Talbot with the Main Monastirian (Fairbridge 1958); the Pamlico (Fairbridge 1958) or Suffolk Scarp (West 1968) with the Late Monastirian; and the Princess Anne with the Epi-Monastirian. The Upper and Lower Normannian beaches in western France may represent the Late and Epi-Monastirian, respectively (Guilcher 1969). In Britain, it has been suggested that the Tyrrhenian is represented by the Boyn Hill terrace, the Clava shell beds, and the so-called '100-foot beach' of Scotland. The Taplow and Upper Flood Plain stages of the Thames and the '50-foot beach' of Scotland have been attributed to the Main and Late Monastirian periods (Zeuner 1952a, 1959, Charlesworth 1957). In Morocco, the Ouljian strandline is the equivalent of the Epi-Monastirian. Numerous other attempts have been made to identify the counterparts of the classical Mediterranean strandlines in, for example, South Africa, India, Japan, Australia, New Zealand, South America and around the Black Sea (Charlesworth 1957, Chatterjee 1961, Ward 1965).

The Cretaceous and Tertiary

High strandlines have traditionally been ascribed to the early Pleistocene or to the Pliocene, partly because of the belief that they could not have survived subaerial denudation over a much longer period. Unless these terraces were raised to their present positions, however, they cannot be reconciled with the Pleistocene palaeoclimatic and palaeosea-level records. Although a great deal remains to be determined about the degree to which pre-Pleistocene sea levels have contributed to the development of rock coasts, it is clear that sea levels have been substantially above and below today's level in the Cretaceous–Tertiary periods. No interpretation of the cliffs and planation surfaces of these coasts, therefore, can be complete without consideration of the possible role of these ancient sea levels (see, for example, Nunn 1984).

Several major transgressions and regressions took place in the Cretaceous, although little is known of their order of magnitude (Cooper 1977, Mörner 1980a). Sea level reached its maximum in the late Cretaceous: 150, or possibly more than 300 m, above today's (Vail *et al*. 1977, Vail and Hardenbol 1979). The Campanian–Maestrichtian transgression, which was followed by an extensive withdrawal of the sea at the end of the Cretaceous, was the longest and largest event in the Cretaceous–Tertiary periods. It was widespread and probably eustatic, possibly as a result of tectonic activity (Hallam 1963, Grasty 1967, Russell 1968, Tanner 1968, Cooper 1977), as there is no unequivocal evidence of any permanent ice at that time (Barron *et al*. 1981). A complex series of fluctuations was also superimposed upon this dominant event (see articles in *Cretaceous Res.* 1980, vol. 1).

There have been many attempts to determine changes in sea level in the Tertiary, although there is presently little agreement on their magnitude or occurrence (Berggren and Van Couvering 1974, Arthur 1979). Bandy (1968) used a planktonic organism *Globigerina pachyderma* to trace variations in the distribution of cold water in the Tertiary. Polar planktonic faunae expanded into temperate regions in the late Miocene and middle Pliocene. This was probably associated with the growth of ice masses and falling sea levels, although probably not on the scale of the Pleistocene. Glenie *et al.* (1968) traced the position of the sea during the Eocene—Pliocene at Adelaide, and from the late Cretaceous to the Pliocene in Victoria, Australia. Stratigraphic evidence suggests that there was a sea level fluctuation in the Palaeocene in Victoria. A second oscillation began in the Eocene, attaining its maximum level in the Miocene. A transgression occurred in New Zealand from the middle Eocene to the end of the Oligocene. Sea level in southern Australia may have been below its present level in the early part of the Pliocene, although it had risen to well above today's level by the late Pliocene.

There are major differences between Glenie *et al.*'s Australasian data and the global scheme discussed by Vail and Hardenbol (1979). The latter workers found that after being more than 300 m above the present level in the late Cretaceous, sea level became fairly stable from the early Palaeocene to the Oligocene (Fig. 7.1). Sea level may have been between 225 and 275 m above today's level in this period, but fell below on several occasions, particularly at about 60, 49.5, and 40 Ma BP. A rapid fall to more than 100 m below the present level occurred at about 29 Ma BP, in the late Oligocene, although the period of stability in South Carolina may have continued into the early, and possibly the middle, Miocene (Colquhoun and Johnson 1968).

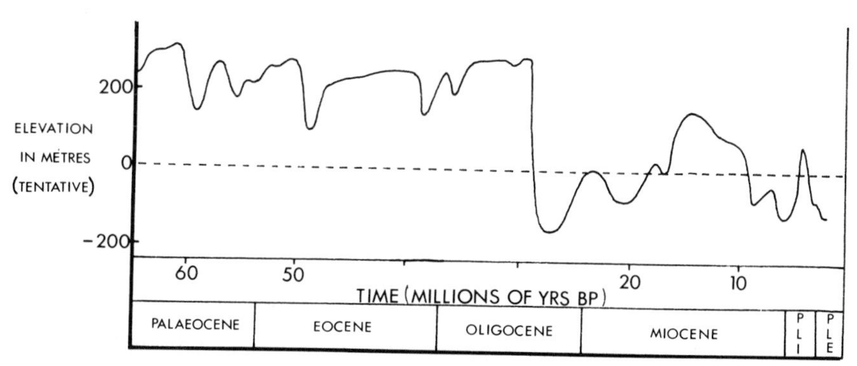

Figure 7.1. Changes of sea level in the Tertiary (Vail and Hardenbol 1979).

There is substantial, although possibly fortuitous, agreement between Glenie *et al.*'s and Vail and Hardenbol's sea level curves for the period following the Oligocene, and with Colquhoun and Johnson's (1968) data from South Carolina for the post-middle Miocene. Glenie *et al.* and Vail and Hardenbol recognized a rise in sea level in the Miocene, which may have been the equivalent of the late Miocene transgression in the southern United States. This transgression was responsible for the Orangeburg Scarp, between 61 and 76 m above present sea level (Colquhoun and Johnson 1968, Alt 1968). Vail and Hardenbol considered that the sea rose to more than 100 m above its present level in the middle of the Miocene, although the rise was interrupted by a fall of between 50 and 75 m at about 22 Ma BP. Other regressions occurred at about 10.8 and 6.6 Ma BP, to about 100 m below today's level. Bandy (1968) also found that the later Miocene was a period of ocean cooling, presumably associated with the growth of ice sheets and falling sea levels. It was also at this time that the Mediterranean became cut off from the world ocean, inducing a salinity crisis and the deposition of evaporites. This could have been the result of a glacio-eustatic fall in sea level of between 40 and 70 m (Adams *et al.* 1977). Further evidence for this fall is provided by facies changes in Atlantic and Caribbean coastal sediments, and by faunal–floral changes of late Miocene age in Alaska and California, and in the Atlantic, Pacific, and Southern oceans. Tanner (1968) has also referred to a Mio–Pliocene fall in sea level of about 75 m, although this is much more gradual than the drop envisioned by Vail and Hardenbol.

Following the late Miocene low, the sea may have quickly risen to almost 100 m above its present level in the early Pliocene (Vail and Hardenbol 1979). An abrupt fall in sea level, however, began about 4.2 Ma BP. In the coastal plain of the United States, Blackwelder (1981) found evidence of falls in sea level between 22 and 19 Ma BP, at the end of the middle Miocene, and between 6.5 and 5 Ma BP; these events are in fair agreement with the scheme of Vail and Hardenbol. Blackwelder also confirmed the occurrence of a major transgression in the early Pliocene, when the sea rose to 43 m above its present level in South Carolina (Colquhoun and Johnson 1968). An abrupt fall in sea level about 4.2 Ma BP may also correlate with ocean cooling in the middle part of the Pliocene (Bandy 1968, Vail and Hardenbol 1979).

On Gran Canaria in the Canary Islands, transgressions took place at about 8, 4.3, and possibly 3 Ma BP, and regressions around 9.6, 5, and 3.8 Ma BP (Lietz and Schmincke 1975). These estimates are in generally good agreement with Vail and Hardenbol's data, which refer to the onset of the transgressive–regressive phases rather than to their culmination.

Vail and Hardenbol's (1979) data do not support the contention that the sea was close to its present level for any significant period in the Tertiary since the early Miocene. In Florida, however, middle Pliocene marine and terrestrial vertebrates have been found in sandy gravels which are only about 2 or 3 m above present sea level. If, as several studies have indicated, the area has been quite stable since the late Miocene, then the sea must have been close to today's level between about 7 and 4 Ma BP (Webb and Tessman 1967). On the basis of estimates of ice volume, it has been calculated that early Pliocene sea levels were about 7 m above today's on the coastal plain of the United States (Blackwelder 1981), and at about the same level at 3 Ma BP (Mercer 1968a). In southeastern Virginia, Pliocene or early Pleistocene shorelines are found between 30 m above and 3 m below the present sea level (Oaks *et al*. 1974).

In a foreword to the *proceedings* of a symposium on Tertiary sea levels, Tanner (1968) proposed that there had been a more or less steady fall in sea level of between 70 and 100 m since the middle of the Miocene. Pitman (1979) calculated changes in sea level which could have resulted from changes in the volume of oceanic ridges. The results indicate that there was a steady fall in sea level from 85 to 15 Ma BP. It fell from more than 300 m above its present level in the late Cretaceous to about 50 m above in the late Miocene. The rate of fall was greatest in the Palaeocene, the early Eocene, and the Oligocene. Hallam (1963) agreed that there had been a progressive withdrawal of the sea from the continents since the Cretaceous. Much of the recent work which has been discussed, however, suggests that the pattern of sea level change in the Cretaceous and Tertiary was very complex. In particular, it has been suggested that Cretaceous sea levels may have changed by different amounts and at different times around the world (Eardley 1964, Mörner 1980a). A crude trend of falling sea level can be identified in many studies, but it is obscured by the oscillations superimposed on it. Much more information is necessary before Cretaceous and Tertiary sea level changes can be confidently determined.

The Last Interglacial and the Early Wisconsin/Weichselian Glacial Period

Uranium series disequilibrium dating methods have provided a fairly reliable record of sea level maxima within the last 150 ka. Fossil corals have been dated using ^{230}Th and ^{234}U in the ^{238}U series (half-lives of 75 and 250 ka, respectively), and ^{231}Pa in the ^{235}U series (half-life 32 Ka). Dates obtained from fossil shells are generally less reliable (Kaufman *et al*. 1971). The use of these and other isotopes has been discussed by Broecker (1965), and a useful summary has been provided by Bowen (1978).

Until fairly recently, tectonically mobile areas were thought to be unsuitable for investigations of eustatic changes in sea level. It is now recognized, however, that stairways of elevated reefs in such areas as New Guinea, the Ryukyu Islands, and Barbados permit the most detailed resolution of sea level changes if the tectonic history of the area can be reliably determined. This is because the sea level record in more stable regions is vertically compressed and partially submerged. On the Gower Peninsula in south Wales, for example, amino acid analysis suggests that the raised beaches formed in the last interglacial period include some reworked material from the penultimate interglacial (Davies 1983).

Numerous studies have shown that sea level in the last interglacial period was between 3 and 10 m higher than today's. Indeed, the strandlines associated with this period are so widespread and conspicuous that they provide an important reference level in many parts of the world. The high interglacial sea level between about 130 and 120 ka BP was confirmed by studies conducted in the Pacific and Indian Oceans (Veeh 1966, Thurber *et al.* 1965), and in Florida and the Bahamas (Broecker and Thurber 1965, Osmond *et al.* 1965). On Barbados, careful examination of an emerged reef complex led to the recognition of two later sea level maxima, in addition to the earlier, higher level (Broecker *et al.* 1968). Other studies on Barbados have confirmed the occurrence of these three maxima (Mesolella *et al.* 1969, 1970, Matthews 1973, Steinen *et al.* 1973). Nevertheless, there has been some disagreement on the precise elevations of the later maxima, which occurred at about 103 and 82 ka BP. Broecker *et al.* (1968) placed them about 10 to 16 m below the earlier interglacial maximum, and similar values have been reported from Haiti (Dodge *et al.* 1983). Stearns (1976), however, considered that the difference was only about half that amount, whereas Fairbanks and Matthews (1978) and Harmon *et al.* (1983) suggested that it was much greater. Strandlines corresponding to the maximum interglacial level have been identified throughout the Caribbean and the western Atlantic, but the later stands which were below today's are absent, or more probably submerged, in the more stable regions (Land *et al.* 1967, Cant 1973, Moore and Somayajulu 1974, Neumann and Moore 1975, Schubert and Szabo 1978, Szabo *et al.* 1978, Harmon *et al.* 1978a, 1983).

A spectacular series of terraces has been studied on the Huon Peninsula of Papua New Guinea. At least 20 reef complexes, consisting of coral with subordinate deltaic gravels, extend up to 600 m above present sea level. Chappell (1974a,b) compared the palaeoecology of each reef with modern coral associations, and he devised an uplift correction curve to separate the effects of tectonic and eustatic mechanisms. The New Guinea eustatic curve (Chappell 1974a,b, 1983a, Bloom *et al.* 1974), as shown in Figure 7.2, is similar to those from Barbados (Broecker *et al.* 1968) and the Ryukyu Islands (Konishi *et al.* 1970, 1974). Further evidence from Atauro near Timor, the New Hebrides, and eastern Australia has confirmed the suggestion that the interglacial maximum actually consisted of two similar peaks, at about 135 and 119 ka BP, separated by a minor regression

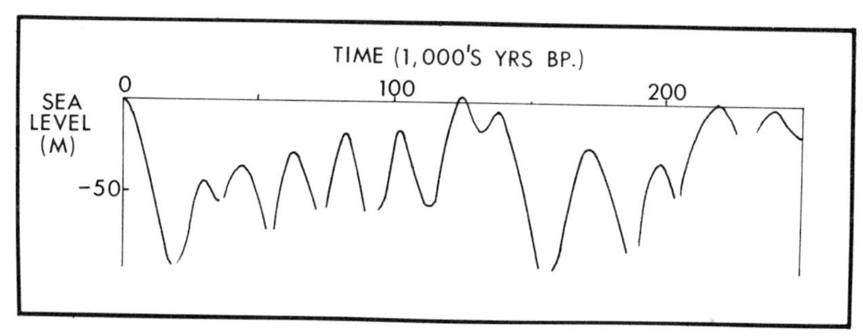

Figure 7.2. Sea level curve on the Huon Peninsula of New Guinea for the last 250 ka, assuming uniform uplift before 120 ka BP (Chappell 1974b).

(Bloom *et al.* 1974, Moore and Somavajulu 1974, Marshall and Thom 1976, Neef and Veeh 1977, Chappell and Veeh 1978a).

Pre-Sangamon/Eemian Levels

Far less is known about high sea levels which preceded the maximum of the last interglacial period. Much of the middle and all of the early Pleistocene lies beyond the limits of the uranium series disequilibrium dating methods which are most commonly used. The identification of ancient shorelines from these periods must therefore depend upon the interpretation of less reliable sedimentary and morphological evidence. Furthermore, the older the strandline, the more probable that it has been affected by tectonism, and the less justifiable is the assumption that rates of movement have been uniform since they were formed.

Uranium series dating and other evidence suggest that high sea levels occurred in the period between 220 and 180 ka BP. It is very difficult to determine their approximate elevation, however, because of the possible role of tectonic movement, and the questionable validity of the assumptions which are necessary to separate the effects of tectonic and eustatic influences. Several investigators have suggested that sea level in the penultimate interglacial period, of undetermined age, was between 10 and 30 m above its present level (Land *et al.* 1967, Biberson 1970, Ward *et al.* 1971, Stearns 1971, 1978), but Fairbridge (1961a) placed it much higher than this. In Britain, the Hoxnian/Holsteinian level has been estimated at between 20 and 30 m above today's level (Stephens and Synge 1966, Guilcher 1969, West 1972, Mitchell 1977). Fossil insect assemblages, however, suggest that the Hoxnian period was no warmer than at the present time, and cooler than in the Ipswichian and Cromerian interglacials (Coope 1977). The sea was about 2 m above its present level in Bermuda at about 200 ka BP (Harmon *et al.* 1983), but most recent studies show that in most areas it did not surpass its present level between about 220 and 150 ka BP. There is some agreement that the sea was from 20 to 30 m below its present level between 185 and 180 ka BP, but estimates vary from about 7 m above to as much as 20 m below in the preceding 15 ka. Evidence from New Guinea, Atauro near Timor, Mauritius, and Madagascar suggests that the sea was very close to its present level by about 220 ka BP, but it is estimated to have been about 32 m below today's level in Barbados, and about 20 m below today's level in Bermuda (Labeyrie *et al.* 1969, Montaggioni 1974, Chappell 1974a,b, 1983a, Chappell and Veeh 1978a, Marshall and Launay 1978, Harmon *et al.* 1978a, 1983, Fairbanks and Matthews 1978, Cronin *et al.* 1981). There is also some evidence of a period of high sea level between about 350 and 300 ka BP, which in New Guinea between 336 and 320 ka BP was about the same as today (Bender *et al.* 1973, 1979, Schubert and Szabo 1978, Chappell and Veeh 1978a, Chappell 1983a).

It is difficult to distinguish any globally consistent pattern for sea level stands before 400 ka BP. Several dates, however, cluster around 400 ka BP, and a transgression appears to have terminated on Atauro at about 400 ka BP (Ward *et al.* 1971, Ward 1973, Lietz and Schmincke 1975, Chappell and Veeh 1978a, Pillans 1983). A number of isolated He/U coral dates reported by Bender *et al.* (1973, 1979) from Barbados have not been confirmed from elsewhere, although they do provide some support for a high stand which could have been responsible for a 570-ka old reef on Curaçao, and one of similar age on Atauro (Schubert and Szabo 1978, Chappell and Veeh 1978a).

The age and the sea levels responsible for even older shorelines cannot be reliably determined using the techniques presently available. It has been proposed that certain shorelines correspond to specific interglacial periods, but without reliable dating of these strandlines, the correlations must remain essentially speculative (Ward 1965, Hoyt and Hails 1967, Oaks *et al.* 1974, Stearns 1978, Paskoff 1980, Cronin 1980).

Wisconsin/Weichselian Sea Levels

Many workers have proposed that the sea was close to, or even above, today's level in at least one of the interstadials of the last glacial period (Zeuner 1952a). This conclusion was based upon morphological and sedimentary evidence, and on the radiocarbon dating of peat, wood, shells, and other materials, and less frequently on ^{230}Th and ^{231}Pa uranium series dating of speleothems and coral (Shepard 1963, Milliman and Emery 1968, Hoyt *et al.* 1968, Osmond *et al.* 1970).

The occurrence of high interstadial sea levels has been contested. Much of the evidence can be attributed to contamination of radiometrically dated material (Mörner 1971a, Butzer 1975, Chappell and Veeh 1978b). Radiocarbon dating is very sensitive to small amounts of contamination in shells and other marine carbonates, and to a lesser extent in organic materials. If the true age of a sample is infinite (>60 ka BP), only 1% contamination is sufficient to provide an apparent age of 38 ka BP; 2% will produce an age of 32.5 ka BP; and 5% will give 24.8 ka BP (Bowen 1978). Radiocarbon dating of even slightly contaminated material from a strandline of the last interglacial could therefore suggest that it was formed during an interstadial of the last glacial period.

Uranium series dating of reefs in the Caribbean, the western Pacific, and New Guinea has shown that sea level fluctuations were superimposed upon a general decline in level in the last glacial period. Five sea level maxima, at intervals of about 20 ka, have been identified in New Guinea, since the peak of the last interglacial (Veeh and Chappell 1970, Bloom *et al.* 1974, Chappell 1974a,b, 1983a, Chappell *et al.* 1974, Chappell and Veeh 1978b). Similar records have been obtained from the New Hebrides (Neef and Veeh 1977), the Ryukyu Islands (Konishi *et al.* 1970, 1974), and Barbados (Stearns 1976). Estimated sea levels from 45 to 35 ka BP are generally about 37 to 45 m below today's, and between 40 and 45 m below from 31 to 28 ka BP. There is also evidence of a high sea level stand about 20 m below its present level between 65 and 55 ka BP (James *et al.* 1971, Chappell 1974a,b, Chappell *et al.* 1974, Konishi *et al.* 1974, Stearns 1976, Neef and Veeh 1977) (Fig. 7.2).

If the sea had risen to anything like its present level during the last glacial period, there should be abundant evidence of major climatic amelioration and ice retreat. In Britain, fossil insect assemblages suggest that summer temperatures were as high as today in the Windermere interstadial, and even briefly higher than today in the early part of the Upton Warren interstadial complex, about 43 ka BP (Coope 1977). There is also some evidence of a rise in temperature in North America at various times in the Wisconsin (Stuiver *et al.* 1978, Alley 1979, Heusser *et al.* 1980), and the glacial margins in the Huron and Erie basins are known to have retreated on several occasions (Dreimanis and Goldthwait 1973). Pollen spectra, and the oxygen isotopic record from deep-sea cores, speleothems, and ice sheets, however, generally show that there could not have been sufficient retreat or disintegration of the ice masses to raise the sea to its present level (Dansgaard *et al.* 1969, Broecker and Van Donk 1970, Epstein *et al.* 1970, Shackleton and Opdyke 1973, Thom 1973, Harmon *et al.* 1978b, Gascoyne *et al.* 1980, 1981). The fairly high sea levels which have been reported at about 60, 40, and 30 ka BP may correspond to the interstadial periods of Ontario. These are the Port Talbot I, whose age has not been reliably determined but may be about 60 ka BP; the Port Talbot II, between 48–42 ka BP; and the Plum Point, at about 30 ka BP. There is also some agreement with the interstadials of northwestern Europe; the Hengelo between 40.5 and 36 ka BP, and the Denekamp at 30 ka BP (Bowen 1978). There does not, however, appear to have been a corresponding rise in sea level during the Moershoofd interstadial between 50 and 43 ka BP. The minimum age of the Odderade interstadial is 58 ka BP, but it is not yet possible to correlate it, or indeed the earlier Brørup and Amersfoort interstadials, with the high sea levels which occurred before 60 ka BP. Even if high interstadial sea levels do correspond to minimum ice volumes, these levels must have been much lower than today's, at least on the basis of the climatic–geologic record (Mörner 1971a).

Low Pleistocene Sea Levels

For the obvious reason that the evidence is usually submerged, much less is known about low levels in the Pleistocene than high levels. It is difficult to obtain reliable dates from submerged sediments, because of the alteration of samples. Most estimates of low sea levels, therefore, are based upon the presence of terraces and terrestrial and near shore sediments on the continental shelf, and approximations of the amount of water stored in the ice sheets during the glacial periods. Maclaren (1842) calculated that the growth and decay of ice sheets caused sea level to vary by as much as 213 m; Tylor (1869) obtained a figure of about 180 m. Several workers have suggested that the maximum lowering occurred before the last glacial period (Daly 1934, Donn *et al*. 1962), although Fairbridge (1961a) considered that glacial sea levels were progressively lower throughout the Pleistocene. He proposed that sea level minima in the earlier glacial periods were very close to the relatively high sea level of today (Fairbridge 1971). Levels of between about 80 and 130 m below today's have generally become accepted as being representative of the amount of sea level lowering in the last glacial period. There is, however, some evidence that much greater depths were attained in some parts of the world. Around the Huon Peninsula in New Guinea, sea level was about 150 m below today's in the last and in the penultimate interglacials (Chappell 1983a). In the Arafura Sea between Australia and New Guinea, shallow water coral in a terrace 200 m below present sea level was probably formed in the penultimate glacial period (Jongsma 1970). Terrigenous sands at a depth of 170 m on the Brazilian shelf may have been deposited at the beginning of a mid-Wisconsin transgression (Kowsmann and Costa 1974). Terraces off southern Australia, Baja California, and on the Bering Shelf, and the floors of submerged caves in Honduras and the Bahama Bank, are at even greater depths (Pratt and Dill 1974). Sediments and terraces on several seamounts in the northeastern Pacific are also indicative of very low glacial sea levels (Schwartz 1972). The oxygen isotopic record from deep sea cores also suggests that sea level fell to about 165 m below its present level at the maximum of the last glacial period (Shackleton 1977a), and even lower in previous glacial periods.

The Postglacial Period

The introduction of radiocarbon dating in the late 1950s and early 1960s led to the formulation of three basic perceptions of the way in which the sea arrived eustatically at its present position (see for example, papers in *Quaternaria* 14, 1971):

1. It rose at a diminishing rate in the Holocene, reaching its present level asymptotically (Shepard 1963, Jelgersma 1966);
2. It reached its present level between 5 and 3 ka BP, and has been essentially stable ever since (Godwin *et al.* 1958, McFarlan 1961);
3. It reached its present position between 5 and 3 ka BP, and then either gradually declined or fluctuated above and below today's level (Daly 1934, Fairbridge 1961a).

Because of the effects of innumerable local factors, the reliability of 'eustatic' sea level curves drawn by grouping data from several areas is very low (eg. Godwin *et al.* 1958, Fairbridge 1961a, Shepard 1963, Curray 1965). Even when data are collected from a single site, the elevation and age of former shorelines cannot be ascertained with sufficient accuracy to determine the precise manner in which the sea reached its present level.

Tidal gauge records, analysed by Fairbridge and Krebs (1962) and Mörner (1973) are oscillatory, and provide support for the contention that Holocene sea levels have fluctuated in response to changes in climate and glacial activity (Mörner 1971b). Walcott (1975), however, has noted that if Holocene oscillations were eustatic, then there should have been synchronous fluctuations of the ice sheets. He argued that this does not appear to have been the case, and in some areas it may have been the ground, rather than the sea, which oscillated up and down. Mörner (1971c) recognized the possibility of isostatic oscillations, but emphasized that most sea level curves converge at about 7 to 6.5 ka BP, rather than diverging through time as would presumably be the case if they reflected differences in isostatic movement.

The Gulf and Atlantic Coasts of North America

In the Mississippi Delta region and in south-central Louisiana, the sea may have reached its present level between 5 and 3 ka BP. (McFarlan 1961, Coleman and Smith 1964). In northwestern Florida, Holocene sea levels are also at their highest level today. It is not known whether this level was reached asymptotically, or following a prolonged stillstand in this area

(Schnable and Goodell 1968), although it is believed that the present level in southwestern Florida was attained asymptotically (Scholl *et al.* 1969).

Nearly all the sea level curves for the Atlantic coast suggest that the sea reached its present level asymptotically. Milliman and Emery's (1968) data have often been used to represent changes in sea level on the Atlantic coast (MacIntyre *et al.* 1978). Curray's (1965) curve for the Gulf coast, and Milliman and Emery's for the Atlantic, differ by as much as 50 m over portions of the curves. Milliman and Emery's curve shows a much lower sea level between 16 and 11 ka BP than Curray's, but it is higher from 25 to 18 ka BP. On the South Carolina shelf before 10 ka BP and in Bermuda at 6 ka BP, however, sea level appears to have been much higher than was proposed by either Curray or by Milliman and Emery (Harmon *et al.* 1978a, Blackwelder *et al.* 1979). Blackwelder *et al.* believed that the South Carolinian shelf has been stable since the beginning of the late Wisconsin period, but tilting of the Delaware and Long Island coasts could explain why Milliman and Emery's sea level curve is too deep at the glacial maximum (Dillon and Oldale 1978). The deepest part of Milliman and Emery's curve was defined by data derived from regions where the shelf has subsided. The curve was modified by Dillon and Oldale to prevent the postglacial sea level from falling below 100 m. Shallower depths were also proposed as a result of a study in southeastern Massachusetts (Oldale and O'Hara 1980). Relative sea level was about 70 m below its present level in this area at 12 ka BP, although further south it was not much below 30 m (Blackwelder 1980).

There has been an abrupt retardation in the rise in sea level on the Atlantic coast in the late Holocene. Reported rates for the last few thousand years have ranged between 0.1 and 1.4 mm yr^{-1}. This has been interpreted as evidence of subsidence and differential warping of the shelf (Stuiver and Daddario 1963, Newman *et al.* 1980a, Blackwelder 1980). Relative sea levels rose most rapidly along the northern Atlantic coast. The New England shelf has probably subsided as a result of glacial rebound to the north, although it may have risen in earlier postglacial time. In the Boston area for example, the crust rose between 14 and 6 ka BP and has been subsiding ever since (Kaye and Barghoorn 1964). Subsidence appears to have begun in the vicinity of New York City, moving later into southwestern Maine, towards the centre of Wisconsin glaciation (Fairbridge and Newman 1968). Dillon and Oldale (1978) compared this pattern to the collapse of a peripheral bulge around an area of glacial rebound.

The pattern continues northwards into the Canadian Maritime Provinces (Grant 1970, 1980). Areas well beyond the ice margin experienced a brief period of rebound, followed by continuous submergence associated with the eustatic rise in sea level, and possibly with forebulge collapse. Areas at or near the margins of the ice, as in southern New Brunswick and Nova Scotia, Prince Edward Island and Newfoundland, emerged until uplift waned, and then began to submerge; this transition occurred about 7 ka BP in this area (Scott and Medioli 1982). It may have been during this period of stability or slow submergence following initial uplift that the 6 m Micmac terrace of the St Lawrence was formed (Goldthwait 1911). Areas such as western Newfoundland, which were further under the ice and therefore greatly depressed, experienced uplift until quite recently, when stability or subsidence finally allowed the transgression to begin. In the Northumberland Strait between Prince Edward Island and New Brunswick, four terraces were formed during the postglacial transgression. The area appears to have lain between the zones of positive and negative crustal movements in the Maritimes, although it has been subsiding for the last 7 ka BP (Kranck 1972). Further north, in the northern Gulf of St Lawrence, the land has still not completed the uplift or emergence phase.

Hillaire-Marcel and Occhietti (1977) analysed several hundred radiocarbon dates from eastern Canada and New England to show the interplay between ice position, isostatic rebound, and eustatic recovery. Relative sea level curves show a strong inflection in the peripheral areas of glaciation. Emergence in the eastern part of Hudson Bay has resulted in the formation of nearly two hundred raised beaches, ranging in age from about 8.5 ka BP to the present time (Hillaire-Marcel and Fairbridge 1978, Hillaire-Marcel 1980). Relative sea level in eastern Baffin Island reached its Holocene maximum at about 8 ka BP, when local ice readvance may have caused isostatic depression or crustal stability (Andrews *et al.* 1972, England and Andrews 1973, Pheasant and Andrews 1973, Andrews 1980). No evidence has been found of this transgression in southern Ellesmere Island, but a transgression may have occurred about 5 ka BP (Blake 1975). A beach formed at that time in the Queen Elizabeth Islands is now more than 25 m above sea level (Blake 1970, 1976). Emergence also dominated the Holocene history of sea level change in Greenland, although submergence has occurred within the last thousand years in the west (Washburn and Stuiver 1962, Donner and Jungner 1975).

Northern and Northwestern Europe

A number of detailed studies have been made in the coastal plain of the Netherlands. It has usually been reported that the sea rose to its present level asymptotically in this area. Jelgersma (1966, 1980) showed that sea level rose rapidly in the Netherlands and in the adjacent North Sea before 6 or 7 ka BP, gradually slowing down as it approached its present level. Her data were essentially derived from the radiocarbon dating of peat lying on top of an inclined Pleistocene surface. It was assumed that peat formation was initiated by the rising groundwater table, induced in turn by marine transgression; this assumption has been contested (Van de Plassche 1981). It should also be noted that although the Netherlands curve of Jelgersma has frequently been used to approximate global eustatic changes, the area is subject to the downwarping associated with its position in the North Sea (Zagwijn 1974).

A similar sea level curve has been devised for the northeasterly coast of France (Mariette 1971). Paepe (1971), however, recognized three 'Dunkerkian' trangressions in Belgium within the last 2.2 ka, and archaeological and geological evidence also suggest that the sea on the northern and western coasts of France has been above its present level on several occasions in the Holocene (Delibrias and Guillier 1971, Delibrias *et al*. 1971, Bourdier 1971). Ters (1975) found that there have been seven transgressions and six regressions on the Atlantic coast in the last 8.25 ka, although the sea never rose above its present level in that time. Gabet (1971) considered that the sea on the mid-Atlantic coast reached today's level at about 2 ka BP.

Relative sea levels fell almost exponentially from a maximum marine limit of 220 m in southeastern Norway, where isostatic uplift was very rapid (Hafsten 1983). In Svalbard and northern and western Norway, however, Holocene emergence was interrupted by the equivalent of the Tapes transgressions (Donner 1980). In western Norway, an early Boreal high sea level stand was succeeded by a Boreal regression and a late Boreal transgression, which was responsible for an extensive Tapes beach ridge (Hafsten and Tallantire 1978, Hafsten 1983). The Tapes transgression reached a maximum elevation of about 26 m above present sea level between 7.5 and 6 ka BP in northern Norway, and 11 to 12 m on the southwestern coast. In northern Finland, uplift of the land has been pronounced, but in the southern part of the country, rapid marine regression was first interrupted by the Ancylus transgression and later by the Littorina (Eronen 1983). In the Helsinki area of the Baltic, the Littorina transgression reached 35 m above today's level at 7 ka BP. Mörner (1969, 1971d,e) investigated the relationship between upheaval of the land and recession of the ice on the west coast of Sweden and the Kattegatt Sea. His eustatic sea level curve is oscillatory when the isostatic influence is removed, although the amplitudes are much smaller than Fairbridge's, and its greatest elevation at about 3.5 ka BP is only about 0.4 m above today's level. Mörner's curve can be correlated with similar records from the Baltic coast of southern Sweden (Berglund 1971) and from southwestern Sweden (Digerfeldt 1975). Koster (1971) also recognized sea level oscillations within the last 2 ka on the coast of northern Germany. De Jong and Mook (1981), however, have argued that apparent sea level oscillations are generated by irregularities in the ^{14}C time scale. The concentration of ^{14}C in the atmosphere varies sinusoidally with time because of changes in the magnetic properties of the solar wind, and changes in the Earth's geomagnetic field (Barbetti and Flude 1979, Stuiver and Quay 1980, Stuiver 1980). De Jong and Mook showed that radiocarbon dates from peats, corresponding to a uniform rise in sea level, could suggest alternations of slow and rapid transgression.

At least three distinct regions have been distinguished in Britain: the southeast, which is associated with the tectonically sinking North Sea basin; the southwest or Celtic Sea area, which is fairly stable; and the north, where isostatic recovery has been, and in some regions may still be, active (Mitchell 1977, Jardine 1981). A great deal of data have been collected on Holocene sea levels in Britain, but they cannot be integrated as yet to provide a regional synthesis (Shennan 1983). Many of the conclusions pertaining to particular sites must be considered tentative, and differences in data collection, terminology, and interpretation frustrate efforts to correlate changes in sea level between different regions of the country. Errors in the calculation of the elevation and age of shorelines only allow the identification of a broad sea level band, and they make it impossible to resolve the arguments over whether changes in sea levels were oscillatory or smooth (see papers for example, in *Proc. Geol. Assoc.* 1982, vol. 93).

Tooley (1974, 1979, 1982) made a detailed investigation of the biostratigraphy of the lagoonal, tidal flat, and sand dune zones of Lancashire and adjacent areas in northwestern England. Twelve transgressive overlap and 12 regressive overlap periods were identified in this area. Several similar episodes also occurred in the Humber estuary of eastern England (Gaunt and Tooley 1974). In the Fens and the Lincolnshire marshes, five positive and four negative sea level tendencies have been recognized before 2.5 ka BP, and several less reliable tendencies since that time (Shennan 1982). Recent work in the Norfolk Broads has identified positive sea level tendencies from at least 7.5 to 4 to 5 ka BP, when the sea rose from 19.3 to 5.5 to 6.5 m below its present level. A second period of positive tendencies had began by at least 1.973 ka BP (Coles and Funnell 1981). The estuaries and marshes of Essex experienced positive and negative changes in sea level (Greensmith and Tucker 1973, 1980). Sea level may have been higher than at present in this area within the last 1.5 ka, but this conclusion must be considered tentative because of the paucity of boreholes and other problems with the data base. It has also been suggested that sea level has been oscillating within the last 8.5 ka in the Thames Estuary and elsewhere in southeastern England (Devoy 1977, 1979, 1982). In the Somerset Levels and in Cardiganshire, smooth, nearly exponential sea level curves were constructed from radiocarbon data by Kidson and Heyworth (1976, 1978). They proposed that the apparent similarity between these two areas suggests that they either lay outside the area of rebound induced by Devensian ice loading, or within an area where it was completed at an early stage. It is difficult at present to recognize any oscillations which may have been superimposed on the general rise in sea level in southwestern Britain, possibly because of their small amplitude relative to the potential errors arising from other factors (Heyworth and Kidson 1982).

Postglacial sea levels in Scotland reflect the varying effects of glacio-isostatic and eustatic influences (Sissons 1967, 1976, 1983, Donner 1970, Sissons and Brooks 1971, Jardine 1978, 1982). Raised strandlines rise towards the centre of glacial accumulation, although in Shetland and other peripheral areas there is evidence only of subsidence. Late-glacial sea levels were much higher than today in many areas in Scotland. The late Devensian marine limit was about +50 m in the east, and around +41 m in the west (Jardine 1982). The most detailed investigations have been made in the southeastern part of the country, where they are based upon radiocarbon dating and information from several thousand boreholes. In east Fife, tilted raised shorelines have been dated at between 17.6 and 14.75 ka BP (Andrews and Dugdale 1970, Sissons 1976). In southwestern Scotland, the sea was +25 m in the Allerød interstade, between 12 and 10.7 ka BP (Jardine 1964, 1971), and it was at least +20 m at Oban in western Scotland by 11.6 ka BP (Sissons 1974). Relative sea level fell in the period before about 8 ka BP, when uplift of the land exceeded the eustatic rise. As the two great northern ice sheets melted, however, the eustatic rise eventually outstripped the waning uplift of the land, and the sea rose to form the Main Postglacial Shoreline (Jardine 1964, 1971, 1975, Donner 1970). The subsequent intermittent fall in sea level resulted in the deposition of carse muds in the sheltered parts of the Forth valley, and the formation of several lower shorelines (Sissons 1976, Jardine 1964, 1971, Gray 1974).

Isostatic recovery in northern Ireland began between about 18 and 16 ka BP, with the progressive northerly retreat of the late Midlandian ice sheet. Raised shorelines in the north of Ireland tilt to the west and south, declining in elevation from 20 to 22 m in the north, to below present sea level south of Dublin. Relative sea level in the north had fallen to possibly 30 to 60 m below today's level by about 12 ka BP. Later inundations of the coast were associated with the waning of isostatic movement (Carter 1982, Devoy 1983).

The Southwestern Pacific

The question of whether Holocene sea levels have been higher than today's has been the central theme of numerous studies of elevated strandlines in the southwestern Pacific. Daly (1920a,b, 1934) thought that the fresh-looking terraces which fringe many of the Pacific islands were formed several thousand years ago, when the sea was up to 6 m above its present level. Other workers have emphasized the presence of terraces which are only a few metres above today's level (Wentworth and Palmer 1925, Kuenen 1933, Stearns 1945). Fairbridge (1961a) believed that the Abrolhos terrace is well represented on many Pacific islands. Several contributions dealing with this subject were published in a special issue of *Zeitschrift fur Geomorphologie* (Supplement Band 3, 1961).

Schofield and Suggate (1971) considered that the evidence for a higher Holocene sea level exists in many areas of New Zealand and the southwestern Pacific where tectonic activity can be discounted. For example, reef rock a metre or more above today's sea level has been radiocarbon dated at between 4 and 1.2 ka BP on Bora-Bora and Mopélia in the Society Islands, Jarvis and Starbuck in the Southern Line Islands, Enderbury and Phoenix Islands in the Phoenix chain, and on Ifalik Atoll in the Caroline Islands (Guilcher 1969, Labeyrie *et al.* 1969, Tracey and Ladd 1974). Six transgressions have been identified on the Gilbert and Ellice Islands, occurring at intervals of about 660 radiocarbon years (Schofield 1977, 1980). These transgressions have been correlated with sea level fluctuations recorded in northern New Zealand (Schofield 1960). Further evidence for a high Holocene sea level in New Zealand was provided by Schofield and Suggate (1971), although based on the evidence from the southern part of the country, Suggate (1968) expressed little confidence in Fairbridge's (1961a) oscillating sea level curve.

Tectonic instability played an important role in several parts of the southwestern Pacific. According to Baltzer (1970), the Holocene transgression in New Caledonia may have preceded and been more rapid than on other islands. Rapid transgressions and low maxima can be partly explained by tectonic subsidence in this area (Lalou and Duplessy 1977). The effect of tectonic movements combined with eustatic changes also account for high Holocene levels on the Île des Pins in New Caledonia, Mare in the Loyalty Islands, and Espiritu Santo in the New Hebrides (Launay and Recy 1972).

Many workers have disputed the evidence for high Holocene sea levels. Johnson (1931) considered that the '2-m bench' in the Pacific, is the product of contemporary storm wave action. Shepard and Curray (1967) pointed out that shells could have been carried onto raised shorelines by storm waves, natives, and even birds, and later cemented into the beaches by groundwater. They could therefore provide dates which are much younger than the beaches themselves. A number of workers have discussed the results of an expedition to Guam and to 33 islands in the Caroline and Marshall chains of Micronesia (Shepard *et al.* 1967, Curray *et al.* 1970, Newell and Bloom 1970). None of these workers found any elevated coral reefs which could be unequivocally established as growing *in situ*. Many ridges and low, flat-topped terraces extending up to and even above the high tidal level consist of cemented coral rubble. These terraces were thought to have been constructed by storm waves, possibly generated by hurricanes or tsunami. Although samples provided dates between about 4.5 ka BP and the present time, none of the workers found any convincing evidence that Holocene sea levels in the Caroline and Marshall Islands were higher than at the present time. Furthermore, Bloom (1970) argued that the stratigraphy and morphology of swampy coastal plains in the eastern Carolines are consistent with a submergence of about 6.2 m in the last 6.5 ka BP, rather than with recent emergence. The controversy over the occurrence of high Holocene sea levels in the south Pacific involves a vertical range of elevations of only ±3 m (Bloom 1980). Interpretation is made even more difficult by local tectonic movements, and by the presence of storm and tsunami debris (Easton and Ku 1980). Bloom (1980) pointed out that many islands in the southern Pacific, including Western Samoa, Fiji, and Santo and Efate in the New Hebrides, have a unique tectonic history which has often been ignored. Bloom expressed the hope that the reconstruction of Holocene sea level events in the Pacific would never again be based upon a single radiocarbon date from one island.

Australia and Adjacent Areas

Evidence for Holocene sea levels above today's has been reported from central and northern Queensland, Victoria, Tasmania, and from South and Western Australia. The evidence is apparently absent, however, in New South Wales and northern Australia. Numerous reports have been made of sea levels up to 4.8 m above its present level in northern and north-central Queensland (Hopley 1971, 1978, 1980, Smart 1977, McLean *et al*. 1978, Belperio 1979, Chappell 1983b). In Victoria, evidence of Holocene sea levels has been found up to 3 m above today's level (Gill 1961, 1964, 1965, Ward 1971, Gill and Lang 1982). Similar levels have been identified in Tasmania and the Bass Strait (Davies 1959, Jennings 1961, Gill 1973b), South Australia (Ward 1965) and Western Australia (Teichert 1950, Fairbridge 1948, 1950, 1961a, Wyrwoll 1977).

No conclusive evidence for high Holocene levels has been found in New South Wales and southern Queensland (Hails 1965, Hails and Hoyt 1971, Thom and Chappell 1975). High levels in southeastern Australia have not been confirmed by the radiocarbon dating of shells, wood, and freshwater peats, and evidence such as aboriginal kitchen middens, which has been used to support their occurrence elsewhere, has been contested (Hails 1965, 1968, Langford-Smith and Hails 1966, Thom *et al*. 1969, 1972). Evidence of high Holocene sea levels is also lacking in the southern and south-central parts of Queensland (Hopley 1978). A chernier plain north of Rockhampton, for example, provided no evidence of the sea level oscillations identified by Fairbridge and Hopley (Cook and Polach 1973).

On New Guinea, the sea was still several metres below its present level by 6 ka BP. This is in fair agreement with the situation in the eastern Carolines, where the sea has risen by more than 6 m in the last 6.5 ka, but not with the Marshall Islands, where it may have been above its present level at 6 ka BP (Bloom 1970, Tracey and Ladd 1974, Chappell and Polach 1976). Sea level may also have been higher than today in this period in Indonesia (Fujii *et al.* 1971).

The Indian Ocean

Beach ridges, shell deposits, notches, emerged reefs, bioerosional borings and other evidence suggest that Holocene sea levels have been higher than today in the Indian Ocean (Taylor and Illing 1969, Geyh *et al.* 1979). On the Laccadives, most storm beaches were formed either in the last five hundred years, or from 3 to 2 ka BP. This could reflect changes in sea level, although it may also be attributed to variations in storminess (Siddiquie 1980). McIntire (1961) attributed the occurrence of several features up to 2 m above present sea level on Madagascar to local diastrophism, geological structure, and contemporary storm wave action.

West Africa

Holocene sea levels higher than today have been identified in west Africa north of Dakar in Senegal, but only in a few areas south of Gambia (Faure 1980). Maximum sea levels in Morocco, Senegal, and Mauritania, were several metres above the present level (Gigout 1959, 1971, Faure and Élouard 1967, Faure *et al.* 1980). There is evidence in several areas of a second high level, probably separated from the first by an intervening period when the sea was below today's level (Davies 1971, Einsele *et al.* 1974). Hoyt (1967) found no signs of high levels in southwestern Africa, but they have been reported near Cape Town (Fleming 1977).

The Mediterranean

Evidence of Holocene sea levels 2 to 4 m above today's has been found throughout the Mediterranean (Butzer 1975, Bortolami *et al.* 1977, Erinc 1978). The evidence in the western basin, however, comes from deltaic regions, or from volcanically or seismically active zones (Flemming 1969). Flemming believed that there has been no net change in sea level in the last two thousand years, but Pirazzoli (1976) argued that the archaeological evidence between Marseilles and Naples suggests that the sea has risen gradually to its present position.

South America and the Caribbean

Holocene relative sea levels may have been particularly high in Argentina, partly at least, because of tectonic movements of the land (Richards and Broecker 1963, Urien and Ottman 1971). Sea levels in Brazil were probably oscillatory, reaching up to a few metres above its present level on several occasions (Van Andel and Laborel 1964, Delibrias and Laborel 1971, Fairbridge 1976a,b, Suguio *et al.* 1980, Martin *et al.* 1980). Emerged shorelines have been dated in several areas on the west coast of South America, although uplift played an important role in these areas (Paskoff 1980). In the Bahamas, Lind (1969) found that the oscillating sea level curve was broadly similar to Fairbridge's, although he considered this to be coincidental. Lind's data contradict Newell's (1961) observation that the sea in the Bahamas and elsewhere in the western Atlantic is presently at its highest Holocene level.

North Pacific

High strandlines of Holocene age are common features in much of the northern Pacific. They are particularly high in Taiwan, where they are up to several tens of metres above present sea level (Pirazzoli 1978). The sea level record is oscillatory, but this is an area of severe crustal instability (Fujii *et al.* 1971). On the tectonically active Ryukyu Islands, sea level stabilized between 3.5 and 1.7 ka BP, when it was less than a metre above its present level (Koba *et al.* 1982). High Holocene sea levels have been reported from the east China Plain, where it reached its present position about 2 ka BP, and from Japan (Iseki 1978). Radiocarbon dating of shell middens suggest that Holocene sea levels have been oscillating in a similar way in Japan and Brazil (Fairbridge 1976a,b, Taira 1980). Much of the northwestern Pacific, however, consists of large, unstable blocks which can move independently of each other (Pirazzoli 1978). Similarities between the sea level events within this area, or with other distant regions, must therefore be largely coincidental.

A tundra-covered bench, less than a metre above present sea level, can be traced for 2,000 km across the Aleutians (Powers 1961). Black (1974, 1980) found that the sea rose to its present position about 5 ka BP on one island, and as long ago as 11 ka BP on others. High sea levels also occurred in Alaska between 6 and 5 ka BP (Karlstrom 1968). Frozen peats were buried by a transgressive sea between 4.6 and 2.7 ka BP at Point Barrow (Brown and Sellmann 1966), and at least two other periods of high sea level have occurred within the last two millennia (Hume 1965). Tectonic–isostatic movements are likely to be important in these high latitudes, however, and it is difficult to separate their effect from those of purely eustatic mechanisms (Black 1980).

The postglacial marine limit in British Columbia was at least 200 m near Vancouver and Kitimat on the mainland coast, but only 75 m at Victoria (Clague and Bornhold 1980). The late Pleistocene marine limit was below present sea level on the Queen Charlotte Islands on the outer coast, where there was little or no glacial depression of the land (Clague 1983). Rapid isostatic uplift took place in coastal British Columbia in the late Pleistocene and the early Holocene (Andrews and Retherford 1978). A eustatic rise in sea level or forebulge migration caused a marine transgression in the late Pleistocene–early Holocene on the Queen Charlotte Islands, and in the mid-Holocene on the mainland and on eastern Vancouver Island (Clague 1983). Within the last 5.5 ka the sea has risen to today's level on the inner coast, and fallen in places on the outer coast (Clague and Bornhold 1980). The Holocene sea level on the Santa Monica shelf in California is at its highest level today. The curve is similar to that for the Texas shelf (Curray 1965), although relative to Texas, the Californian shelf may have risen before 10 ka BP, and subsided thereafter (Nardin *et al.* 1981).

Wentworth and Palmer (1925) attributed a bench between 1.2 and 3.7 m above present sea level on the Hawaiian Islands to a sea level about 4 m above today's. Stearns (1935, 1961) identified two low benches in Hawaii. The lowest, only a little above mean sea level, is always awash at high tide; the upper bench, between 0.9 and 2.4 m above mean tidal level, is swept by storm waves. Stearns believed that these benches are probably the result of contemporary wave action, although the upper surface could be the result of a former sea level 2 to 4 m above its present level. Support for a contemporary origin was provided by a detailed study of a fringing reef on Oahu (Easton and Olson 1976). Data from 63 radiocarbon-dated samples suggested that it is unlikely that the sea has been higher than today within the last 3.5 ka.

The Oxygen Isotopic Record

Variations in the isotopic composition of deep sea cores can be used to substantiate sea level records derived from radiometrically dated strandlines. When water evaporates, the isotopes ^{16}O, ^{17}O, and ^{18}O are carried away at different rates. At the beginning of a glacial period, water evaporated from the oceans contains a disproportionately large amount of the lighter ^{16}O isotope. As this water is used for glacial growth, the oceans become slightly enriched with the heavier residual isotope ^{18}O. The analysis is usually made on planktonic and benthic foraminifera, but corals, coccoliths, molluscs, and other organisms have been used. The isotopic composition of the foraminifera depends upon the temperature and the isotopic composition of the sea water. Emiliani (1955, 1966, 1971) thought that isotopic variations are largely the result of changes in the temperature of the water, so that they essentially provide a record of palaeotemperatures. Most other workers, however, believe that the isotopic record is primarily of the fluctuations in the isotopic content of sea water, and consequently of the alternate growth and decay of ice sheets (Olausson 1965, Shackleton 1967, Dansgaard and Tauber 1969, Broecker and Van Donk 1970, Shackleton and Opdyke 1973, Imbrie *et al*. 1973). According to Shackleton (1977a), the oceans were enriched with ^{18}O by about 0.1‰ for every 10 m drop in sea level. This relationship allows the Pleistocene sea level record to be estimated from isotopic data. The record, however, can be obscured by bioturbation of the sediment. The effect of postdepositional mixing is greatest where sediment accumulates slowly, extreme isotopic fluctuations being less likely in cores from areas where sediment accumulation is fairly rapid (Shackleton 1977b). In much of the world's oceans, accumulation rates are less than 0.2 mm yr^{-1}, but several areas are known where the rates are high enough to provide a meaningful record of isotopic variations (Shackleton 1977a,b).

Shackleton and Opdyke (1973) defined isotopic stages at particular depths in an excellent core from the Pacific Solomon Rise (V28–238), placing the isotopic record within the framework of palaeomagnetic reversals (Fig. 7.3). They defined 22 stages, the first 16 corresponding to those of Emiliani (1966). These stages were determined by reference to the position in the core of the Brunhes–Matuyama magnetic boundary, and then by extrapolation along the core, assuming a constant rate of sedimentation. The dates corresponding to the boundary of each stage are somewhat older than those of Broecker and Van Donk (1970), and as much as 20% older than Emiliani's (1966). A number of cores now extend to the base of the Pleistocene, including: V28–239, (Fig. 7.4), which was taken from close to the site of V28–238 in the Pacific (Shackleton and Opdyke 1976); V28–179 (Fig. 7.5) from the equatorial Pacific (Shackleton and Opdyke 1977), and V16–205 (Fig. 7.6) from the tropical Atlantic (Van Donk 1976).

Shackleton (1977a) found that the isotopic record is basically similar in more than 60 deep sea cores from around the world. Van Donk (1976) has compared his record from core V16–205 from the tropical Atlantic, with the records from Caribbean core V12–122 (Broecker and Van Donk 1970, Imbrie *et al*. 1973), Pacific core V28–238 (Shackleton and Opdyke 1973), and Caribbean core P6304–9 (Emiliani 1966). The records were

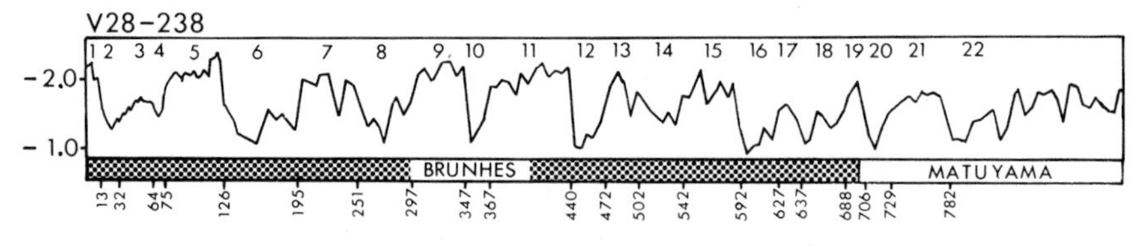

Figure 7.3. Implied changes in sea level for about the last 900 ka obtained from deep sea core V28-238 from the Solomon Rise in the Pacific Ocean. Stages are identified below the upper axis. The vertical axis represents the deviation per mil from the Emiliani Bl standard, 0.29‰ must be added to these values for comparison with the PDB standard (Shackleton and Opdyke 1973).

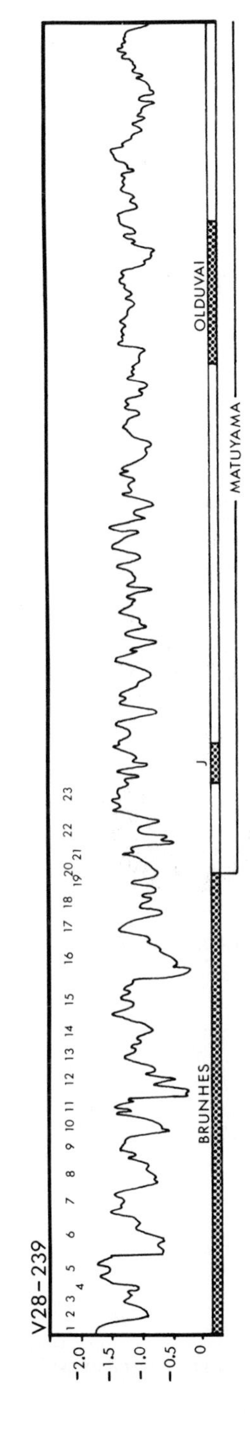

Figure 7.4. Implied changes in sea level for the entire Pleistocene from core V28-239 from the Solomon Rise. The deviation on the vertical axis is referred to the PDB standard (Shackleton and Opdyke 1976).

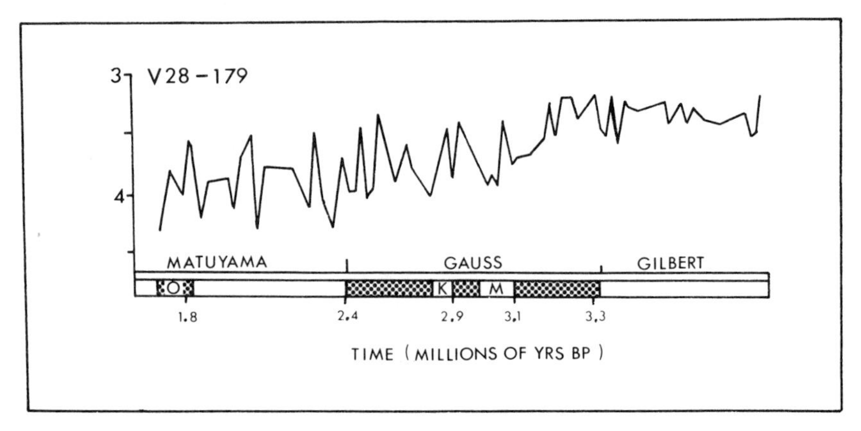

Figure 7.5. Implied changes in sea level for the Pleistocene (>3.5 Ma) from core V28-179 from the equatorial Pacific, north of New Guinea. The deviation on the vertical axis is referred to the PDB standard (Shackleton and Opdyke 1977). Higher values than in V28-238 and V28-239 are the result of the use of a different species of foraminifera.

found to be similar in several ways, although there are also some significant differences. Van Donk (1976) found that the sea was as high or higher than at present on 12 occasions within the last 1.9 Ma. A particularly prominent high level appears in his record at about 1.4 Ma BP, and others occur in stages 5, 9, and 15. He considered that periods of glaciation equivalent to the Wisconsin maximum, however, occurred on only four occasions, all within the last 750 ka. Before this, there appears to have been much less glacial ice. The records of Emiliani (1966) and Shackleton and Opdyke (1973) show that with the exception of stage 14, each glacial period seems to have been of roughly the same intensity throughout the Brunhes. The isotopic values of the interglacials are also quite similar in these records, although stages 3 and 17 are lower, and 1 and 5 are higher than the others

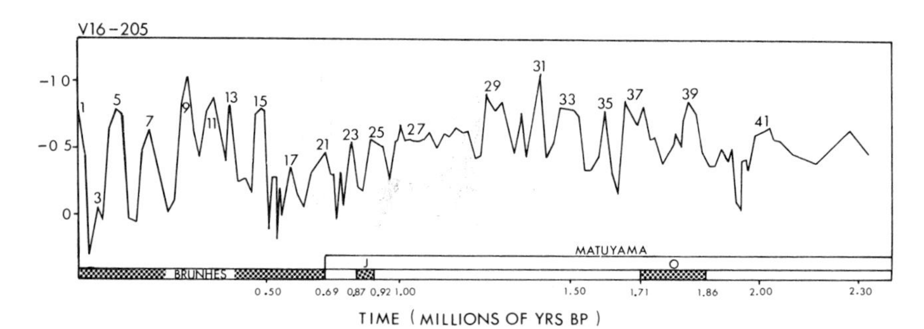

Figure 7.6. Implied sea level changes for the last 2.3 Ma from core V16-205 from the tropical Atlantic. The deviation on the vertical axis is referred to the PDB standard (Van Donk 1967).

(Emiliani and Shackleton 1974). Some of the differences between these records may be the result of the distortion of Van Donk's Atlantic record because of a very low rate of sedimentation (5.5 um yr^{-1}) (Shackleton 1977b); or possibly because of $CaCO_3$ dissolution cycles, which have increased the amplitude of the isotopic fluctuations in the Atlantic and reduced it in the Pacific (Erez 1979). Mörner (1981a), however, has argued that the differences between the records reflect geoidal changes in the water mass distribution over the Earth, which influence the isotopic composition and the micro-organisms in the sea.

Deep sea sediments also contain a palaeoclimatic record which is expressed by variations in their microfaunal composition. Temperature curves can be based upon changes along a core in the relative abundance or size of temperature-sensitive foraminifera (Malmgren and Kennett 1978). The record has been extended to what may be the base of the Pleistocene by correlating and overlapping core records with reference to magnetic reversals, and changes in the coiling direction of *Globorotalia truncatulinoides* (Ericson *et al*. 1964, Ericson and Wollin 1968, 1970, Wollin *et al*. 1971). A number of glacial and interglacial periods have been identified from Atlantic and Caribbean cores. The records show that a warm period occurred between about 150 and 100 ka BP, and a period of generally warm but fluctuating conditions between about 380 and 190 ka BP (Ericson and Wollin 1968). Variations in the Sr concentrations in two planktonic foraminifera from deep sea cores in the Indian Ocean suggest that the warmest climate in the last 300 ka was at about 125 ka BP, and that it was about the same as today from 220 to 200 ka BP (Cronblad and Malmgren 1981). Warm conditions within the Wisconsin have been identified at about 65, 40, and 25 ka BP (Wollin *et al*. 1971). The isotopic and faunal methods have been compared using Caribbean core V12–122. Imbrie *et al*. (1973) derived a number of mathematically defined faunal indices which describe the average salinity and the summer and winter temperature of the sea surface. They found that for this core, the broad chronological patterns derived from the isotopic and faunal methods are generally similar.

There is general agreement between the major elements of sea level curves based upon radiometrically dated strandlines and those inferred from the oxygen isotopic record. On Barbados, for example, (and possibly in Taranaki, New Zealand), each isotopic interglacial stage back to between 640 and 680 ka BP is represented by at least one reef (Bender *et al*. 1979, Pillans 1983). Numerous dated shorelines around the world testify to the high sea level implied by the isotopic values in stage 5, and in particular in the interglacial substage 5e. Shackleton and Opdyke (1973) showed that there is excellent correspondence between the occurrence of emerged reefs in Barbados and New Guinea (Broecker *et al*. 1968, Veeh and Chappell 1970) and high sea levels implied by the isotopic data of core V28–238. If it can be assumed that the isotopic data really do provide a reliable estimate of ice volume and sea level, then they confirm that sea level was higher than today in substage 5e, and that somewhat lower stands occurred at about 100 and 80 ka BP. Further confirmation of the occurrence of the peak of stage 5 at about 125 ka BP was provided by oxygen isotopic analysis of shells and reef crest coral on the terraces of Barbados (Shackleton and Matthews 1977, Fairbanks and Matthews 1978). The isotopic data from deep sea cores also indicate that there could not have been a mid-Wisconsin interstadial rise in sea level to a position close to its present level. Neverthless, the isotopic data of Emiliani (1966) and Shackleton and Opdyke (1973) do appear to underestimate the amplitude of the well-documented fluctuations in sea level which occurred between 115 and 15 ka BP (Chappell 1974a, 1983a).

According to Emiliani's (1966) and Shackleton and Opdyke's (1973, 1976) isotopic data, the sea in substage 5e was higher than at any other time in the middle and upper Pleistocene. Many workers have proposed that the sea was above its present level in the 'penultimate' interglacial period. In southern England, the Hoxnian level, which may in fact correspond to isotopic stages 11 and 13, rather than to stage 7 (Bowen 1978), may have been about 25 m above today (Stephens and Synge 1966, West 1972, Mitchell 1977). Similar levels between 250 and 222 ka BP have been reported from Morocco (Biberson 1970), Tonga (Stearns 1971), Hawaii (Stearns 1978), and Alaska (Hopkins 1973). Other workers have recognized strandlines corresponding to the penultimate interglacial, represented by isotopic stage 7 (251 to 195 ka BP, according to Shackleton and Opdyke 1973), which are only a few metres above the present level of the sea (Butzer 1975, Cronin *et al.* 1981, Harmon *et al.* 1983). Nevertheless, estimates of levels 7 to 32 m below today's, at various times between 220 and 180 ka BP (Labeyrie *et al.* 1969, Fairbanks and Matthews 1978, Harmon *et al.* 1978a, Marshall and Launay 1978), seem to be more consistent with the isotopic record, which suggests that the sea did not reach its present position in stage 7 (Shackleton and Opdyke 1973, Van Donk 1976).

Shackleton and Opdyke's (1973) data indicate that sea level in stage 9 (347 to 297 ka BP) was only a little above today's. Van Donk's isotopic record, however, suggests that it was considerably higher than at present. Reefs formed at this time on Barbados (Bender *et al.* 1973), on La Blanquilla off the coast of Venezuela (Schubert and Szabo 1978), and on Atauro near Timor (Chappell and Veeh 1978a). On the Huon Peninsula of New Guinea, the sea appears to have been close to its present level from about 336 to 320 ka BP (Chappell 1974b, 1983a).

Stage 11 (440 to 367 ka BP) is represented by strandlines on Réunion and Madagascar (Montaggioni 1974), on Barbados (Bender *et al.* 1979), in New Zealand (Pillans 1983), in Gippsland, Australia, and South Carolina (Ward *et al.* 1971), on Gran Canaria (Lietz and Schmincke 1975), on Oahu (Ward 1973), and on Atauro (Chappell and Veeh 1978a). As sea level was about the same (Shackleton and Opdyke 1973) or a little higher than today (Van Donk 1976), strandlines of this age, which are up to 29 m above sea level in South Carolina (Cronin *et al.* 1981), on Gran Canaria, Oahu, and Madagascar, were probably uplifted.

Other high sea levels, corresponding to stage 13 (502 to 472 ka BP) have been identified on Barbados (Bender *et al.* 1973, 1979); to stage 15 (592 to 542 ka BP) on Curaçao (Schubert and Szabo 1978), Barbados (Bender *et al.* 1979) and Atauro (Chappell and Veeh 1978a); to stage 17 (647 to 627 ka BP) on Barbados (Bender *et al.* 1979); to stage 19 (706 to 688 ka BP) on Atauro (Chappell and Veeh 1978a); and to stage 21 (782 to 729 ka BP) in Gippsland, Australia and South Carolina (Ward *et al.* 1971).

Deep sea records are further substantiated by variations in the isotopic composition of the Greenland and Antarctic ice sheets. Because the ^{18}O concentration of polar ice decreases with falling temperature, the $^{18}O/^{16}O$ ratio is largely determined by the temperature of the snow when it is deposited on the ice sheet. Ice core records are basically similar from Camp Century, Greenland, and from Byrd Station and elsewhere in Antarctica (Dansgaard *et al.* 1969, 1971, Johnsen *et al.* 1972, Epstein *et al.* 1970, Lorius *et al.* 1979), and they are also generally consistent with the records from deep sea cores. The Greenland record shows that the main cooling took place in the early Wisconsin, about 73–59 ka BP. Temperatures appear to have been more stable in the Middle Wisconsin (59–32 ka BP), and lower in the Late Wisconsin, when minimum glacial temperatures were attained.

Similar records have been obtained by radiometric dating and oxygen isotopic analysis of speleothems. Duplessy *et al*'s. (1971) record from southern France shows that there was a gradual rise in temperature between 130 and 120 ka BP. Warm conditions persisted with little variation until about 97 ka BP, when there was a sharp decline in temperature. Analysis of speleothems in Mexico, Bermuda, Kentucky, Iowa, and Alberta suggest that warm periods occurred between 190 and 165 ka BP, from 120 to 100 ka BP, and at 60 and 10 ka BP. Cold periods were identified from 95 to 65, and from 55 to 20 ka BP (Harmon *et al.* 1978b).

The Trend of Falling Sea Level

Many workers believed that glacio-eustatic fluctuations were superimposed upon a general fall in sea level in the Pleistocene (Zeuner 1952a,b, Butzer and Cuerda 1962, Hoyt and Hails 1967, Butzer 1971, 1976, Sparks and West 1972, Stearns 1978). Fairbridge (1961a) placed the Aftonian interglacial sea level about 100 m above today's, the Yarmouthian about 50 m above, and the Sangamon maximum about 18 m above. The concept of progressive glacio-eustatic lowering of the high interglacial sea levels, however, cannot be reconciled with the Pleistocene climatic record. The radiometric dating of emerged reefs suggests that interglacial sea levels were similar or only slightly above today's, at least within the Brunhes polarity epoch (eg. Bender *et al.* 1979). The temperatures of the interglacial periods appear to have been similar throughout the Pleistocene, according to such evidence as Zagwijn's (1975) pollen record, but the glacial periods may have become progressively cooler (eg. Kerr 1981a). If all the ice presently on the Earth melted, sea level would rise by between 60 and 80 m, assuming that the change faithfully reflected the loss of ice volume. The deep sea core records and other evidence indicate that this did not occur during the Pleistocene. It is possible that sea levels fell in the early Pleistocene because of tectono-eustasy, or because of changes in the position of the poles (Zeuner 1959, Fairbridge 1971). Many workers, however, believe that strandlines which are much higher than the maximum of the last interglacial period either predate the Pleistocene, or must have been uplifted after formation.

The growth of ice sheets has affected sea levels for at least 10 to 15 Ma, and possibly for considerably longer. The east Antarctic ice sheet had reached or even surpassed its present size by at least 12 to 15 Ma BP, although it may have been growing in the previous 10 or more million years. The west Antarctic ice sheet had reached its present size by 7 Ma BP, and by 3 Ma BP, major continental glaciation had begun in the northern hemisphere (Rutford *et al.* 1968, Mercer 1968a, Goodell *et al.* 1968, Berggren and Van Couvering 1974, Stump *et al.* 1980, Kerr 1981b, Woodruff *et al.* 1981). The fall in sea level during the latter part of the Tertiary probably reflects the progressive growth of these ice sheets, which was not reversed during the Pleistocene. Earlier falls in sea level, as at the end of the Cretaceous, may have been tectono-eustatic, involving an increase in the volume of the ocean basins. It is also possible, however, that the Earth had a significant ice budget at that time. Tertiary ^{18}O isotopic records suggest that glacio-eustatic sea level fluctuations could have occurred as far back as the Eocene, and even throughout much of the Cretaceous (Matthews and Poore 1980).

Mechanisms of Sea Level Change

Markedly different opinions have been advanced to explain the way in which the sea attained its present level. There is a strong geographical component to variations in Holocene sea level curves around the world.

The sea seems to have risen in the last 6 ka in large areas of the northern hemisphere, and fallen in much of the southern hemisphere. Pirazzoli (1977) has examined more than 700 radiocarbon dates covering the last 2 ka BP. He concluded that submergence was dominant on the western side of the north Atlantic, in western Europe, and most of the Mediterranean, and in areas adjacent to former ice caps. Emergence has characterized the shores of the Indian, Pacific, and Arctic oceans. Newman *et al.* (1980b) used up to 3,000 published radiocarbon dates to plot variations in Holocene sea levels. Glacio-eustatic uplift was identified as a persistent feature in an area extending from the western Canadian archipelago to Nova Zemlya. A second emergent area, extending from Japan and Taiwan to New Guinea, is probably associated with the boundary between several tectonic plates. Submergent areas include the northeastern United States, where the submergence might be the result of the downwarping of a passive continental margin, collapse of an ice peripheral bulge (Daly 1934), or sedimento-isostasy induced by the deposition of glacial outwash. In some other areas, the direction of vertical earth movement has been reversed within the Holocene.

This and other work clearly demonstrates that no areas can be regarded as being truly 'stable'. There are probably no places where relative changes in the level of the land and sea can be simply interpreted as reflecting eustatic changes in sea level. All sea level curves record changes in relative sea level, and they are relevant only to the areas from which the evidence was obtained. Attempts are now being made to provide an explanation for disparate and apparently contradictory sea level records. It is now recognized that the sea may have been above its present level, in for example Australia, while it was lower in western Europe, and that differences of this sort must have occurred not only in the Holocene, but also at other times in the past.

Several local factors can induce variations in sea level, although the effects are likely to be rather small on rock coasts, particularly when of short duration. These include changes in the temperature and salinity of the oceans, atmospheric pressure, wind stress, the coriolis effect, and river outflow and the hydraulic head in estuaries (Fairbridge 1966). Schofield (1980) argued that the maximum postglacial sea level was attained earlier in northern New Zealand than in the equatorial Pacific, reflecting the delay involved in the mixing of glacial meltwater with more saline sea water. The sea level evidence for this assumption has been disputed (Gill 1971), and no relationship has been found between surface layer densities and mean annual sea level in the English Channel and the North Sea (Rossiter 1962). Possibly of greater significance is the occurrence of maximum and minimum tidal generating potentials at intervals of about nine centuries. The last maximum occurred about 517 BP and the last minimum at

1400 BP, reflecting changes in the average generating force of about 6% (Pettersson 1914).

In addition to glacial eustasy, which involves changes in ocean water volume, major changes in the level of the ocean surface can occur as a result of (*a*) geoidal-eustasy, involving ocean mass/level distribution; and (*b*) tectono-eustasy, involving ocean basin volume (Mörner 1983).

Geoidal-eustasy

Geodetic measurements made from Earth-orbiting satellites have confirmed that the ocean surface is uneven, consisting of a series of humps and depressions. The geoid, or the geodetic sea level, is an equipotential surface of the gravitational and rotational potentials. Vertical and horizontal changes in the configuration of the geoid could, in extreme cases, induce quite rapid changes of up to 200 m in sea level. This would happen if the New Guinea 'hump' (+76 m) interchanged with the Maldive Islands 'depression' (−104 m), causing a 180 m regression in the former area and an 80 m transgression in the latter (Mörner 1976). If the configuration of the geoid is unstable (Mörner 1976, Newman *et al.* 1981), then eustatic curves derived from global data are meaningless, as eustatic changes in the ocean level would not produce parallel deformations of the geoid (Mörner 1981a). Changes in the geoidal configuration, therefore, provide a mechanism to which short-term fluctuations in sea level can be referred.

Many mechanisms can generate changes in the geoid. Geoidal changes resemble a precession cycle, and they can be matched by changes in the atmospheric ^{14}C production cycle and by palaeomagnetic events and excursions. This suggests that they have a common origin, possibly associated with changes in the core/mantle coupling and interface. Any relative movement of an irregular core/mantle topography could generate migrational deformations of the geoid (Mörner 1976, 1980b, 1981b).

The gross form of the geoidal ellipsoid could also be deformed by polar shift, and by changes in the tilt of the axis of spin (Mörner 1976, 1980b). Centrifugal forces generated by the Earth's rotation cause its geoid to swell at the equator and to be depressed at the poles. Any change in the rate of rotation, therefore, produces differential variations in sea level, distributed in a circumferential pattern around the Earth. Isostatic adjustment or eustatic sea level changes related to the growth or decay of ice sheets would alter the moment of inertia and the angular velocity of the globe.

Small changes in the Earth's rotational velocity have been inferred from the records of ancient astronomers (Munk and Revelle 1952, Fairbridge 1958, 1961a, Rossiter 1962), although it is not known whether the rotation rate has retarded cyclically or linearly. Glacial conditions may have prevailed throughout the Precambrian because of a rotation rate 2 to 2.5 times greater than today (Hunt 1979). The geological record suggests that sea level could have fallen by about 240 m in the equatorial regions since the Cretaceous, and risen by a similar amount in the Arctic. Eardley (1964) attributed these changes to a reduction in the Earth's rotation caused by the drag of lunar tides, although they could also be explained by the expansion of the Earth, or by the shift in mass associated with the formation of ice caps. Continental drift and the growth and decay of ice sheets could induce a shift in the position of the geographical pole (Munk and MacDonald 1960, Fairbridge 1971). Weyer (1978) considered the question of whether the load of the ice sheets in the last glacial period was sufficient to have produced detectable changes in the location of the geographical pole. Polar shift would cause sea level to change by variable amounts in different quarters of the globe (Jardetzky 1962). The sea level in coastal regions which moved closer to the equator would rise as they slipped under the equatorial bulge. Alternatively, sea levels would fall in those areas which shifted away from the equator. Theoretically at least, changes in sea level could be considerable. If for example, the poles shifted by only 1°, relative sea levels along the meridian of polar movement and at the 45° latitudes could change by up to 373 m.

Tectono-eustasy

World-wide sea level fluctuations could be caused by sedimento- and tectono-eustatic factors, as well as by the growth and decay of ice sheets (Fairbridge 1961a, 1966). Sea level would rise because of a progressive reduction in the volume of the ocean basins, caused by the deposition of sediments from the land. The effect, however, is likely to be quite small, because contemporary mean accumulation rates are only about 1 to 2 mm per century in deep water far from the shore, and perhaps five to six times greater near the coast. Tectonism may be responsible for major changes in the shape and the capacity of ocean basins (Worsley *et al.* 1984). Hallam (1963) considered that the relative relief of the Earth has been increasing since the Cretaceous. Local tectonic movements caused sea level fluctuations in the Tertiary about a regressive trend related to increases in ocean volume induced by subsidence and other major epeirogenic movements. Menard (1969) has discussed the mechanisms responsible for the subsidence of the oceanic crust. Grasty (1967) noted that periods of world-wide regression appear to have coincided with orogenic episodes and epochs of folding. Transgressions may occur when low orogenic activity, and hence reduced horizontal compression, causes the continents to subside. Conversely, increased horizontal compression during orogenic episodes causes the continents to rise, with the result that sea level falls. Major sea level changes can only occur if orogenic events occur simultaneously in different areas. Grasty suggested that the regression of the Cretaceous sea could have been associated with the collision of India and Asia, which occurred at about that time. It is generally accepted that the ocean floor has undergone considerable subsidence in the Cretaceous, and particularly in the Cenozoic. This could have been sufficient to have caused sea level to fall by more than 1,000 m. It actually fell by much less, possibly because of the release of water which was retained in the basaltic layers of ancient platforms (Rezanov 1979; see also Revelle 1955, and Menard 1969).

Epeironesis, orogenesis, and ocean floor spreading are closely related within the framework of plate tectonics (Jacoby 1972, Hallam 1981). Slower periods of ocean floor spreading appear to be associated with the peak phases of continental orogenies (Rona 1973). Mid-oceanic ridges are much larger than continental mountain ranges, and their growth or decay could cause sea level to rise or fall by as much as 500 to 650 m (Fairbridge 1961a, Hallam 1963, Russell 1968, Menard 1969, Flemming and Roberts 1973, Hays and Pitman 1973, Pitman 1979). Transgressions may be the result of the expansion of oceanic ridges associated with the rapid spreading of the sea floor, and net orogenic quiescence. Alternatively, regressions could result from ridge contraction associated with slow sea-floor spreading and orogenic compression of the continental crust (Rona 1973). Valentine and Moores (1970) suggested that regressions correspond to the assembly of continents, and transgressions to their fragmentation. Pulses in the rate of sea floor spreading may have been responsible for transgressions and regressions in the middle and upper Cretaceous. Rapid transgressions and much slower regressions in this period suggest that mid-oceanic ridges expand rapidly, but subside more slowly until the onset of a new expansive episode (Cooper 1977). In the post-Cretaceous period, changes in sea level could have been partly induced by suturing and continental underthrusting in the Alpine–Himalyan belt, and by movements on the Darwin Rise. It has been suggested that a 10 per cent reduction in ocean depth and a eustatic rise in sea level of about 350 m in the middle to late Miocene was the result of the uplift of the southern mid-Atlantic ridge by about 500 to 1,000 (Flemming and Roberts 1973). Alternatively, transgressions and regressions may be primarily the result of the tectonic movement of the continental blocks. Jeletzky (1978) has argued that with the exception of the Maastrichtian regression, and possibly a late Santonian/early Campanian regression, sea level changes in the Cretaceous and most of the Tertiary were local, rather than world-wide or eustatic events related to differential subsidence or uplift of the land.

The expansion of the Earth provides a possible driving mechanism for continental migration and sea-floor spreading (Van Diggelen 1976, Owen 1976, Wesson 1978, Van Flandern 1979, Stewart 1981, Vigdorchik 1981). Carey (1975) has reviewed the evidence for the expansion of the Earth. He noted that the major oceans of the world are fairly recent geological features. The Arctic, Atlantic, and Indian oceans only date from the

Mesozoic, and their size has doubled since the Eocene. The Pacific was much smaller before the Mesozoic than it is today. Machado (1967) considered that the pattern of alternating marine transgressions and regressions in the geological record suggests that the gravitational 'constant' has been pulsating. Maximum values appear to have occurred during the Carboniferous and the Cretaceous, and minimum values in the Triassic and at the present time.

The Origin of the Ice Ages

Two mechanisms which could provide a partial explanation for the occurrence of glacial periods have important implications for the interpretation of Pleistocene sea level oscillations.

The Antarctic Surge Theory

Wilson (1964, 1969) suggested that ice ages could be induced by the surging of Antarctic ice. The Antarctic may become unstable when ice thickness is sufficient to cause basal melting. A rapid surge could then cause about a quarter of the Antarctic ice to flow into the Southern Ocean, producing a vast floating ice shelf which would increase the Earth's albedo and so promote cooling. The albedo would further increase as secondary ice sheets developed in the northern hemisphere. The glacial period would end with the eventual breakup of the Antarctic ice shelf and the reduction in the Earth's albedo.

Hollin (1965, 1969, 1972) considered Wilson's theory with regard to its potentially important implications for effecting changes in sea level. The theory suggests that when the Antarctic ice surges, sea level rises very rapidly, possibly by 10 to 20 m in a century or less. The increased albedo, and therefore lower temperatures, could then promote the buildup of ice in the Antarctic and in the northern hemisphere and a slow fall in sea level, possibly by about 120 m in about 50 ka. Sea level would eventually begin to rise again when the Antarctic ice shelf, and then the northern ice sheets, broke up and melted. This would be succeeded by a slight fall in level associated with the continued buildup of the Antarctic ice sheet, preceding the next surge.

The ice surge theory implies that there should be a rise in sea level at the end of each interglacial period, at a time when temperatures are falling. An alternative explanation has been provided by Mercer (1968b, 1981). Lake sediments and inactive solifluction lobes in central Antarctica suggest that at some stage in the Pleistocene, summer temperatures were from 7 to 10°C above those of today. Temperatures this high would result in the rapid disintegration of the west Antarctic ice sheet and recession of the ice shelves. Most of the associated rise in sea level could take place within a century (Thomas *et al*. 1979). Mercer proposed that high sea levels during the latter part of the last interglacial were the result of meltwater flowing into the Southern Ocean, rather than surging of Antarctic ice because of a period of lower temperatures.

Fluctuations of the Antarctic ice sheet do not appear to conform in a simple way to changes in temperature. It has been proposed that the east Antarctic ice sheet thickened during the interglacials of the northern hemisphere, possibly because fairly moist, warm air was able to penetrate the Antarctic at those times (Denton *et al*. 1971, Hendy *et al*. 1979), although this has been disputed (Drewry 1980). Certainly, variations in Antarctic precipitation would induce long-term changes in sea level (Oerlemans 1981). Denton *et al*. (1971) recognized that the surface elevaion of the interior of the east Antarctic ice sheet may have been controlled by changes in the accumulation rate, or by periodic surging, but they did not consider that deep sea cores from the Southern Ocean provide support for the triggering of ice ages by Antarctic surging. Chappell and Thom (1978) also argued that the best sea level records are not consistent with east Antarctic surges, and the detailed record from Bermuda (Harmon *et al*. 1978a, 1981) shows that a surge could not have happened about 95 ka BP, as suggested by Hollin (1977, 1980).

The evidence for a brief transgression towards the end of the last interglacial is inconclusive on the coast of South Africa (Davies 1981), but the oxygen isotopic and sea level records from New Guinea are consistent with the occurrence of a surge of either the east or west Antarctic ice sheets, at about 120 ka BP (Aharon *et al*. 1980). The sea in this area may have been about 5 m above its present level by 133 ka BP. This high was succeeded by a fall to below today's level, followed in turn by a rise to about 8 m above at 120 ka BP, at a time when temperatures were in the process of falling by about 2°C.

Further support for the surge hypothesis has been provided by Bowen (1980), and by Flohn (1981). Flohn showed that planktonic and benthic microfossils in deep sea cores from the southern Pacific suggest that there was a rapid collapse of the west Antarctic ice sheet in the last interglacial period. He argued that temperatures fell in the middle of substage 5e, although global ice volume did not increase, nor did sea level fall, until several thousand years later. This time-lag may have been the result of a surge of the west Antarctic ice sheet, which caused the sea to rise to about 5 m above its present level, several millennia before the development of large ice sheets in the northern hemisphere.

Variations in Solar Radiation

Several attempts were made in the last century to relate the occurrence of ice ages and glacial periods to astronomical events, but it was Milankovitch

(1938) who first placed the theory on a sound theoretical basis. Milankovitch calculated variations in the amount of solar radiation which has reached the upper limits of the Earth's atmosphere over the last 600 ka. Solar radiation varies in a predictable manner because of:

1. precession of the equinoxes, associated with the wobbling of the Earth's axis;
2. changes in the obliquity of the elliptic, or changes in the angle made by the axis with the plane of orbit; and
3. variation in the eccentricity or shape of the orbit.

These variations have periods of about 19 and 23; 41; 95 to 136, and 413 ka, respectively (Imbrie and Imbrie 1980). The cycles affect the seasonal and geographical distribution of solar radiation. The precession and eccentricity effects produce opposite changes in insolation in the northern and southern hemispheres. When the obliquity increases, insolation increases in summer and decreases in winter in nearly all latitudes, although the annual total increases above 45° and decreases below. Changes in the obliquity are more significant at higher latitudes, but changes in eccentricity and precession are predominant at lower latitudes. Milankovitch's original calculations have been modified and improved by a number of workers (Brouwer and Van Woerkom 1950, Van Woerkom 1953, Sharaf and Budnikova 1969, Vernekar 1972, Berger 1976).

The hypothesis that glacial climate and sea level vary according to perturbations in the Earth's orbit can be tested by reference to the records of palaeoclimatic and sea level indicators. Attempts to assess the effects of variations in solar insolation using numerical energy-balance models have led to the conclusion that the direct temperature response at the Earth's surface is too small to account for major climatic changes (Schneider and Thompson 1979, Suarez and Held 1979, North and Coakley 1979). When the effect of the ice sheets on the climate has been considered, however, it has been concluded that the Earth's orbital variations were sufficient to trigger the initiation and termination of the North American ice sheet (Budd 1981, Budd and Smith 1981).

Support for the astronomical explanation has recently been provided by a number of workers (Chappell 1973, Goreau 1980, Sergin 1980, Imbrie and Imbrie 1980, Idnurm and Cook 1980, Lockwood 1980, Kukla *et al.* 1981). Several investigators have discovered that there is considerable agreement between insolation curves, the oxygen isotopic records from deep sea cores, and the sea level data from Barbados, New Guinea, the Ryukyu Islands, and elsewhere. On Barbados, for example, the approximately 20-ka transgression–regression cycle could reflect the precession of the equinoxes and its influence upon climate (Broecker *et al.* 1968, Mesolella *et al.* 1969, Veeh and Chappell 1970).

The most significant effect of orbital perturbations is in changing the geographical and seasonal distribution of radiation. The greatest problem, therefore, is to identify the areas and the seasons most sensitive to these changes. Rising temperatures and sea levels are probably not synchronous. There could for example be a time lag of about 2 to 7 ka between changes in insolation at high latitudes and sea level fluctuations in New Guinea (Veeh and Chappell 1970). Chappell (1974a, 1981) found that there was an excellent peak-to-peak correspondence between the precession effects and the New Guinea sea level data. Because the precession effects are most significant at low latitudes, the strong relationship supports the suggestion that Pleistocene climates are controlled by variations in the insolation at low rather than at high latitudes (Broecker 1966, Broecker *et al.* 1968). This may be associated with the transfer of heat to the Fennoscandia and Laurentide ice sheets by the Gulf Stream and the north Atlantic current. In addition to its possible effect on the Earth's climate, changes in orbital geometry can directly influence the level and the isotopic composition of the sea, through changes in the Earth's gravity and rotation, which induce geoidal changes in the water mass distribution over the globe (Mörner 1981a).

It is also possible that climatic and sea level changes are the result of variations in the solar 'constant' (Eddy 1977, Scherrer 1979). Several workers have correlated sunspot activity with palaeoclimatic indicators. Fairbridge (1961b, 1966), for example, considered that there is considerable agreement between sunspot, temperature, and sea level records covering the last three centuries. Mason (1976), however, found little evidence to suggest that the total annual solar flux at the top of the atmosphere varies by more than 1%. He doubted that the sunspot cycles cause changes in climate, and Stuiver (1980) found no relationship between climate and solar changes, as recorded by variations in atmospheric ^{14}C. Numerous correlations have been reported between climate and sunspot activity, but many of these are the product of the particular data selected. Certainly at present, there is no adequate explanation of the way in which slight variations in solar activity could significantly modify weather patterns (Williams 1980).

Suspended volcanic material ejected into the atmosphere by eruptions filter out some of the incoming solar radiation, although this could be partly offset by the enhanced greenhouse warming effect, as a result of the increased opacity at the infra-red wavelengths (Pollack *et al.* 1976, Hammer *et al.* 1980). Bloch (1965) proposed that erratic sea level fluctuations in prehistoric and historic times may be the result of changes in the polar albedo induced by terrestrial and volcanic dusting. Schofield (1970) found that there has been some correspondence between fluctuations in sea levels and the frequency of volcanic eruptions in the last few hundred years. A possible relationship between glacial periods and volcanic activity is also suggested by a marked increase in the number of volcanic eruptions in the Quaternary (Kennett and Thunell 1975). Several workers have provided further evidence of this relationship, but the data do not allow any firm conclusions to be made at present (Bray 1974, Bryson and Goodman 1980, Porter 1981).

Models of Sea Level Change

Mention has previously been made of the isostatic deformation of the land, and the changes in relative sea level which result from variations in ice loading (Andrews 1970, Walcott 1970, 1972a, 1980). It has been proposed that over the time-span of the postglacial period, the response of the Earth to applied stresses has been elastic in the short term, with slow deformation involving redistribution of mass in the mantle over longer periods (Brotchie and Silvester 1969, Walcott 1972b, Chappell 1974c).

The sea floor also adjusts to variations in the water load associated with the alternate growth and decay of ice sheets. The concept of hydro-isostasy was discussed by Daly (1925, 1934) and Lawson (1940), and developed theoretically by Bloom (1965, 1967) and Higgins (1969). The width of the shelf determines the degree to which a coast will adjust to changing water loads. Fairbridge and Richards (1967) suggested that the depression of coasts with wide shelves partly explains why they show little evidence of high sea levels in the mid-Holocene, unlike shelfless oceanic atolls or other areas where the shelf is narrow. Bloom (1967) compared the effects of hydro-isostasy on oceanic islands and on 'stable' continental coasts. Flint (1957) and Mörner (1971f) suggested that crustal subsidence could cause sea level to rise by only two-thirds of the amount expected on the basis of the quantity of meltwater returned to the oceans. Bloom (1967), however, showed that this is only true if the ocean floor simultaneously adjusts to the rise in water level. If there is a lag between the rise in level and crustal response, the altitudinal range of the strandlines on a stable continental coast could be as much as one-third greater than the eustatic change in sea level. Bloom and Flint also believed that the sea floor would rise when water was removed to nourish the ice sheets, although Kaitera (1966) and Mörner (1971f, 1972, 1976) considered that this effect would be small. Hydro-isostatic downwarping may explain about one-quarter of the submergence which occurred on many coasts in the last 6.5 ka (Bloom 1971).

Hydro-isostatic adjustments could account for discrepancies between the relative sea levels within and between shelf areas. Thom and Chappell (1978), for example, attributed differences in Holocene sea levels along the northern Great Barrier Reef Province of Australia to hydro-isostatic influences. Hopley (1977) found no evidence of hydro-isostatic subsidence in the outer shelf in the last 6 ka in a part of northern Queensland, although the systematic variation in sea level perpendicular to the coast throughout this area has been attributed to isostatic influences (Thom and Chappell 1978, Chappell *et al*. 1982).

Several interesting attempts have been made to model changes in sea level which are consistent with the decay of the last glacial ice sheets. Walcott (1972b) considered changes in relative sea level in the last 20 ka, using a viscous flat Earth model. Calculations were made of the movements of the Earth's surface emanating from eustatic rise in sea level, isostatic compensation, and the elastic deformation of the Earth. He concluded that postglacial events can be explained without assuming that there was a substantial eustatic rise in sea level in the last 6 ka. A weakness of Walcott's approach is that he separated the problems associated with ice and water loading. He recognized three distinct regions:

1. regions under the decaying ice experienced rapid uplift;
2. those peripheral to the ice sheets became submerged; and
3. far from the ice, coastal regions tilted and emerged.

Cathles (1975) used a more realistic spherical visco-elastic or Maxwellian model of the Earth, but his results generally support those of Walcott. His model considers the deformation of the solid surface of the Earth by ice melting and hydro-isostatic factors. It shows the effects of uplift in areas of melting ice, and the sinking of the loaded ocean basins as meltwater flowed back into the sea. Australia was initially uplifted as a result of the temporary continental storage of mantle material squeezed out by the loading of the ocean basins. Cathles (1980) found that if the viscosity of the mantle significantly increases with depth, a peripheral bulge can develop as regions are loaded, and a peripheral trough as they are unloaded. The reverse would occur if the viscosity is constant with depth.

Peltier and others developed a theory and model to consider glacial isostatic adjustments, which depend upon the viscosity of the mantle and an accurate reconstruction of glacial history (Peltier and Andrews 1976, 1983, Peltier *et al*. 1978, Peltier 1980). The version employed by Clark and co-workers considers deformation of the ocean floor by glacial and water loading: geoidal perturbations caused by the attraction of the ocean to large ice sheets and the redistribution of matter within the Earth. Relative sea levels are raised by the gravitational attraction in the vicinity of large ice sheets. It has been claimed that the presence of strandlines up to 85 m above present sea level in Hudson Bay, for example, can be explained by this factor alone, without recourse to the effects of isostatic compensation. As an ice sheet melts, weakening of the gravitational attraction may cause an initially rapid emergence of areas near to the ice margins, and an unexpectedly high rise in sea level elsewhere (Clark 1976). The model is based upon the assumptions that the viscosity of the mantle is constant, that sea level rose by 75 m between 18 and 5 ka BP, and that there has been no eustatic change in sea level in the last 5 ka BP. Farrell and Clark (1976) used the model to determine changes in sea level produced by the melting of the Fennoscandia and Laurentide ice sheets. They found that if enough ice melted to produce a uniform rise in sea level of 100 m around the world, the sea would actually rise by as much as 120 m in the south Pacific, and by less than 100 m in the north Atlantic. There is also a zone in the north Atlantic where there would be almost no change in sea level, and an area near Norway and Greenland where it could fall by more than 100 m. A thousand years after the disappearance of the ice, a forebulge would migrate towards the former position of the ice, causing the northerly transfer of water in the Pacific, and the formation of raised strandlines in the south Pacific. Clark *et al*. (1978) and Clark (1980) recognized six zones. Emerged strandlines are unlikely to be found in the two zones where submergence is dominant. In the other four zones, emerged shorelines can occur at great distances from the former ice sheets. The zones are as follows (Fig. 7.7):

(*a*) Zone I consists of regions which were beneath the ice sheets. Immediate emergence occurred in this zone because of elastic uplift and reduction in the gravitational attraction of the ice on the ocean waters. Emergence has continued in this zone because of the viscous flow of material beneath the mantle.

Areas peripheral to the ice sheets fall into a transitional zone, characterized by initial emergence following shrinkage of the ice, and later submergence as a collapsing forebulge migrates towards the ice sheets.

(*b*) Zone II is an area in which submergence was related to the flow of mantle material into the uplifted areas, causing collapse of the forebulge.

(*c*) Zone III is an area of initially rapid submergence, which slowed down until it was succeeded by slight emergence of less than a metre, several thousand years ago.

(*d*) Zone IV experienced continuous submergence of between 1 and 2 m in the last 5 ka BP, despite the assumption in the model that ocean water volume was constant during that time.

(*e*) In zone V, initial submergence gave way to slight emergence (1 to 2 m), when meltwater was no longer being added to the oceans.

(*f*) Zone VI consists of all the continental margins, other than those which are adjacent to zone II. Mid-Holocene emergence occurred in this zone, although it was assumed that ocean water volume was not increasing at that time. This was because the increased water load on the ocean floor forced mantle material to flow into the area beneath the continents.

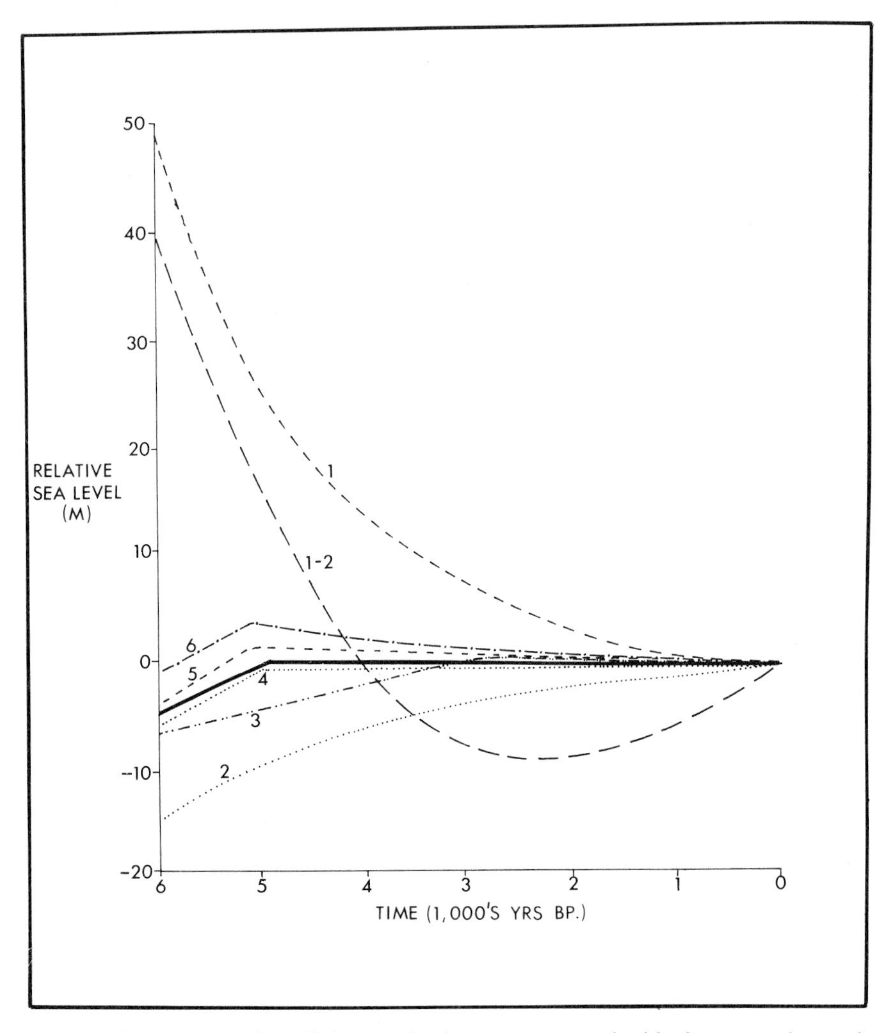

Figure 7.7. Representative relative sea level curves compared with the assumed eustatic change in sea level (solid line) for the last 6 ka. The numbers on each line refer to the sea level zones defined by Clark and Lingle (1979).

The model predicted emergence of about 1.5 to 2 m for Australia, the south Pacific, South Africa, and the southern part of South America. Discrepancies between model predictions and actual sea level data appear to be substantial in the area of the proglacial forebulge, and on the east coast of North America. Peltier (1980) slightly extended the basic theory of the model, providing modifications which gave a better fit to the data from areas such as Hudson Bay, although not to the forebulge regions.

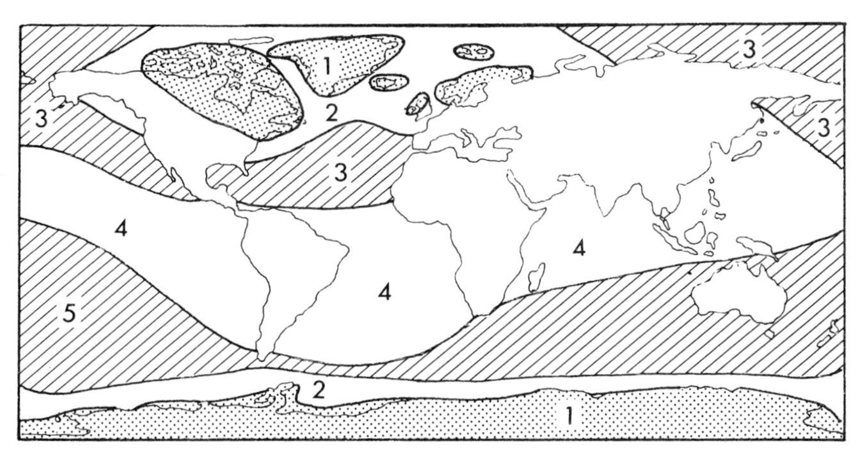

Figure 7.8. Distribution of the sea level zones of Clark and Lingle (1979), assuming that northern ice sheet disintegration was complete by 5 ka BP, and that eustatic sea level subsequently rose by 0.7 m because of melting of Antarctic ice. Stippled areas are those that experienced continuous emergence, whereas shaded areas are those where Holocene sea levels have been higher than at present.

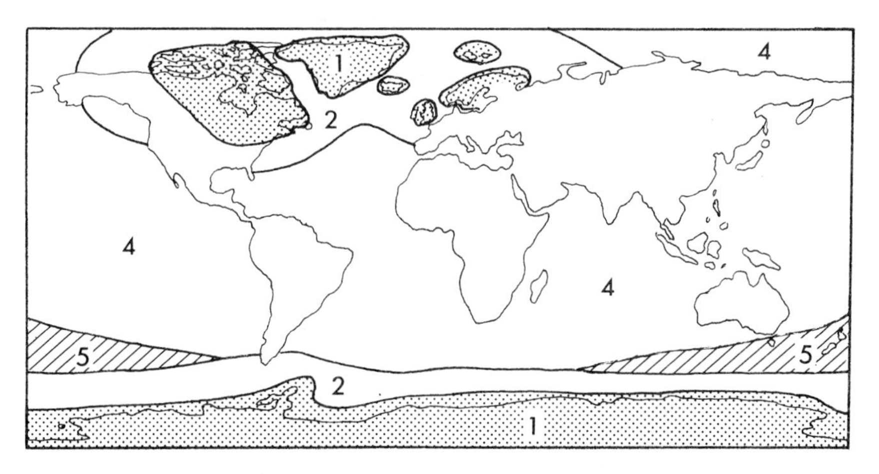

Figure 7.9. Distribution of the sea level zones of Clark and Lingle (1979), assuming that ice disintegration was complete by 5 ka BP in the northern and southern hemispheres. Stippling is used as in Figure 7.8.

Clark *et al*. (1978) did not consider the effect of changes in the Antarctic ice sheet in the last 18 ka. Further work has suggested that melting of the west Antarctic ice sheet could have been responsible for about a quarter of the total global rise in sea level. If only about 3% of the west Antarctic contribution was made after 5 ka BP, few mid-oceanic islands could have emerged. Alternatively, if the west Antarctic ice sheet remained about the same size in the last 6 ka BP, many islands would have emerged in the southern hemisphere. The distribution of the zones defined by Cl[illegible] [illegible] models (Clark and Lingle 1977, 1979), as shown in Figures 7.8 and 7.9.

Each of the models discussed here is based upon assumptions regarding the rheology of the Earth, and the distribution and size of the ice sheets during and following the maximum of the last glacial period (Peltier and Andrews 1983). Mörner (1981a,b) believes that glacial loading models cannot explain global variations in relative sea level. He has argued that glacial loading in Fennoscandia was largely compensated locally, and that differences in sea level records around the world are the result of geoidal eustatic changes. The differential loading and the geoidal models show that changes in sea level emanating from the return of meltwater to the oceans cannot be uniform around the world. The evidence for Holocene sea levels above today's in some parts of the world can now be reconciled with concurrent levels below present sea level in other areas, although the mechanisms responsible remain to be determined. This does not of course, necessarily guarantee that the published sea level data for specific regions are correct. At present, global theories and the interpretation of Holocene sea level changes are based upon a good deal of questionable field evidence. The ability of any model to represent Holocene events cannot be fully evaluated, therefore, until reliable sea level data are available from around the world.

II. LANDFORMS

8

Coastal Cliffs

It has been estimated that sea cliffs are present around about 80% of the world's oceanic coasts (Isakov 1953, Emery and Kuhn 1982) (Fig. 8.1). There are few systematic descriptions of the form and development of sea cliffs in the literature, despite their morphological importance and the aesthetic contribution they make to the littoral environment. Information has to be gleaned from a variety of sources, most of which are primarily concerned with other aspects of the coastal environment. The literature on shore platforms, for example, generally contains only a few brief comments on the form of the cliffs behind. In part this reflects the difficulty of investigating precipitous or heavily vegetated coastal slopes, which are often in remote areas. It is therefore impossible at present to provide a globally representative survey of the morphology, processes, and evolution of coastal cliffs. Much remains to be done, in measuring slope form in different morphogenic environments and in assessing the effect of a myriad of morphological and geological factors on cliff development. The form and evolution of marine cliffs, which are dominated by large mass movements, and the occurrence of notched and corroded limestone cliffs in low

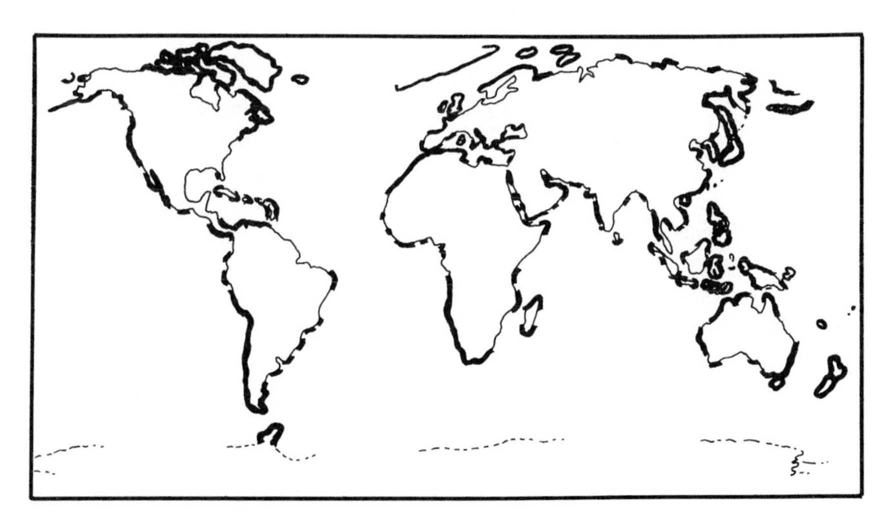

Figure 8.1. Generalized distribution of sea coasts backed by cliffs (black) (after Isakov 1953, and Emery and Kuhn 1982).

latitudes, are discussed elsewhere (see Chapters 6 and 10, respectively), and will not be considered here.

Global Characteristics

Certain types of marine cliffs are more characteristic of some parts of the world than others, although it is difficult to classify them on the basis of climate, wave regime, or other aspects of the morphogenic environment. The profiles of marine cliffs are the result of the interplay of such factors as geology, climate, wave and tidal regime, vegetation, nearshore water depth, the type and amount of beach material at their base, the topography of the cliff-top area, and changes in the relative level of the sea. Variations in local factors, such as rock structure or lithology, can produce greater differences in cliff morphology within the same morphogenic regions than between different regions. Nonetheless, the importance of differences in climate and wave regime around the world (Davies 1964, 1972) does permit some generalizations to be made.

Davies (1972) suggested that, in general, the gradients of marine cliffs are greatest in the vigorous wave environments of the temperate latitudes. Lower cliff gradients in the tropics reflect the prominence of chemical weathering, whereas mechanical weathering and mass movement play a similar role in high latitudes. In the mid-latitudes of the northern hemisphere, storm waves frequently attack the coast, producing notches or overhangs at the base of steep cliff faces where the structure and lithology are suitable. In particularly resistant rocks, waves have not accomplished much erosion in the few thousand years in which the sea has been at its present level. In these cases, steep wave-cut cliffs only occupy the lower portions of coastal slopes.

In the tropics, the low velocity and variable direction of the winds, and the presence of protective coral and algal reefs generally provide wave conditions which are not conducive to the formation or maintenance of steep marine cliffs. Vigorous wave action is experienced on some tropical coasts. Cliff recession induced by storm wave attack can be associated with onshore monsoons, as in Malaysia (Nossin 1965, Swan 1968), with the passage of tropical cyclones (McIntire and Walker 1964), and with the trade winds in exposed areas of the Caribbean.

In discussing cliff formation it is necessary to distinguish between the humid and the arid tropics. Vegetated coastal slopes, on which only the lowest few metres are bare, are more common in the hot, wet tropics, whereas in arid regions steep, bare cliffs are more common (Bird and Hopley 1969, Bird 1970b, Tricart 1972, Faniran and Ježe 1983). In the humid tropics, chemical weathering plays an important role in coastal development, partly because of the climate, but also because wave action is fairly weak in most areas (Edwards 1958, Tricart 1959, 1962). In most humid tropical regions, as western Malaysia (Swan 1971), the Indonesian Archipelago and New Guinea (Bird and Hopley 1969), and in El Salvador, the Pacific coast of Columbia, and in Brazil (Tricart 1972), perennially weak wave activity produces limited cliffing, and the coastal slopes are usually densely vegetated. Steep coastal tracts are often restricted to headlands and other exposed areas (Swan 1965). Cliff recession in the crystalline and sedimentary rocks of the western parts of the Ivory Coast, for example, is very slow, and the vegetated cliffs are usually convex, although there are steep, bare cliffs on some of the more exposed headlands (Tricart 1957, 1972, Buckle 1978). Rocky coasts are more common on small islands than on continental coasts, possibly because of the small quantities of alluvium available on islands for the buildup of protective beaches or coastal plains (Swan 1971, Tricart 1972). Cliffs are particularly susceptible to slumping and other forms of mass movement in rocks which weather rapidly, such as granites, gneisses and lower micaschists.

Although wave attack is generally rather weak in the hot, dry tropics, weathering associated with such mechanisms as salt crystallization, alternate wetting and drying, and corrosion facilitates cliff formation, particularly in exposed areas. Vegetational growth is inhibited by salt spray and high rates of evaporation, and cliff faces are therefore predominantly bare. Steep cliffs have been cut in dolerite and columnarly jointed basalt near Dakar, Senegal, for example (Buckle 1978), and in a variety of sedimentary rocks along the arid coast of Pakistan (Snead 1969).

A broad morphological distinction between the wet and dry tropics is generally justified, but the form of coastal cliffs in the tropics actually depends upon a number of factors, including whether the rocks are weathered or not, the degree of shelter of the site, and the rock type (Swan 1971). In some types of rock cliffs are quite stable. In the volcanic breccias of El Salvador, the cliffs are still aligned along the original fault scarps because weathering is retarded by high rock permeability and steep slopes (Tricart 1972). In eastern Thailand, steep cliffs have developed in non-crystalline rocks, such as shales and conglomerates, as well as in limestones, but they are more gentle in granites and other crystalline rocks which have been subjected to deep chemical weathering and denudation (Consentius 1975). Weathering of coastal rocks provides conditions which are suitable for the wave erosion of otherwise-resistant rock types, or rocks which are in sheltered environments, although in some areas, small, locally derived waves can be effective mechanisms for cliff formation in weathered and unweathered rocks (Swan 1971). On the Singapore Islands, for example, deep weathering is a vital precursor to the development of active cliffs in resistant rocks, although weathering is unnecessary on the less resistant formations. A further objection to a strictly morphogenic classification of tropical cliffs is the occurrence of steep, bare cliffs on limestones throughout the wet and dry tropics. Benches, notches, and visors are commonly produced by chemical or biological action in these rocks (see Chapter 10), although they can also be found on steep cliffs cut in other rock types (Russell 1963, Tricart 1972).

Most coasts in high latitudes are in low-energy wave environments. Wave action is generally weak because of sea ice, and in some places coastal configuration, although locally derived waves reach the shore when the immediate offshore area is ice-free (Hume and Schalk 1967; McCann and Owens 1969, 1970; King 1969). There are few true marine cliffs in these areas, and most steep coastal slopes, as in eastern Greenland and northeastern Baffin Island, are probably the result of glacial oversteepening, although weak wave action can induce landslides and other forms of cliff collapse (Bird 1967, John and Sugden 1975). Even in some more vigorous wave environments, contemporary wave action has only slightly modified cliffs which were cut in the Tertiary or in a Pleistocene interglacial, when sea level was fairly constant for a considerable period and sea ice was less extensive (John and Sugden 1971). The cliff-abrasion coasts of northern Norway, for example, are probably the result of Tertiary faulting. In the Pleistocene, material loosened from the cliff by waves, frost, and mass movement in the interglacials was removed by ice during the glacial periods (Klemsdal 1982). In the southern Canadian Arctic, few cliffs are being cut at the present time, other than those in limestones. Limestone cliffs are usually steep (Bird 1967, Taylor 1980), but similar forms have developed in Precambrian crystalline rocks in the exposed parts of southeastern Baffin Island (Miller *et al.* 1980), and in folded volcanic and metamorphic rocks in northern Labrador (Rosen 1980). Further south, in southeastern Hudson Bay, steep structural cliffs have formed on the limbs of cuestas consisting of sedimentary rocks with a basaltic cap (Guimont and Laverdiere 1980). Wave erosion plays an active role in these areas, despite the presence of protective coastal ice for the greater part of the year.

In the polar regions frost weathering assumes an important role in the development of coastal cliffs; if frost weathering is effective at the cliff foot overhanging cliff profiles are produced. It has been suggested that convex cliff profiles are associated with rocks which are moderately resistant to weathering, and vertical profiles with rocks which are particularly susceptible to frost shattering (Flores Silva 1952, Tricart 1969b). Unless wave action is vigorous, however, the accumulation of weathered fragments will progressively bury the cliff in a debris apron (Bird 1967, Howarth and Bones 1972, Taylor 1980).

Geology and Cliff Type

The variety of cliff profiles found in nature is testimony to the numerous factors which determine the efficacy of the processes sculpturing marine cliffs. The form of a cliff profile is a function of the relative effects of marine and subaerial processes, and the amount of time in which they have operated. Geological factors partly determine the relative efficacy of these processes, and they usually explain variations in cliff type along a coast. Although structural and lithological factors assume an important role in determining the form of coastal cliffs, it is impossible to classify them according to rock type. Nevertheless, within a single morphogenic region, in which such factors as the tide, wave regime, and climate can be considered to be essentially uniform, there is usually some relationship between cliff scenery and rock type.

Several attempts have been made to derive a classification of cliff profiles, using rock structure and lithology as the basic criteria (Guilcher 1958a, Small 1970). Emery and Kuhn (1982) have provided a simple classification of sea cliffs, based upon the relative importance of marine and subaerial erosion, and the effects of variations in rock hardness on the efficacy of these erosive agencies (Fig. 8.2). In homogeneous rocks, steep cliffs tend to develop in areas where marine processes are dominant, whereas convex profiles form if subaerial mechanisms are most important. Where subaerial and marine processes are effective, the cliffs may be convex in the upper portions and steep at the base. The shape of the cliff profile in homogeneous rocks is therefore largely a function of the climate and the wave environment. Strata which differ greatly in their resistance to erosion, however, often superimpose marked variations on these basic forms. Thus, despite the dominance of subaerial agencies, a steep cliff can develop in fairly weak materials if resistant cap rocks prevent the formation of a convex slope; alternatively, where marine processes are more powerful than subaerial processes, the upper part of the cliff can be convex because of the presence of an outcrop of weaker material. The most comprehensive survey of the relationship between cliff form and geology has been provided by Steers (1964), for the coast of England and Wales, and later for Scotland (1973). Although this shows some relationship between lithology and cliff form, his examples emphasize the dominant role of structure in accounting for differences in coastal scenery between similar lithological units.

Igneous coasts are usually quite different from other rock coasts, but they also differ greatly between themselves. Granites, for example, are usually resistant to erosion in extratropical regions. In coastal Maine, steep granite cliffs were cut by glaciers during periods of low sea level. They have been little modified by contemporary wave action, and glacial striations can still be traced down to the water level (Johnson 1925, Shepard and Wanless 1971). The effects of marine and subaerial processes on granites are concentrated along the joints, forming great cuboidal blocks, as on Land's End, England, where high castellated cliffs have formed along shear

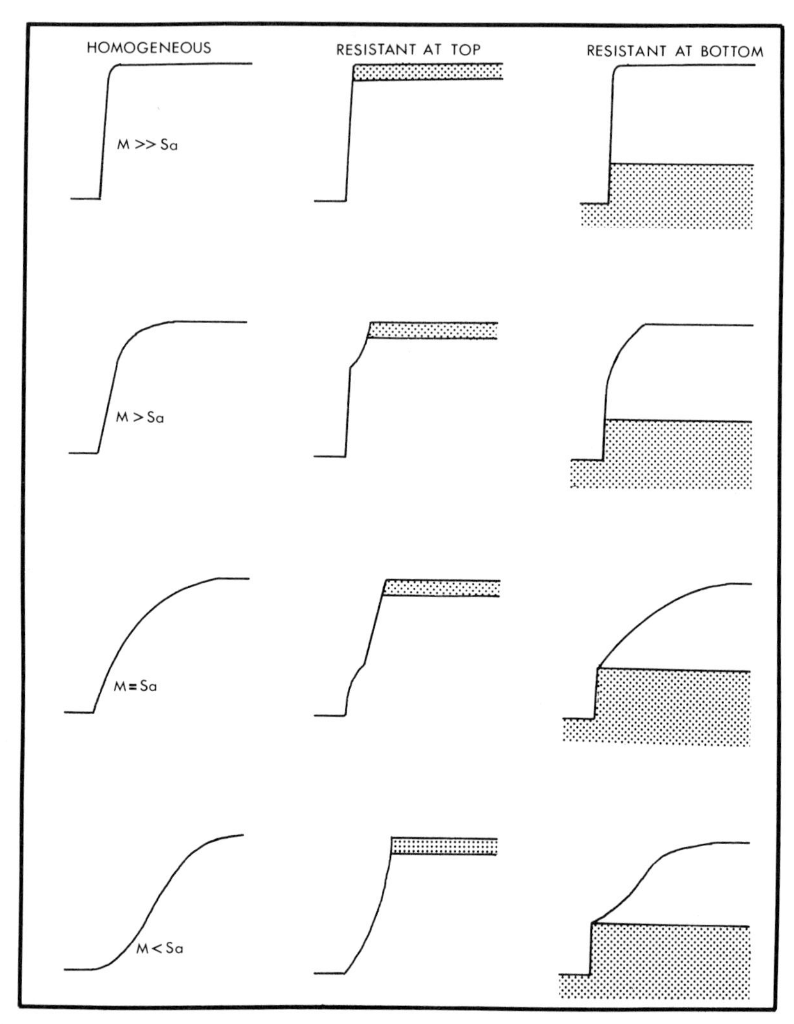

Figure 8.2. Cliff profiles according to variations in rock resistance and in the relative efficacy of marine (M) and subaerial (Sa) processes. The more resistant rock outcrops are shaded (after Emery and Kuhn 1982).

faces coincident with the vertical joint planes (Arber 1949). Similar cliffs have been described in Victoria, Australia (Hills 1971), but because of structural and compositional differences, castellated cliff scenery is poorly developed in the coastal granites of Peterhead in eastern Scotland (Steers 1962a). The western islands of Scotland have excellent examples of coasts formed in lava flows, dykes, and sills. The dykes often form great cliff buttresses, and occasionally re-entrants along the cliffs. Lava flows usually form high, precipitous cliffs which are resistant to erosion. Differences in basaltic coastal scenery are explained in some cases by compositional differences, flow contacts which are susceptible to subaerial and marine erosion, and the presence of columnar jointing (Moign and Moign 1970, Thomson 1979). In western Skye, cliffs more than 300 m high are composed of at least 25 separate lava sheets (Steers 1962a). In southeastern Iceland, notches, overhangs, and caves coincide with the presence of weaker basalts and with tuffaceous material deposited between the basaltic flows (Bodere 1981). High lava cliffs plunge into the sea on Easter Island (Paskoff 1978a). In Northumberland, northern England, the Whin Sill dolerite produces high cliffs in several areas. Tuffs and agglomerates form irregular cliffs, fortified in places by dykes. Near St Andrews in Scotland, a number of irregular stacks have formed from old volcanic necks which pierce the local sedimentary rocks.

Classification of cliffs in sedimentary rocks on the basis of purely lithological criteria is even more difficult than for igneous rocks. The cliff scenery of igneous rock coastlines is strongly influenced by structural factors, but in sedimentary rocks such factors usually dominate. Furthermore, many sedimentary formations consist of alternations of strata of variable resistance to erosion; coastal scenery is then partly determined by the resistance, thickness and position of the weakest members of these alternations (Trenhaile 1971). Limestones in south Wales provide an excellent example of the inability of simple lithological classifications to predict the characteristics of rock coasts. Along only a few kilometres of the western-facing coast of the Vale of Glamorgan, Wales, coastal scenery changes dramatically from the sometimes massively bedded, essentially homogeneous Carboniferous Limestone tract between Porthcawl and Ogmore, to the Liassic limestones, shales, and mudstones of Nash Point (Trenhaile 1972). Thick beds, and the general lack of weak interbedded shales, make the old Carboniferous limestones resistant to mechanical wave action, and lapies and other corrosional forms are common. The younger Liassic limestones, which are thinly bedded and densely jointed, are undermined by erosion of the friable black shales. Mechanical wave action is dominant, and corrosional features are less well developed. The old Red Sandstone coasts of Caithness and Arbroath in Scotland are also quite different from the lithologically similar coast of south Pembrokeshire, Wales. One of the main reasons is that the sandstones in Pembrokeshire have been subjected to Amorican folding, whereas the same rocks in Scotland have very low dips (Steers 1962a). Only perhaps one lithology in Britain permits fairly confident prediction of the coastal scenery associated with it: the homogeneity of the chalk results in very steep or even vertical cliffs in southeastern England and at Flamborough in northeastern England, as well as in northern France (Prêcheur 1960, May 1971). Jointing and faulting provide zones of accelerated wave erosion (So 1965), which [illegible] for the presence of arches, caves, and geos in these [illegible] essentially horizontal in most areas, coastal [illegible] where the rocks have been folded. Strengthen- [illegible] by folding, however, accounts for the formation of the [illegible] series of chalk stacks off the Isle of Wight (Steers 1962a). In the [illegible] France, cliff retreat in the chalk is largely the result of small [illegible] and falls following heavy rainfall. These are induced by the presence of joints which are partly enlarged by corrosion (Guilcher 1953).

The presence of slaty cleavage, foliation, and schistosity in metamorphic rocks introduces further complicating factors which frustrate attempts to derive a lithological classification. Metamorphic rocks have often been subjected to intense pressure and folding. The inner beds of small synclines running at right angles to wave action are usually the first to be eroded by the sea, particularly if the beds can slide seawards. Synclines are also unstable when the sea is eroding parallel to the axis of folded rocks. Once the limb of an anticline has been removed, the next limb, which is also part of the following syncline, will be quickly eroded (Arber 1911). Anticlines are generally more stable; although caves can be worn into the heart of the fold, the crest normally remains (Arber 1911, Steers 1962a).

There is a complex relationship between the shape of cliff profiles and the dip of the bedding and joint planes in the rock. Even the weakest, mechanically intact, unweathered rock could theoretically withstand the pressures generated within vertical cliffs well over a thousand metres in height. These heights are never attained, because the critical height is limited by the presence of joints, faults, and other mechanical defects, rather than by the strength of the rock itself (Terzaghi 1962). A. Young (1972) also pointed out that the slopes of marine cliffs can be in short-term rather than in long-term equilibrium, and are therefore unrelated to rock strength.

There is a tendency for cliffs to be very steep in rocks either horizontally or vertically bedded, whereas slopes are generally more moderate in rocks which dip seawards or landwards. If the dip is seawards, profiles are often coincident with individual bedding planes, and they are therefore quite smooth. Irregular profiles can form in seaward-dipping rocks along the cross-joint surfaces perpendicular to the dip. When the dip is landwards, a large number of beds can be exposed, and differential erosion often produces a very irregular cliff face. Other geological and non-geological factors provide numerous exceptions to these generalizations. Cliff profiles in rocks with a low seaward dip, for example, can be sawtoothed—a series of steps, each associated with a resistant stratum.

Terzaghi (1962) developed the theory of slope stability in hard, jointed, and unweathered rocks. The resistance of rock to shear is dependent upon such factors as the continuity, spacing, and other aspects of the joint pattern. In massive rocks, the joint pattern can be fairly continuous, and their properties are similar to those of stratified rocks. If the joint pattern is random (as in granites, marbles, and some limestones and dolomites), the critical angle for slopes on hard, massive rocks is about 70°, assuming that seepage pressures are not acting on the joint walls. On stratified sedimentary rocks, however, if the joint pattern is rectangular, the steepest stable slope is a function of the angle of friction, the dip and its direction, and the spacing and offset of the bedding and joint planes (Terzaghi 1962). Consideration of these variables can account for cliff profiles which are neither coincident nor perpendicular to the bedding planes, as well as for those common situations in which the profiles are determined by this factor. If the rock is horizontally bedded, there can be no slippage, and the critical slope is vertical. If the rocks are dipping, the critical slope will depend upon the orientation of the cross joints and the bedding planes relative to the slope. A. Young (1972) has conveniently summarized Terzaghi's theory for cliff forms produced by various combinations of rock dips, angles of friction, and joint patterns (see Chapter 6). The theory shows that under certain circumstances, vertical and seaward-sloping cliffs can be formed in horizontal seaward- or landward-dipping rocks.

Whether a portion of a cliff profile fails because of its intersection with a fault plane depends upon the dip of the plane, its orientation relative to that of the cliff, and the resistance to sliding along the fault plane. The resistance to sliding depends upon the condition of the rock along the fault, which may have experienced physical or chemical change (Terzaghi 1962).

A fault coast forms along the scarp which separates a raised block of land from one which has been depressed below sea level. De Martonne (1909) defined fault coasts, and cited examples in Greece, southern Italy, and Japan. Several cliff coasts in New Zealand are thought to have resulted from faulting (Cotton 1916, 1968), and many of the high, steep cliffs on the arid coast of Pakistan are attributable to fault scarps (Snead 1969). Faulting has played an important role in western Scotland, where many steep cliffs in lavas and sills are fault controlled. In Skye, fault coasts in fairly sheltered waters have only been very slightly modified by wave action. These cliffs can be regarded as fault-line scarps (Steers 1962a). In northern Norway, sea cliffs are also primarily the result of faulting, although they have been kept steep by the glacial removal of debris (Klemsdal 1982).

Shepard and Wanless (1971) have described several fault-induced cliff lines in North America. A broken block of land, for example, has dropped below sea level on the northeastern side of San Clemente Island, off California, forming a straight cliff without a shelving bottom. The absence of a bordering shelf also implies that the steep Santa Lucia Mountain coast of central California is originally the result of faulting. Some of the cliffed coasts of Alaska are the result of faulting associated with earthquakes. In Hawaii, cliffs can be produced by faulting induced by vulcanicity, although on Molokai Island, the coast may have retreated up to 5 km from the original fault scarp.

In the humid tropics, some of the vegetated cliffs are still aligned along the original faults (Tricart 1972). In arid northern Chile, the cliff which extends for about 800 km along the coast is more than 1,000 m high in places, with slopes of up to 80°. It is still attacked by the waves in some places, but has been abandoned elsewhere. It has been suggested that the structure is: a true wave-cut cliff formed during a Pliocene transgression when there was also prolonged coastal subsidence (Mortimer and Saric 1972, 1975); a fault scarp which was eroded in Plio-Pleistocene time (Rutland 1971); or the result of upper Miocene fault scarps which retreated during a major transgression in the middle to upper Pliocene (Paskoff 1978b). It has apparently been little modified by Quaternary sea level fluctuations. Most of the higher calcareous cliffs in Malta are tectonically induced, although stratigraphy and lithology have determined their subsequent evolution (Paskoff and Sanlaville 1978).

Composite Cliffs

In many areas, cliffs with steep, flat-topped profiles have been cut into coastal plateaux (Arber 1911). Other cliff profiles, however, consist of a fairly gentle slope element above a steep, wave-cut face (Fig. 8.3). The origin of these cliff profiles has been the subject of debate for more than a century. They are particularly common on coasts formed in resistant, ancient massifs, but they are not limited to those areas (Guilcher 1958a). They are widely distributed throughout western Britain and Ireland, and they have also been reported from eastern Canada (Hetu and Gray 1980), Brittany, and southern France, and on many oceanic islands in the southern hemisphere. Indeed, it may be the prevalent cliff profile in areas composed of resistant materials.

A variety of terms have been employed to describe this type of cliff profile; they include 'slope over walls' (Whitaker 1911), 'coastal slopes' (Challinor 1931, 1948, 1949), 'marginal scarps' (Leach 1933), 'two-storied cliffs' (Fleming 1965) and 'composite cliffs' (Orme 1962, A. Young 1972). The terms 'hog's back' (Arber 1911) and 'bevelled' cliffs have frequently been used, although the nomenclature lacks precise definition and the distinction between them is poorly defined. Wood (1959, 1962) used the term 'bevel' to encompass seaward slopes which only occupy the upper parts of cliffs, as well as those which almost extend down to the cliff base; this terminology has been adopted by King (1972). Most workers, however, have used either the shape or the length of the upper slope to

Figure 8.3. Composite cliff in landward-dipping rocks at Coombe Martin, Devon, England.

distinguish between bevelled and hog's back cliffs. Wilson (1952) used 'hog's back' to describe cliffs which have rectilinear upper slopes, whereas 'bevel' was used to describe upper slopes which are curved; Steers (1981) employed a similar definition. A. Young's (1972) attempt to provide a more precise definition of hog's backs and bevels is a formalization of the implied definitions of Orme (1962) and Small (1970). Bevelled cliffs were defined as those in which the upper seaward slopes are shorter than the lower free faces, whereas hog's backs are those in which the seaward slopes are longer. Savigear (1962) has objected to the use of these terms; unless a genetical distinction can be made, there seems to be little purpose in using morphologically descriptive terms which are so poorly defined. In this discussion, Orme's term 'composite cliff' will be used to describe the cliff profiles, and Savigear's 'seaward' or 'seaward-facing' slopes will be used to refer to the upper sloping surface.

Arber (1911) thought that vertical cliffs can develop from composite or hog's back cliffs where vigorous wave action has been able to remove the upper, seaward slopes. Arber (1949) has also proposed that the presence of hog's backs, bevels, or vertical flat-topped cliffs, reflects the relative importance of marine and subaerial erosional processes. A vertical, wave-cut cliff with a flat top forms in exposed areas where there are no controlling structural planes. Arber thought that hog's backs develop in rocks which dip landwards and are sheltered from vigorous wave action; but they can also form in rocks which dip steeply seawards. Similarly, she attributed bevelled cliffs to situations in which there is a rough balance between marine and subaerial erosion, in the absence of dominant structural planes; but it is difficult to reconcile this explanation with her statement that bevelled slopes may nevertheless occupy the greater part of the cliff height. In parts of Victoria, Australia, marine cliffs pass into composite cliffs and then into degraded bluffs as one moves from a high-energy section of the coast into a more sheltered area (Bird 1977).

It has been suggested that composite cliffs, are essentially contemporary features. Challinor (1931, 1948, 1949) believed that they are the result of subaerial processes operating at the same time as wave erosion at the cliff base. Although Guilcher (1958a) also considered that convex cliffs are subaerially eroded hill slopes trimmed by wave action at their base, he recognized that similar profiles could be the result of subaerial denudation during a period of low sea level. Fisher (1866) first suggested that composite cliffs are the product of alternations of periods of marine and subaerial erosion. Most workers now accept his explanation, and consider these cliffs to be vestiges of sea level changes induced by the climatic fluctuations of the Pleistocene period. The first stage in the two-cycle theory in its simplest form is the cutting of a steep cliff by wave erosion. The cliff is then abandoned when sea level falls during a glacial period; subaerial processes then progressively degrade it, covering it in debris. When the sea level rises, renewed wave attack steepens the base of the cliff, leaving the gently sloping, regolith-covered remnant above (Davis 1912, Balchin 1946, Cotton 1951a, Fleming 1965, Cotton and Wilson 1971).

Savigear (1952) made a classic study of a cliff at Pendine in south Wales which was progressively abandoned as a spit grew eastwards. He suggested that when a steep wave-cut cliff is abandoned, parallel retreat of the free face initially takes place. This surface is gradually replaced from below by a slope of about 32°, which declines in steepness if removal is impeded. With unimpeded basal removal, regolith-covered slopes of 32° may retreat in a parallel fashion. Kirkby (1983, 1984) devised a mathematical model for slope development based upon Savigear's data, incorporating the long-term effects of landsliding processes. It was found that at assumed periglacial rates of erosion, 100 ka or more would be necessary for the formation of convex seaward slopes. Orme (1962) has also considered the slopes of cliffs abandoned by wave action. He thought that since the lower segments of abandoned cliffs are quickly mantled and protected by accumulating debris, they must retain much of their original steepness. As the debris slopes rise up the cliff faces, which are partly degraded by subaerial processes, however, the buried cliffs assume convex profiles. Above the highest levels attained by the waste, gradients decline to between 20 and 35°. In southern Britain and in other northern latitudes, frost action and solifluction may have accelerated the degradation of the upper slopes, and head or glacial drift frequently mantle the lower slopes (Guilcher 1950, Cotton 1951a, Wood 1959, Orme 1962, Fleming 1965, Davies and Stephens 1978).

The uniformity of the upper slopes suggests that some composite cliffs have a simple two-cycle origin, all morphological evidence of previous stages of development having been destroyed during the last sequence of marine and subaerial erosion (Savigear 1962, Fleming 1965, Cotton 1968). Many cliffs may be polycyclic, retaining the vestiges of a number of sea level fluctuations. Seaward slopes frequently consist of a series of straight or convex segments, which can be traced for some distance along a coast and along the flanks of adjacent valleys. Savigear (1962), for example, recognized as many as four or five slope elements in parts of northern Cornwall and Devon. It has been suggested that each of these upper slope segments may represent a period of cliffing following one of subaerial denudation; an explanation which is similar to that proposed for polycyclic valleys (Savigear 1956, 1962, Te Punga 1957, Orme 1962). The presence of raised beaches fronting degraded cliffs which laterally merge into the seaward slopes indicates, that the composite cliffs in Carmarthen Bay, Wales, may also be the product of several phases of marine and subaerial erosion (Wood 1959, 1962). The slopes appear to have been cut into and rejuvenated by marine erosion on several occasions, separated by intervals long enough to allow subaerial processes to reduce the steep cliff to a gentler slope.

A mathematical model has been devised to simulate the development of a marine cliffs through two interglacial–glacial cycles (Trenhaile, in preparation). The model was based upon the assumption that a vertical cliff, cut by waves during an initial interglacial period, undergoes parallel slope retreat above the accumulating scree of the following glacial, low sea level period. A decay function was used to represent the depth-dependent decline in the rate of erosion of a rock face beneath scree. Intervals of marine and subaerial erosion were commensurate with the last glacial period. Rates of 1.0, 0.1, and 0.01 m per century were used for subaerial erosion on bare rock faces, and 0.01 and 0.001 m per century for wave erosion at the cliff base. Four basic forms were produced (Fig. 8.4):

(*a*) *Vertical cliffs* developed at the end of the first cycle under conditions of slow weathering and fast wave erosion. This form usually reappeared at the end of the second cycle.

(*b*) *Two-storied cliffs* developed at the end of the first cycle if slow

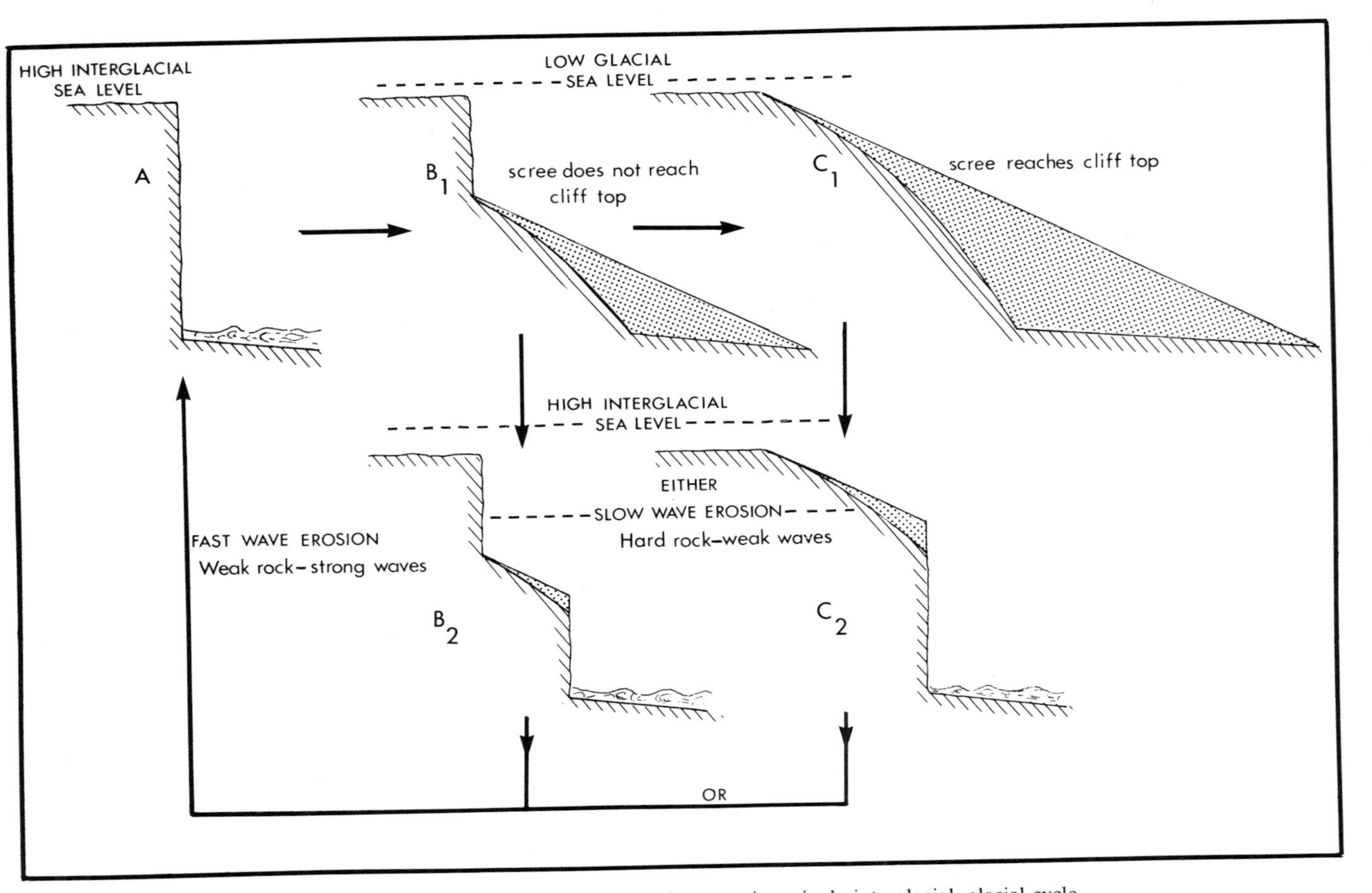

Figure 8.4. Model of coastal cliff development in a single interglacial–glacial cycle.

weathering was combined with slow wave erosion. This form was the result of situations in which the scree failed to reach the cliff top by the end of the first glacial period (Fig. 8.4, profile B_1). It consisted of two vertical rock faces separated by a seaward-facing slope (Fig. 8.4, profile B_2). It occasionally reappeared at the end of the second cycle, but was usually replaced by type (*d*).

(*c*) *Composite or hog's back cliffs* formed at the end of the first cycle when moderate to fast weathering was combined with fast or slow wave erosion (Fig. 8.4, profile C_2). This form either reoccurred at the end of the second cycle or was replaced by type (*b*).

(*d*) *A three-storied cliff* developed from type (*c*) at the end of the second cycle. It consisted of three vertical segments separated by two seaward-facing slopes. Additional steps, each resembling a former wave-cut surface, would correspond to further interglacial–glacial cycles.

Alternative suggestions have been made to account for the development of composite cliffs. Reid *et al.* (1910) considered that the seaward-facing slope in northern Cornwall may represent a denuded newer Pliocene shoreline, although they also recognized the possibility that it could be the result of the rounding of the cliff edge under Arctic conditions. Gullick (1936) suggested that the seaward slopes of Land's End are the product of wave erosion when the sea was higher than today, but Balchin (1937) found no evidence of this planation level. Cotton (1952) envisaged very high, steep cliffs developing as the result of the cumulative erosional effects of waves operating at a variety of levels. A similar explanation has been put forward for the high composite cliffs in the Old Red Sandstones and quartzites of south and southwestern Ireland. Guilcher (1966) thought that they are the result of successive periods of wave attack during Pleistocene periods of high sea level. Wave erosion was assisted by the formation of pressure release joints, and by glacial action during the last glacial period. The lower parts of these composite cliffs are steep and undercut, but the gentler slopes above, which developed primarily as a result of wave erosion, were eventually degraded by subaerial processes.

Some workers have emphasized the role of geological factors. Flett and Hill (1912) suggested that seaward slopes on the Lizard of western Cornwall are the result of landslipping related to the presence of dykes. Leach (1933) attributed the seaward slopes of southern Pembrokeshire, Wales, to subaerial processes, but he considered that their gradients are determined by the inclination of the cleavage planes in the thinly bedded strata, and by joints and dips in the more massive rocks. Balchin (1937) noted that composite cliffs in northern Cornwall are associated with seaward-dipping rocks which strike parallel to the coast, whereas vertical cliffs are usually found where the coast cuts across the strike. Wilson

(1952) also related the occurrence of composite profiles in northern Cornwall to rock joints and dips, although near Tintagel they may be produced by the stripping of inclined fault planes. In the southwestern part of New Zealand and in some other areas, hog's back cliffs are associated with monoclinal flexures (Cotton 1968). Arber (1949, 1974) suggested that seaward slopes are the result of geological structure, and particularly, rock dip. She noted that the seaward slope is a long, straight escarpment surface on landward-dipping rocks, and a dip slope on rocks which slope steeply seawards. The ambiguity of Arber's (1949) statements on the origin of composite cliffs is reflected in the different interpretations of her views that have been made by several workers. Savigear (1962) and A. Young (1972) considered that Arber supported Challinor's (1931) one-cycle theory, and justified this conclusion by her statement that a subaerial seaward slope can coexist above a steep, wave-cut form. As Cotton (1951a) has noted, however, Arber also implied that the two slope elements do not develop contemporaneously, although she did not say so explicitly.

Some composite cliffs are simply the one-cycle product of differential erosion, brought about by the juxtaposition of lithologies with markedly different resistance to erosion. This is the case, for example, on the Isle of Purbeck in southern England, where Purbeck Limestones have rapidly weathered above a vertical cliff of massive Portland Stone (Small 1970), and on Bull Nose, west of Barry in south Wales, where a concave, gulleyed slope in weak shales lies above a vertical cliff cut in more resistant limestones (Trenhaile 1969). It has been shown that a cliff must assume a composite profile in seaward-dipping rocks, if slippage along the bedding planes is resisted by cohesion between these surfaces. The gradient of the seaward slope in this case is determined by the height of the vertical cliff at its base, the amount of cohesion, the angle of friction between the bedding planes, and the density and dip of the rocks (Terghazi 1962) (see Chapter 6).

No single explanation can account for composite cliff profiles found in different rock types and locations. Cotton (1967) has suggested that a two-cycle development is characteristic of resistant rocks, whereas on weaker rocks, cliff recession may have kept pace with rising sea levels during the postglacial transgression. The upper and lower slopes of composite cliffs could have developed contemporaneously in some cases: where wave erosion is slow in relation to subaerial weathering, as perhaps in some tropical areas; where weak rocks overlie more resistant lithologies; or where the seaward slopes coincide with structural planes.

In northern Cornwall, composite profiles may have developed as a result of the stripping of fault planes dipping seawards at about 45°, and the subsequent undercutting of these structural slopes by wave action. This is essentially a one-cycle concept, but a similar result could be obtained by supposing that there was then a glacial fall in sea level, parallel slope retreat under subaerial attack, and the accumulation of debris at the degraded foot of the cliffs. The cliffs would then have been rejuvenated by wave attack when the sea rose again in postglacial times (Wilson 1952). This two-cycle theory differs from those discussed previously, in that the composite cliff is formed during an interglacial, and then exhumed and modified by contemporary wave action.

Air photographs of siltstone and mudstone cliffs near Tokyo, taken over a six-year period, suggest that composite profiles are created and destroyed quite quickly in some cases (Horikawa and Sunamura 1967). Nevertheless, it has been argued that in the general case, a low, regolith covered seaward slope cannot retreat as rapidly as a steep, wave-cut cliff, which must therefore progressively encroach upon the slope above (Cotton 1951a, Savigear 1962, A. Young 1972). If the seaward slope is about 30°, however, the rate of subaerial downweathering perpendicular to the surface, which is necessary for parallel retreat without encroachment, is only one-half the rate of cliff retreat, and only one-third if the seaward slope is 20°. In most areas where there are high composite cliffs, wave erosion is so slow that it cannot be assumed that the free face is necessarily encroaching, and thereby obliterating the seaward slope above. Such low rates of weathering, however, could not have produced composite cliffs within postglacial time.

It is imprudent to confer polycyclical significance to breaks in the seaward slopes of composite cliffs without a detailed investigation of structural and lithological variations along their profiles. Savigear (1962) accepted that some breaks of slope are induced geologically, and this seems to be the explanation for many of the slope facets near Coombe Martin (Fig. 8.3), and elsewhere along the northern Devon coast. The relationship between geological factors and breaks of slope may not be obvious on regolith-covered slopes. The persistent reoccurrence of similar slope facets over many kilometres (Savigear 1962) may reflect a common polycyclical history, but more work is required before this possibility can be confirmed.

The idea that composite cliffs are contemporary features is inherent to the one-cycle theory. The two-or multicyclic theory suggests that these cliffs are essentially relics of the Pleistocene which have been slightly trimmed or freshened by contemporary wave action. Without precise dating of associated beach and other deposits, attempts to determine the age of these cliffs must inevitably be speculative.

Most workers have suggested that cliff cutting occurred during interglacial intervals of high sea level, whereas degradation took place when sea level was low during the cold glacial periods (Arber 1949, Orme 1962). Agar (1960) for example, believed that the 30 to 40° slopes of the upper cliff in northern Yorkshire, England, are relics of the last interglacial period. Cotton (1951a) originally agreed that the seaward slopes were formed during interglacials, but he later suggested that the alternations of marine and subaerial erosion could have been accomplished during the stadials and interstadials of the last glacial period (Cotton 1967). Modern evidence suggests that interstadial sea levels were probably too low for effective wave erosion at the base of contemporary sea cliffs (Chapter 7).

Periods of cliff cutting were followed by the burial of the fossil cliff in debris during the last cold period (Savigear 1962, Wood 1962, Orme 1962, Fleming 1965, Davies and Stephens 1978). Near Naples, Italy, aeolian red sands mantle the lower slopes of flysch cliffs which are eroded by salt action. The sands contain volcanic ash in places, and are attributable to the last cold phase of the Würm. With the subsequent postglacial rise in sea level, the sand has been partly excavated, and a marine abrasion platform has formed at the base of the rejuvenated cliffs (Baggioni 1975). Many cliffs, of all types, are fronted by raised beaches—for example, on both sides of the Bristol Channel and in many other areas. Amino acid analysis suggests that the raised beaches of southwestern Britain contain material from the middle Pleistocene, reworked and added to in the last interglacial period (Andrews *et al*. 1979, Davies 1983). This implies that the cliffs are very old features which have been modified, albeit slightly in many cases, by contemporary wave action at their base.

Composite cliffs, more than 300 m in height in parts of northeastern Skye, have developed in sedimentary rocks which have been intruded by doleritic sills and cut by minor dykes (Richards 1969). The cliff profiles are dominated by long, steeply inclined seaward-facing slopes, sometimes containing vertical sections corresponding to outcrops of lava or dolerite. The intermittent undercliff, which is up to 12 m in height, appears to be a fossil feature in this area. It is out of the reach of the waves in some places and heavily obscured by debris. Cemented beach deposits, apparently covered by till, occur at the base of the undercliff and within sea caves. Richards proposed that the occurrence of the 'preglacial platform' in this area marked the first phase in the formation of the long, seaward slopes. Lower-level transgressions were then thought to have cliffed the base of these slopes. These high cliffs are therefore of some antiquity, being much older than the postglacial transgression, and owing little to contemporary erosion since the sea reached its present level.

In Victoria, Australia, some degraded cliffs were rejuvenated when the sea returned to its present level, but others, because of local uplift or the deposition of protective beaches, were not. In sheltered areas, bluffs may have been temporarily rejuvenated during an episode of higher Holocene sea level (see Chapter 7), which would have increased the depth of the nearshore waters, and the incoming wave energy in these areas. They have subsequently reverted to bluffs. There may therefore be two types of degraded cliff in Victoria: those which were cliffed and degraded during the last interglacial–glacial sequence, and those which were cliffed several thousand years ago. Unfortunately, although shells have been found in beach sands and gravels at the base of a bluff which is now about 1 m above high spring tides, no precise dating has been attempted (Bird 1977).

Plunging Cliffs

Most wave-cut cliffs are fronted by shore platforms or beaches, but some plunge directly into deep water (Cotton 1951b, 1952, 1968, 1969). Plunging cliffs are particularly common around southern hemisphere basaltic islands (Daly 1927, Davis 1928, Hinds 1930, Cotton 1967). The resistance of these rocks naturally inhibits cliff erosion, and on some tropical islands further protection is provided by the growth of coral reefs (Davis 1928, Cotton 1951b). Furthermore, deep water conditions can persist at the cliff base because large amounts of material cannot be transported alongshore in these areas (Cotton 1967). Nevertheless, plunging cliffs are also found on the resistant rocks of continental coasts (Johnson 1925, Baulig 1930, Guilcher 1958a, Clague and Bornhold 1980). Wave reflection associated with a deep water clapotis, and the lack of abrasive materials at the water level, inhibit erosional processes and help to preserve the plunging condition (see Chapter 1). Hydrostatic pressures, however, can eventually produce caves, which coalesce to form a narrow platform; breaking waves can then eventually destroy the plunging cliff face. Plunging cliffs can also be destroyed as a result of the accumulation of material at their base, which also causes waves to break before reaching the cliff. This material can be derived from longshore movement or subaerial cliff weathering.

The formation of plunging cliffs usually requires rapid drowning. If the rate of cliff recession and sediment accumulation were slow in relation to the rate of drowning, a plunging condition could persist long after sea level had become stable. Plunging cliffs were probably cut during Pleistocene interglacial periods when sea level was high, although some recession could have occurred at lower levels during glacial periods (Daly 1927, Hinds 1930, Guilcher 1948, Cotton 1967). Rejuvenation of these cliffs occurred in the early postglacial period, and was followed by rapid drowning during a subsequent period of rapidly rising sea levels (Davis 1928, Cotton 1967). Plunging cliffs can be the result of local submergence, but others, such as those associated with fault scarps, structural escarpments and fiords, are not primarily the result of wave erosion (Cotton 1967).

Cliff Erosion

Most investigations of coastal cliffs by physical geographers and geologists have been concerned with their morphology and with the role of long-term phenomena. Johnson (1919), for example, applied Davisian theory to the development of rock coasts. His model predicted that the slope of coastal cliffs would be reduced as the width of the shore platforms increased at their base, causing subaerial processes to become dominant over marine. Models such as Challinor's (1949), however, which suggest that shore platforms retreat in a parallel manner (Chapter 9), imply that the cliff must also undergo parallel slope retreat with the persistence of steep gradients.

A few Japanese coastal workers have investigated the short-term effects of coastal dynamics in several areas east of Tokyo. Horikawa and Sunamura (1967) and Sunamura (1973) have plotted the mean annual erosion rate of coastal cliffs against the compressive strength of the rocks at the cliff base. They claimed that the plots showed a tendency for erosion to increase when the strength of the rock declines. The data were not subjected to statistical analysis, and it appears that the trends of the estimated best-fit lines are determined in each case by a single data point, representing a rock whose compressive strength is much less than the average for this area. Although there may be a relationship between these factors, the correlation is certainly statistically insignificant in those studies. In any case, the compressive strength is of uncertain value as a measure of the resistance of a rock to wave action, and its use does not take into account the presence of joints, faults, and other structural weaknesses.

Trenhaile (see Chapter 9) examined the relationship between platform width, rates of cliff erosion, and combinations of rock dip and strike in sedimentary rocks consisting of alternations of variable resistance to erosion. The model predictions were satisfactorily tested against field data from Japan, Wales, and Canada. Fastest erosion is associated with horizontal rocks, and moderately dipping beds which strike perpendicularly to the rock face. Slowest erosion occurs in vertical strata which strike obliquely or parallel to the cliff face. This model essentially formularizes and extends a proposal made by Everard *et al*. (1964). Horikawa and Sunamura (1967) and Sunamura (1973, 1982a) have also calculated the minimum or critical wave height at the cliff base which is capable of causing erosion. Variations in the amount of cliff recession which occurred over periods of several years can be explained by differences in the frequency of waves of greater than the critical height. On a crenulated coast southeast of Tokyo, cliff recession was fastest where the trend of the shore is normal to the prevailing wind direction. These workers have derived several equations to predict rates of cliff erosion (see Chapter 9). Sunamura (1977) found that:

$$dx/dt \propto \ln(f_w/f_r),$$

where dx/dt is the horizontal rate of erosion, f_w is the wave force and f_r is the rock resistance. This can be written as:

$$dx/dt \alpha [\ln(\rho g H/S_c) + C],$$

where S_c is the compressive strength of the rock at the cliff base, H is the wave height at the cliff base, and C is a dimensionless constant. This expression can also be used to calculate the minimum or critical wave height capable of causing erosion (H_{crit}):

$$H_{crit} = S_c \cdot e^{-C} / \rho g.$$

Although these equations are based upon the assumptions that the erosive force of the waves is proportional to the wave height, and that the rock resistance is proportional to the compressive strength, they do represent a first attempt to quantify some important aspects of coastal erosion using variables which can be measured in the field or laboratory. Sunamura (1982a) has tested this erosional model in two field areas in Japan. The results suggest that only the larger but more infrequent waves cause erosion, and variations in the frequency of these larger waves therefore account for variations in the amount of erosion recorded over a number of years. Many other factors need to be considered, however, for an equation which will reliably predict rates of cliff recession.

Several workers have suggested that high cliffs retreat more slowly than low cliffs because they produce more debris, which must be removed before further erosion is possible (Shepard and Grant 1947, Kawasaki 1954, Williams 1956). Sunamura (1973), however, found no relationship in a study area in Japan, possibly because of the effect of other factors. The type of material at the cliff base can also influence rates of cliff recession. Robinson (1977c) found that the width of shore platforms, and therefore presumably the rate of cliff recession, progressively decreases where sandy beaches, bare rock, boulder beaches, and talus cones occupy the cliff foot. In south Wales, wide platforms are common where the cliff foot is bare, and to a lesser extent where there are pebble beaches, but they are narrow where there are boulder beaches. No relationship could be discovered using Japanese data, although this may reflect the lack of consideration of such factors as the amount and potential mobility of the material at the cliff foot (see Chapter 9). Sunamura (1982b) has provided a further expression, based upon a wave flume experiment, which relates the erosion at the foot of a cliff to the amount of beach material present.

Other models which provide expressions for rates of cliff retreat were primarily designed to determine changes in shore platform morphology through time (discussed in detail in Chapter 9). One equation (Trenhaile and Layzell 1980, 1981) is similar to the equations of Horikawa and Sunamura (1966) and Sunamura (1977). It provides a value for the annual amount of erosion, R, at the high tidal level, which is assumed to be the cliff base:

$$R = A\ F_n \tan \alpha_{n-1},$$

where A is the erodibility factor, the ratio of deep water wave energy to the amount of energy required to erode a unit depth of undercut; F_n is the tidal duration value, or the amount of time each year that the waves operate at the cliff foot; and α is the slope of the intertidal platform. A second model (Trenhaile, in press, Trenhaile and Byrne 1986) provides an expression for the time ($T_{n,x}$) needed to undercut the cliff to the point of collapse, and to remove the debris:

$$T_{n,x} = (W)(x)(C_1 + HC_2),$$

where W is the intertidal platform width; H is cliff height; C_1 is a constant equal to $\tan \alpha/UT_r$; C_2 is a constant equal to $\tan \alpha/ST_r$; and T_r is the tidal range. U is the amount of cliff undercutting per year, and S is the amount of debris removed each year. These models also use variables which can be measured in the field.

Many papers mention rates of cliff recession in various parts of the world, but it is still difficult to discern trends related to such factors as rock type and wave and tidal conditions. Several techniques have been used to determine rates of erosion. They include sequential photography, terrestrial and aerial (Shepard and Grant 1947, Trenhaile 1969, Shepard and Wanless 1971); morphological evidence (Gill 1973a); direct measurement using cliff-top stakes or metal pins driven into the cliff face (Hodgkin 1964, Trenhaile 1972, Pinckney and Lee 1973, Bird *et al.* 1979); repeated survey (May 1971); erosion of archeological sites, such as ancient cliff top forts, or Second World War pill-boxes (Trenhaile 1969); old maps (Keatch 1965); dated inscriptions (Emery 1941); and the use of the microerosion meter (Trudgill 1976b). In general, techniques which seem most accurate were only used to measure erosion over very short periods of time. Many of the values are crude estimates, and many others are of uncertain reliability. Reported rates vary from virtually nothing up to 100 m yr^{-1} (see summaries in Sunamura 1973, and Kirk 1977). Very rapid rates of erosion can result from vulcanicity placing unconsolidated materials into high wave-energy environments (Richards 1960, Shepard and Wanless 1971, Mogi *et al.* 1980). Rates of 75 to 100 m yr^{-1} have been recorded on Surtsey and Heimaey, and up to 6.6 m yr^{-1} in the soft pyroclastic deposits of Krakatau, Indonesia, although the erosion rate is only a few centimetres per year on more resistant lava flows (Moign and Moign 1970, Norrman 1980, Bird and Rosengren 1984). Maximum rates of 1 to 2 m yr^{-1} have been reported for weak sedimentary rocks such as mudstones and siltstones, and values of less than 0.01 m yr^{-1} for many hard rock coasts. In most cases, actual rates of recession are far lower than these maxima (Horikawa and Sunamura 1967, Rudberg 1967, Sorensen 1968, Trenhaile 1972, Emery and Kuhn 1980); thus, in southern California, old photographs show that virtually no erosion has occurred in 50 years, even in fairly weak rocks (Shepard and Grant 1947). The persistence of raised beaches and the presence of composite cliff profiles in many parts of the world testify further to the generally slow rates of erosion of rock coasts. It should additionally be emphasized that even if we knew the contemporary rates of recession of coastal cliffs in different rocks types and in different morphogenic environments, there is no guarantee that they are representative of the mean rates of development since the sea reached its present level.

9

Shore Platforms

Shore platforms are conspicuous elements of rock coasts in many parts of the world. They have been defined as horizontal or gently sloping surfaces, produced along a shore by wave erosion (*Glossary of Geology*, Am. Geol. Inst. 1972). This definition, which might be better applied to the term 'wave-cut platform', assumes that shore platforms are necessarily the product of wave action, which is not always true. The term 'shore platform', which has no genetic connotation, will therefore be used to refer to rock surfaces of low gradient, within or close to the intertidal zone. The term 'wave-cut platform' will be used to refer to a specific category of shore platform, rather than as a synonym for all platforms.

Although it is difficult to generalize about platform geometry, two types of platform have frequently been distinguished. Gently sloping platforms (1 to 5°), which are sometimes called ramps, usually extend from the cliff–platform junction to below the low tidal level, without any major break of slope or abrupt seaward terminus (Fig. 9.1). Subhorizontal platforms, which can be supra-, inter- or subtidal, generally terminate in a low-tide cliff or ramp (Wright 1967, Sanders 1968a) (Fig. 9.2). Geological and other local factors produce sloping and horizontal platforms along a stretch of coast, although the former are most common in macrotidal environments, and the latter in meso- and microtidal regions (Trenhaile 1972, 1974a, 1978, 1980).

Until quite recently, the literature on shore platforms was dominated by Australasian workers, and their descriptions and interpretations of the platforms in Australia and New Zealand have an important influence on the treatment of shore platforms in coastal texts. The older work was largely descriptive, and mainly concerned with the relative significance of mechanical wave erosion and subaerial weathering processes. Important genetic criteria were usually thought to include the degree of shelter from wave action afforded by coastal sites, and the elevation of the platforms relative to specific tidal levels (although the platforms were rarely surveyed).

The sea's ability to produce intertidal shore platforms was first commented on more than a century ago (Ramsey 1846, Dana 1849). From the beginning, mechanical wave erosion was usually emphasized, based on the vigorous wave environments of the northern Atlantic (De la Beche 1839, Ramsey 1846, Davis 1896, Fenneman 1902). The possible role of

Figure 9.1. Shore platform in conglomeratic sandstone at Stubborn Head, Minas Basin, Bay of Fundy, Nova Scotia. Platform gradient is between 2.5 and 3.5° in this macrotidal area.

Figure 9.2 Subhorizontal platform with low tide cliffs in aeolianite near Sorrento, Victoria, Australia.

mechanical wave action in Australasia, and elsewhere in warm temperate and tropical swell wave environments, has however been a contentious issue, and it has often been accorded a secondary role to subaerial weathering. Even in northern storm wave environments, it has been recognized that subaerial weathering aids wave erosion, and may be a vital precursor to platform development in some areas (Lyell 1873, Green 1887, Reid 1890, Jukes-Browne 1893, Geikie 1903).

The first section of this chapter is particularly concerned with the following questions:

1. Are different processes responsible for cliff retreat and platform formation in different environments?
2. If the answer to (1) is yes, do these processes develop a zonally distinct platform morphology?

An important distinction should be made between the processes which accomplish cliff erosion and those which planate the residual surface at its base. This distinction is rarely made, although the mechanisms which cause cliff retreat are not necessarily those which planate the platform.

Many Australasian workers have acknowledged the efficacy of wave erosion in exposed regions, but it has been argued that because of the variability in wave intensity, and in the levels at which waves operate, subaerial weathering is essential for the formation of subhorizontal surfaces (Wentworth 1938, 1939, Hills 1949, 1971, Hawley 1965, Gill 1967, Sanders 1968a, 1970). It has not been generally appreciated that there is no consensus on the way in which weathering produces horizontal platforms. Critical examination of the various theories shows that most accord a suprisingly major role to mechanical wave erosion, although this is rarely explicitly stated. Those workers who deny the competency of wave erosion to produce horizontal platforms usually attribute their formation to subaerial cliff weathering, or to the modification by weathering, of rough, sloping wave-cut ramps.

Cliff Weathering

Dana (1849) emphasized the role of cliff weathering in the formation of shore platforms in northern New Zealand, but the concept did not gain prominence until the later contributions of Bartrum (1916, 1926, 1935). As Bartrum (1926) noted, his theory is not identical to Dana's, although many workers have subsequently overlooked several crucial differences (see, for example, Edwards 1958). Bird and Dent (1966) made the distinction between wave erosion of unweathered rocks, and the washing away of weathered materials by wave action (Fig. 9.3*a* and 9.3*b*). This distinction, which is characteristic of the thinking which has promoted the

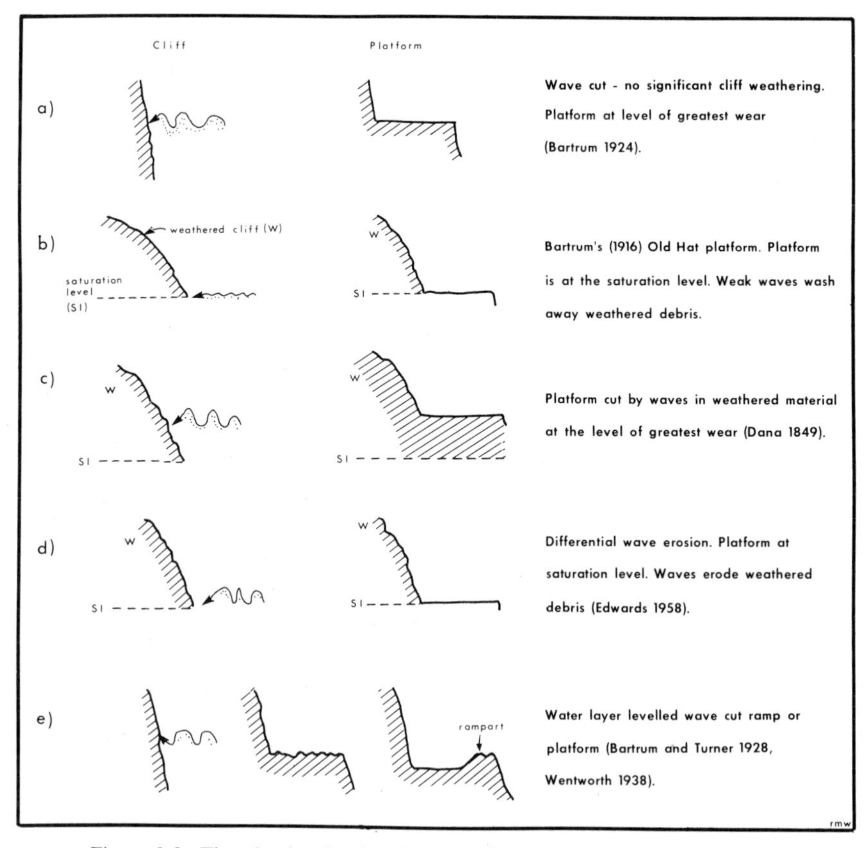

Figure 9.3. Theories for the development of subhorizontal shore platforms.

wave erosion versus weathering debate for more than a century, is misleading, because it recognizes only the two extremes of a spectrum of platforms on which wave erosion and weathering have operated to varying degrees. The question which needs to be asked is not, is the cliff weathered? but, Is platform development independent of wave action to the extent that its morphology is unrelated to the distribution of wave energy within the tidal range? This distinction can be best illustrated by considering the various theories of platform development in weathered cliffs.

Wave-cut Platforms in Weathered Rock

Dana (1849) proposed that platforms are cut by waves at the level of maximum wear, in cliffs which are weathered down to the point at which the rocks are saturated with sea water (Fig. 9.3*c*). Bartrum (1926) concurred with Dana, but only for those situations where wave action is fairly vigorous.

Weathered Debris Removed by Weak Waves

In contrast to Dana (1849), Bartrum (1916) considered that the platforms he termed 'Old Hat', were not the result of wave attack concentrated upon a definite zone of weathered rock. He agreed that coastal cliffs are weathered down to the level at which they become permanently saturated with sea water, but he believed that the platforms develop at this saturation level, rather than at the level of maximum wave erosion (Chapter 2) (Fig. 9.4). Bartrum proposed that they form about 0.5 m below the high tidal level. They develop in impermeable rock, free from open joints, in very sheltered areas where waves are weak and only competent to wash away the weathered debris (Bartrum 1916, 1926, 1935) (Fig. 9.3*b*). Bell and Clarke (1909) were actually the first to attribute the development of benches just below high tidal level to subaerial cliff weathering and the removal of the weathered debris by wave action. They ascribed the presence of upstanding irregularities on the platforms to their greater resistance to wave erosion, compared to that of the platform in general. As Bartrum (1926) has remarked, their interpretation is inconsistent with the suggestion that the function of wave action is to wash away the debris produced by weathering. Bell and Clarke's thesis therefore appears to be closer to Dana's wave erosional theory than to Bartrum's proposal. Although many other workers have cited Bartrum in emphasizing the role

Figure 9.4. Mill or Kaiaraara Island at Russell, Bay of Islands, northern New Zealand. The notched side of this island faces towards the Pacific.

of weathering in the development of platforms, only a few (Bird and Dent 1966, Healy 1968a) have strictly adhered to his interpretation regarding the impotence of mechanical wave erosion.

McLean and Davidson (1968) have suggested that shore platforms in sandstones and mudstones on the Gisborne coast of New Zealand are the result of mass movement on the cliff, and the removal of the debris by waves. Although they argued that wave erosion is not the dominant process, it is difficult to understand why mass movement should take place if not for the oversteepening of coastal slopes by the waves. This is supported by the tendency in this region for mass movement to occur in exposed areas. In most cases, shore platforms are absent or poorly developed on coasts where there is large-scale mass movement (Wright 1969).

Another type of weathered platform has been described by Wentworth (1939). He believed that subhorizontal benches on limestones on Oahu, Hawaii, are formed by solution associated with fresh water, which is progressively inhibited downwards by lime-saturated sea water. It is probable, however, that chemical or biochemical processes associated with sea water play a substantial, if not dominant, role in their formation or modification. These surf platforms and notches, which are common along the coral and aeolianite limestone coasts of warm climatic regions (Fairbridge 1948, 1950), are discussed in greater detail in Chapter 10.

Differential Wave Erosion

Another group of workers (Edwards 1958, Gill 1967, Russell 1971, R. W. Young 1972, Bradley and Griggs 1976, Sunamura 1978a) agreed that platforms develop where the rocks become saturated by sea water, but suggested that this is accomplished by wave erosion of the weak weathered material lying above this level (Fig. 9.3*d*). Fairbridge (1952) and Mii (1962, 1963) also appear to have subscribed to this theory, although Fairbridge considered that the saturation level, and therefore the platform, are closer to the low than to the high tidal level. A novel interpretation has been provided by Russell (1971), who was apparently unaware of the Australasian literature. He argued that on the western coast of the United States, because the water table is lowered as the cliff retreats, platform surfaces develop at successively lower elevations landwards; this has not, however, been confirmed elsewhere. In some fairly exposed areas in southern Victoria, subhorizontal, mid-tidal platforms have developed in aeolianite, which, in its unmodified state, is too permeable and porous for weathering to take place as a result of pools of spray and splash (Fig. 9.2). Hills (1949) and Gill (1973a) have attributed the formation of these platforms to the removal of the weak dune rock above an indurated zone. According to Hills, induration is the result of the downward percolation of fresh rain water, and according to Gill, to contact with sea water.

Platform Morphology and Cliff Weathering

Dana's (1849) theory suggests that platforms develop in weathered rocks at the level at which wave erosion is most effective, rather than at the saturation level. In some environments, the elevation of the two surfaces may happen to be similar, but Dana's theory suggests that platform morphology is determined by wave action, whereas Bartrum's implies that it is determined by the form of the saturation-level surface. Even if a platform is produced by wave erosion through the exhumation of the saturation level from beneath weathered rocks (Edwards 1958), its morphology would probably be partly related to wave and tidal conditions. It is unlikely that the platform surface would be perfectly coincident with the saturation level as implied by Bartrum's theory, because wave action which is capable of eroding weathered material must also exert some influence on the form of the erosion surface.

Platform Weathering

Bartrum and Turner (1928) first described weathering processes associated with alternating wetting and drying, which smooth and lower shore platforms (see Chapter 2) (Fig. 3*e*). If the tidal range is small, secondary lowering could significantly alter the relationship between platform elevation and specific tidal levels. Johnson (1938) stressed that the modification of wave-cut shore platforms should not be confused with the formation of platforms by weathering alone. Hills (1949) also implied that a distinction should be made between water layer levelled and Old Hat platforms. He suggested that Wentworth's (1938) term 'water level weathering' should be reserved for weathering related to sea level, or to the groundwater in the vicinity of the coast, as are the platforms of the Old Hat type, rather than for weathering associated with the surfaces of water in platform pools. The elevation of water layered surfaces was originally conceived as either being determined by the size of the irregularities on wave-cut platforms (Bartrum 1935, Hills 1949), or by the occurrence of any surfaces where spray and splash can lodge (Wentworth 1938). More recently it has been inferred that platform surfaces are reduced to the same level of permanent saturation as Bartrum's Old Hat platforms (Bird and Dent 1966, Bird 1968, Sanders 1968a, Davies 1972, Abrahams 1975, Takahashi 1977). Several workers also consider that cliff and platform weathering operate together, although Bartrum (1916) only considered weathering on platforms which were originally wave-cut.

There are therefore several theories on the way in which shore platforms are produced or modified by weathering. Bartrum's Old Hat theory relegates wave erosion to a very minor role, but the others infer that mechanical wave erosion assumes an important, if rarely recognized, role in the development of shore platforms in weathered rocks.

Weathering: The Evidence

Water Layer Levelling

It will be argued that subhorizontal platforms can be cut by wave action, although this does not preclude the possibility that water layer levelling smooths, levels, and lowers some rugged wave-cut surfaces, where the rocks are suitable. Bartrum (1926) emphasized that platform surfaces in horizontally bedded rocks are usually the result of differential wave erosion rather than weathering. Davies (1972) and others, however, considered that weathering operates best on rocks with low angles of dip. There is an obvious danger, therefore, that fairly smooth surfaces largely produced by the wave erosion of horizontal strata, but subsequently slightly modified by secondary subaerial weathering, could be mistaken for water layered surfaces (Fig. 9.5). It should also be noted that water layer levelling may in some cases make platforms more, rather than less, irregular (Johnson 1938).

Seaward ridges or 'ramparts', often rising a metre or more above the general platform surface, have been attributed to water layer levelling

Figure 9.5. Subhorizontal structural platforms in Jurassic sandstones in Curio Bay, near Invercargill, southern New Zealand.

(Fig. 9.6). Several workers considered ramparts to be residual features which resisted water layer levelling because of their position at the seaward margins of platforms, where they are kept moist by spray and splash (Bartrum 1935, Wentworth 1938, Hills 1949, 1971, Takahashi 1977). It has been suggested that ramparts are characteristic features of volcanic rock coasts, particularly, but not exclusively, in low latitudes (Guilcher *et al.* 1962, Guilcher and Bodere 1975). Ramparts are absent on many platforms, however, and they are often discontinuous where they do occur. Some platforms have a series of ramparts, and single ones are not necessarily restricted to the seaward margins of platforms. The alternative proposal has therefore been made that ramparts are the result of differential erosion related to outcrops of more resistant rocks, and that they have no genetic significance (Johnson 1938, Edwards 1951). Gill (1972a) recognized two types of ramparts on the Otway coast of Victoria: those related to outcrops of massive arkoses, and those owing to ferruginous induration of the rocks preceding the Holocene transgression. The occurrence of ramparts at the seaward edge of platforms would be a natural consequence of their association with resistant beds, which hold up recession of the low-tide cliff

Figure 9.6. Supratidal, subhorizontal water layered platform and rampart in massive arkose at Point Sturt, Otway Coast, Victoria, Australia.

(Gill 1967). This may explain the formation of ramparts in Alaska, where the platforms are cut by hydraulic wave action (Chastain 1976).

Water layer levelling processes appear to be active at all elevations between low tide and the upper limit of spray and splash, and there is little evidence to support the contention that they operate down to a saturation level close to the high tidal level.

Old Hat Platforms

Cliff weathering may produce a distinct class of subhorizontal platform. According to Bartrum (1916), the term 'Old Hat' was first used by Hochstetter in 1864, to describe the form of Mill or Kaiaraara Island, off Russell in the Bay of Islands in northern New Zealand (Fig. 9.4). The general narrowness of the platforms testifies to the hardness of the rocks and the sheltered environment of the bay. The cliffs in this area are weathered, but they are still very resistant and cohesive, although there could be a slow, imperceptible release of granular material. The cliffs are quite steep, and the platforms are usually widest on the seaward side of the island. The undercut cliff or notch on the seaward side of Mill Island may be due in part to the effect of salts, but the presence of fresh rock surfaces in the lower part of the cliff, and the rough, uneven surface of the platform, suggest that wave quarrying played a significant role in the development of the platforms in this area. Bartrum (1935, p. 136) recognized that the steep seaward slope of the Old Hat platforms is 'undoubtedly being cut back by waves at phases of lower water . . .'. If wave action in this sheltered bay is sufficient to erode these resistant rocks at low tidal levels, it is difficult to understand why he thought that their work would be limited to washing away weathered material at higher water levels.

Healy (1968a) attributed the platforms of the Whangaporaoa Peninsula north of Auckland to cliff weathering and the removal of the debris by weak wave action. I cannot agree with this interpretation, which fails to account for the presence on the platforms of large quarried debris, potholes, undercut scarps, freshly quarried surfaces, loosened joint blocks and steep, fresh cliff faces. Bird and Dent (1966) are the only other workers I am aware of who have strictly adhered to Bartrum's Old Hat interpretation. They proposed that some of the platforms on the south coast of New South Wales attain their best development on the sides of islands and headlands which are sheltered from the prevailing south-easterly swell. Abrahams and Oak (1975) did not find any significant relationship between platform development and the degree of exposure to the prevailing winds in part of Bird and Dent's study area. Wide platforms in some types of rock are also found in fairly sheltered areas in cool, storm wave environments where subaerial weathering is not the dominant erosive agent. It is possible that variations in platform width have some genetical significance in parts of southern Australia, although it needs to be clearly demonstrated that they do not reflect variations in the structure and lithology of the rock. In general, particularly near Sydney, the south coast is characterized by steep, actively receding cliffs, and platforms which show the results of quarrying and differential wave erosion.

The field evidence for the existence of Old Hat platforms, *in sensu stricto* (Bartrum 1916), is lacking; there are, however, two main variations on the Old Hat theory which should also be considered (Figs. 9.3*c* and 9.3*d*). Several workers have attributed platform formation to the wave erosion of the weathered material above the saturation level (Edwards 1958, Gill 1967). This proposal, as well as Bartrum's Old Hat theory, are based upon speculations on platform elevations, few of which have been surveyed or precisely related to tidal levels. They are also based upon the assumptions that there is a well-defined permanent level of saturation in the intertidal zone; that there is an abrupt transition between weak weathered rock above this level and more resistant, unweathered rock below; and that the saturation level stays at the same elevation in the cliff as it retreats landwards. They are also dependent on the assumption that the platform surface is coincident with the saturation level, even though this level, if it exists, must vary in time and space according to geological, climatic, and tidal and wave conditions. The lack of evidence for the validity of these assumptions (see, for example, Trenhaile and Mercan 1984), which has been considered in Chapter 2, must cast some doubt on the role of saturation levels in the development of platforms above the low tidal level. The second main variant on the Old Hat theory is essentially attributable to Dana (1849). This supposes that platforms are cut in weathered rock by wave action, at the level of greatest wear. Although weathering in many environments allows platforms to develop where wave action would otherwise be too weak or the unweathered rocks too resistant, Dana's theory proposes that the platforms are cut by waves. These platforms must therefore acquire morphological characteristics which are determined by the distribution of wave energy and the strength of the weathered rock.

A good example of a situation where weathering is apparently an essential precursor to platform development is in a sheltered part of Whangaruru Bay, about 25 km southeast of Russell in northern New Zealand. In this area, in contrast to the Bay of Islands, small rock fragments in the cliff are probably loose enough to be washed away by weak wave action. It is not, however, an Old Hat platform, as the narrow platform is also intensely weathered; the platform surface therefore cannot be coincident with the interface between weathered and unweathered rock, but must be determined by the level of most effective wave action, as proposed by Dana.

Frost and Ice

In high latitudes, frost and ice may play a similar role to chemical weathering and wetting and drying in low latitudes (Trenhaile 1983a). The occurrence of strandflats in narrow fiords and in other areas where wave action is fairly weak has led many workers to propose that frost assumes an important role in the erosion of coastal rocks (Rekstad 1915, Vogt 1918, Nansen 1922, Grønlie 1924, Nordenskjold 1928, Holtedahl 1960). The main emphasis has been on the weathering of the cliffs by frost, although Tietze (1962) suggested that strandflats resulted from frost operating under shelf ice, as it rose and fell with the tides. Nevertheless, the strandflats are widest and best developed on headlands. They are much narrower in the outer parts of fiords, and absent in the inner parts. Nansen (1904) attributed this pattern to the progressive decline in wave intensity in the fiords. He later emphasized, however, the favourable conditions created by frost action around the ice foot in sheltered fiords, where there are narrow, rugged ledges at about the high tidal level (Nansen 1922). He attributed the formation of strandflats to the combined effects of waves and particularly frost action. Gelifraction is an essential mechanism for the development of strandflats and contemporary shore platforms in the fissile and fractured rocks of Spitsbergen (Moign 1974a,b, Guilcher 1974a). It may also have played an important role in the formation of the Younger Dryas Norwegian Main Line platform (Andersen 1968), although the occurrence of contemporary shore platforms in northern Norway is more closely related to the exposure to wave attack (Sollid *et al*. 1973). In western Scotland, shore platforms may have developed in sheltered areas because of frost action in the Younger Dryas, Loch Lomond stadial, although increased storminess would also have aided wave erosion (Sissons 1974, Gray 1978, Dawson 1980). Frost action has also been active on the shore platforms in the Gulf of St Lawrence, in Hudson Bay, and on the Shantarskie Islands in eastern Siberia (Corbel 1958, Gubkin 1969, Guilcher 1981, Brodeur and Allard 1983, Allard and Tremblay 1983). Even in the fairly mild climate of southern England, chalk wave-cut platforms were damaged by frost during a severe winter (Williams and Robinson 1981).

The coastal environment may be particularly suited for frost action and temperature dependent wetting and drying. This has been discussed in some detail in Chapter 5, but it is worth briefly reiterating some of the reasons here:

(*a*) Spray, splash and groundwater flow may allow high levels of saturation to be attained in coastal cliffs. In the intertidal zone, critical levels of saturation could be attained as a result of tidal inundation (Trenhaile and Mercan 1984).

(*b*) Most studies show that rocks which are saturated by solutions with 2 to 6% of their weight in salts are more susceptible to frost damage than those which contain fresh water.

(*c*) More freeze–thaw cycles and more rapid changes in temperature can occur in the intertidal zone than above the high tidal level.

Although the precise mechanisms are still poorly understood, field evidence suggests that frost and temperature-dependent wetting and drying are very effective on coasts in high latitudes. We do not know, however, under what conditions, and how frequently, critical saturation levels are attained in the cliff and in the intertidal zone (Trenhaile and Mercan 1984). Nor do we have any data on the magnitude and frequency of temperature fluctuations within coastal rocks. Furthermore, it needs to be emphasized that frost weathering is only really effective in certain types of rock, particularly those which are fine-grained, of low tensile strength, and with numerous lines of weakness; that is, rocks which are also susceptible to mechanical wave erosion. If it is claimed that shore platforms can be produced by frost weathering, as distinct from its role in contributing to the recession of the cliff, then it must be demonstrated that its efficacy ceases, or markedly decreases, at a specific datum which is coincident with the surface of the platform. Numerous engineering and geological investigations imply that this cannot be true. Considerable differences in the susceptibility of various lithologies to frost action make it particularly difficult to understand how it could planate an intertidal zone consisting of an alternating series of rock types (Trenhaile and Rudakas 1981).

The role of shore ice must also be considered in high latitudes. Moign (1974a,b, 1976) believed that drift ice is essentially protective, and she considered that its role in the development of shore platforms in Spitsbergen is negligible. Although shore ice can transport loose material, most workers believe that it is unlikely to be a major process on rock coasts (Martini 1981). Collapse of a melting ice foot can dislodge rocks from the cliff face (Nielsen 1979), although Moign (1974a,b) believed that the overall effect of the ice foot is small. The ice foot can also carry away debris as it floats away from the cliff (Joyce 1950, Corbel 1954, 1958, Howarth and Bones 1972, Nielsen 1979), but Nansen (1922) thought that this was of little significance, because much of the ice foot melts *in situ*. Several workers have suggested that frost and other processes associated with the ice foot can produce terraces (see, for example, discussions in Nansen 1922 and Charlesworth 1957), but Nielsen (1979) believed that they would be small. It has been proposed that the planation of intertidal areas can be accomplished by shore ice (Hansen 1894, 1898—cited by Nansen 1922, Goldthwait 1933). This could be a result of the growth of segregation ice in cracks in the bedrock beneath the ice foot or shore-fast sea ice (Matthews *et al*. 1986). In the South Shetland Islands, subhorizontal shore platforms have been attributed to ice freeze-on, quarrying by impact, and abrasion (Hansom 1983). Fairbridge (1977) even argued that the wide shore platforms of western Europe, southernmost Australia, and other temperate regions were cut by frost and floating ice in the early glacial periods. Apart from the questionable assumption that early glacial sea levels were similar to today's (see Chapter 7), this hypothesis as noted by Dionne (1978), exaggerates the role of ice in the formation of platforms and strandflats in high latitudes. Zenkovitch (1967) for example, was doubtful that ice could abrade in the coastal zone, and Nansen (1904, 1922) considered that floating ice is an insignificant erosive agent. Floating ice accomplishes little erosion in the Canadian Arctic (Bird 1967), and Moign (1974a,b, 1976) rejected it as an effective erosional agent on the shore platforms of Spitsbergen. In James Bay, Canada, ice has little effect on rock platforms, possibly because floating ice is unable to keep abrasives in contact with submerged rock surfaces (Dionne 1980). Ice can dislodge loose rock from the intertidal zone, and is capable of abrading striations in weak rocks in the intertidal zone, but most workers consider that its role is largely protective, preventing storm waves from reaching the shoreline for much of the year.

Wave Erosion

It has been emphasized that it is necessary to distinguish between those processes which are responsible for cliff erosion, and those which lower, smooth, and level the shore platform. The importance of this distinction has often been overlooked. Few workers have actually proposed that platforms are produced by weathering, as opposed to recognizing the important, often essential, role of weathering in *facilitating* cliff erosion. In most environments, weathering is probably the component which allows shore platforms to develop in sheltered areas, and/or where the rocks are otherwise too resistant to wave action. Even in the low wave energy, humid tropical environment of the Singapore Islands, however, shore platforms have been cut into fairly unweathered granites, quartzites, conglomerates, and sandstones, as well as in intensely weathered rocks (Swan 1971).

Several Australasian workers have insisted that horizontal shore platforms can be cut by waves, particularly by the most vigorous storm waves (Bartrum 1924, 1926, 1935, 1938, 1952, Johnson 1931, 1933, 1938, Jutson 1939, 1940, 1949a,b, 1950, 1954, Edwards 1941, 1942, 1945, 1951, Cotton 1963). According to Bartrum (1952), these platforms develop between about 0.5 and 2.5 m above the high tidal level. They are generally backed by steep, often notched and overhanging cliffs, which are actively receding. The platforms are awash during storms, when debris which has fallen from the cliff is rapidly broken up and swept away. Many workers also consider that wave action is important on coasts in high latitudes, despite the protection afforded by ice for a large part of the year. Wave erosion, often aided by frost, has been considered by some workers to be the primary mechanism for the formation of the Norwegian strandflat (Reusch 1894, Richter 1896, Vogt 1907, Rekstad 1915, Johnson 1919, Strøm 1948). Wave erosion may also have played a fundamental role in the development of strandflats and contemporary shore platforms in Spitsbergen (Hoel 1909, Peach 1916, Werenskiold 1953, Moign 1974a,b, Guilcher 1974a), Iceland (Thorarinsson *et al.* 1959), and on the South Shetland Islands and the Antarctic Peninsula (John and Sugden 1971, 1975). Nansen (1904, 1922) recognized that wave erosion and transportation were important for the formation of strandflats in several regions, and their best development is on capes and in other areas where strong wave action is able to work in unison with frost (Nansen 1904, 1922, Moign 1974a,b). Mechanical wave erosion usually dominates in the mid-latitude storm wave environments of the northern hemisphere, although other processes also play an important role in preparing rock surfaces for eventual destruction (Johnson 1919, Everard *et al.* 1964, So 1965, Wood 1968, Trenhaile 1969, 1972, 1978, Wright 1970, Ters and Peulvast 1972, Sunamura 1978). Few investigations have been concerned with determining the mode or efficiency of the wave erosional processes. The presence of fresh, steep, and often undercut cliff faces, fresh rock scars on the platforms, large, angular debris, and other evidence has convinced many workers that wave quarrying is of greatest importance. Further support for the wave erosional hypothesis is provided by the relationships which exist between various aspects of platform morphology and several environmental factors.

Platform Morphology and the Morphogenic Environment

Platform Gradient

Several workers have commented on the possible effect of tidal range on platform morphology (Edwards 1941, Gill 1967). Davies (1972) suggested that a large tidal range encourages the formation of sloping intertidal platforms, and King (1959) and Wright (1967) mentioned that low platform gradients in most of Australasia could be the result of the small tidal range. Apart from a few passing comments, however, most of the century-old debate on the origin of shore platforms has been conducted in almost total ignorance of the fundamental role of tidal range.

A strong, positive relationship has been identified between tidal range and platform gradient in several areas in the storm wave environments of the northern Atlantic (Trenhaile 1972, 1974a, 1978). Using more than 800 surveyed profiles from three macrotidal areas in England and Wales,

Trenhaile (1974a) found a correlation of 0.92 between local means of platform gradient and tidal range. Adding survey data from Gaspé in eastern Québec, where the tidal range is low mesotidal, provided an overall correlation of 0.88 (Trenhaile 1978). The regression equation suggests that platforms become almost horizontal when the tidal range is less than about 2.5 m. The relationship between gradient and tidal range appears to exist in areas outside the north Atlantic as well, despite the effects of different climates and wave regimes (Fig. 9.7). Platform gradients between 0.2 and 0.5° in Japan (Takahashi 1977), where the tidal range is only between 0.4 and 2.7 m, are consistent with this relationship. In central California, platform gradients of about 0.3° occur in an area with a tidal range of about 2 m (Bradley and Griggs 1976). The northerly increase in tidal range could explain why a platform near Lisbon, Portugal, is the most northerly subhorizontal example in Europe (Hills 1972). It is certainly tempting to speculate that the very low platform gradients which are typical of much of Australasia reflect the small tidal range in most of this region (Trenhaile 1974b). Although sloping ramp-like platforms are found in some types of rock in southern Australia and New Zealand (Wright 1967, Hills 1972,

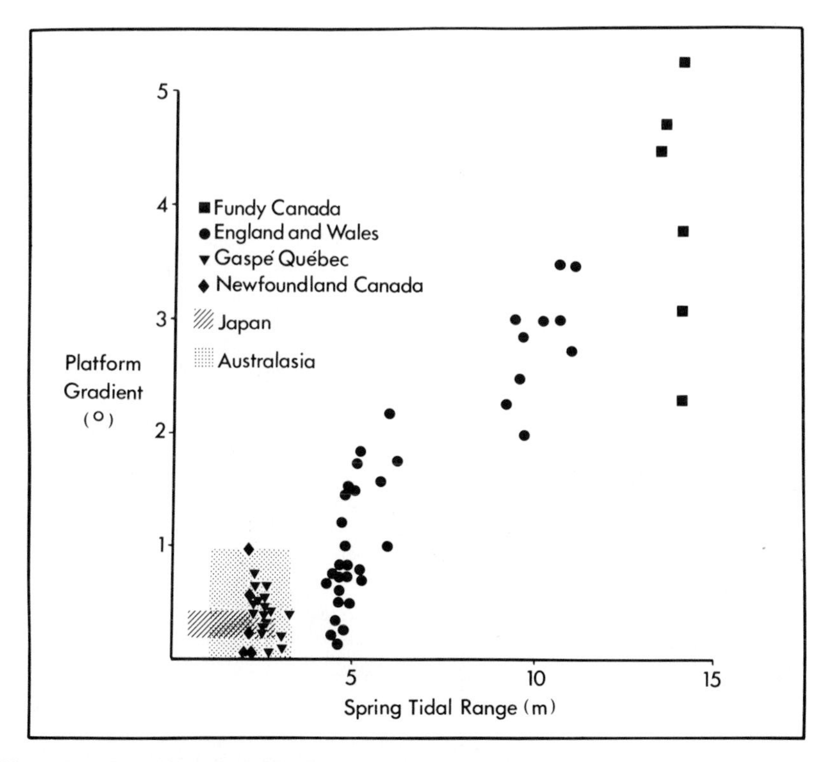

Figure 9.7. Plot of platform gradient against tidal range. Each point is a regional average representing numerous surveyed profiles.

Gill and Lang 1983), most are obviously controlled by rock structure, and their gradients are nevertheless much lower than those in macrotidal areas.

Until quite recently, it was generally believed that shore platforms in cool, wave-dominated environments necessarily slope seawards. Hills (1971) for example, commented that Dana had not encountered any horizontal platforms along the shores of the north Atlantic. This supported the argument that horizontal platforms cannot be produced by wave erosion, and that Australasian platforms are therefore essentially the product of subaerial weathering. The general dominance of macrotidal ranges in the north Atlantic, and meso- and microtidal ranges in Australasia, served to obscure the effects of tidal variations within each region. Differences in platform morphology between these two areas were therefore usually attributed to differences in climate rather than to tidal conditions. In eastern Canada, however, horizontal shore platforms occur in shales, argillites and greywackes in Gaspé, Québec (Trenhaile 1978), and in limestones, sandstones, siltstones, mudstones, and shales in Gros Morne national park in western Newfoundland (Fig. 9.8). The spring tidal range in Gaspé is from 2.25 to 3.5 m, and about 2 m in western Newfound-

Figure 9.8. Subhorizontal platform and structural 'rampart' in sedimentary alternations in Lobster Cove, Gros Morne National Park, western Newfoundland.

land. The morphology of these platforms is very similar to those in Australasia, where the tidal range is about the same. It can be concluded, therefore, that wave-cut platforms are horizontal in vigorous storm wave environments, where the tidal range is small. This conclusion supports the work of Bartrum, Johnson, Jutson, Edwards, and Cotton, who believed that horizontal surfaces can be cut by waves.

It has been suggested that floating ice has produced subhorizontal platforms in the South Shetland Islands (Hansom 1983); it could therefore be argued that platforms in eastern Canada are horizontal because they are the product of frost action and abrasion by floating ice rather than mechanical wave erosion. In the Minas Basin of the Bay of Fundy, however, where frost and ice also operate, platforms in conglomerates, sandstones, shales and basalts have a seaward slope of between 2 and 4.5°, presumably because of a tidal range of between 13.5 and more than 14.5 m (Fig. 9.1). An independent investigation has also shown that within the Bay of Fundy, platform gradient increases to the northeast in accordance with an increase in tidal range (Thomson 1979). Edwards's (1958) paper on the platforms of Yampi Sound in northwestern Australia is the only one which considers Australasian shore platforms in a macrotidal environment. Davies (1972) noted that these platforms 'slope uncharacteristically', presumably in comparison with those in the meso-and microtidal environments of southern Australia. It is notable that weathering has not produced horizontal platforms in this area, which is not surprising when one considers that the tidal range in Yampi Sound is about 10 m.

The relationship between tidal range and platform gradient has important genetic implications. If some shore platforms are produced by chemical weathering, or by frost action, then some explanation must be sought to explain why their gradient varies with the tidal range. If platforms in weathered or unweathered materials are cut or fashioned by waves, then the relationship between tidal range and gradient must be attributable to the way in which wave energy is distributed within the intertidal zone.

The relationship between platform gradient and the total wave energy expended on a coast, as distinct from its distribution within the tidal range, remains to be determined. Mathematical modelling suggests that the gradient of wave-cut platforms is least where wave action is strongest (Trenhaile and Layzell 1980, 1981). The wave energy reaching a coast is influenced by the fetch, and by the offshore slope. In three areas of southern Britain, platform gradient declines logarithmically as the fetch, measured perpendicularly to the shoreline, increases (Trenhaile 1974a), but this relationship does not exist in Gaspé (Trenhaile 1978). The occurrence, in some areas, of steeper platforms in embayments than on the adjacent headlands (So 1965, Wood 1968, Hills 1971, 1972), supports the hypothesis that gradients decline with the strength of the waves. In other areas, platforms are steeper on the headlands than in the embayments (Trenhaile 1972, 1974a, Kirk 1977, Sanders 1968a). This conflicting evidence probably reflects the effect of varying rock strength and hardness, as well as the exposure to wave attack. The offshore slope also affects the rate at which wave energy is expended as it approaches a coast, although no relationship was found between the offshore slope and platform gradient in southern Britain, nor in Gaspé, Québec (Trenhaile 1974a, 1978).

It has been suggested that the gradient of shore platforms progressively declines through time (Bradley and Griggs 1976), until an equilibrium state is attained (Trenhaile and Layzell 1980, 1981). The height of the cliff determines the amount of debris which must be removed to permit continued extension of the platform at its base. For this reason, it may take longer for equilibrium to be attained by platforms backed by high cliffs than by low bluffs. Steeper platforms may therefore be associated with the areas of higher cliffs along a coast. In southern Britain, there is a positive correlation ($r = 0.79$) between platform gradient and cliff height in the Chalk, but a negative one in Liassic limestones and shales. Variations in cliff height seem to have exerted little influence on platform gradients below the hanging valleys of the Seven Sisters in southeastern England (Trenhaile 1974a), and in Alaska (Chastain 1976), presumably because of the effects of numerous other factors.

Platform Width

Platform gradient is a particularly important morphological element because of its effect on the wave energy reaching the cliff base. Width, however, defines the occurrence and the degree of development of platforms, and it is surprising that there has been so little attempt to investigate the factors which determine the width of platforms along a coast. The possible relationship between platform width and tidal range is much less clear than that between gradient and tidal range. Geometrically, the width of essentially rectilinear platform profiles could increase, decrease, or remain constant as the gradient increases with increasing tidal range.

Mathematical modelling suggests that the width increases with the tidal range (Trenhaile and Layzell 1980, 1981, Trenhaile 1983b), but the field evidence is contradictory. Several workers have found that the width of shore platforms does increase with the tidal range (Edwards 1941, Flemming 1965, Wright 1969). A strong positive relationship was found between these variables when composite data were considered from south Wales, eastern Canada, and Japan. When additional data from southern Japan (Takahashi 1977), southeastern England (So 1965) and northeastern England (Robinson 1977c) were included, however, to try to normalize the data with information from areas with intermediate tidal ranges, the correlation was low and insignificant (Trenhaile, in preparation).

A negative correlation between platform width and tidal range is implied by the correlation between width and gradient in Japan, Gaspé (Québec), and south Wales. The width of shore platforms in these areas generally increases as the gradient decreases, which, in view of the relationship between platform gradient and tidal range, suggests that there is a negative correlation between width and tidal range. Platform gradient and width are also negatively correlated in central California (Bradley and Griggs 1976). In areas with a low tidal range, the abrupt truncation of the shore platforms by low tide cliffs can sometimes allow a concomitant decrease in platform width with platform gradient.

The relationship between platform width, gradient, and tidal range is dependent upon the scale of the study. Over large areas where there are significant variations in the tidal range, there is a positive relationship between width and gradient, and between width and tidal range. At the local scale, where tidal range can be assumed to be constant, platform width must vary because of other factors, such as differences in rock resistance and exposure. Theory suggests that if rock hardness decreases or wave intensity increases, platform width will increase as the gradient decreases; and vice versa (Trenhaile and Layzell 1980, 1981). The relationship between platform width and gradient can therefore be negative within local areas (Trenhaile, in preparation).

It is logical to assume that in wave-dominated environments there should be a direct relationship between platform width and the intensity of wave action, but quantification of this relationship is difficult, because of the general lack of wave data from rock coastlines. Several workers have tried to consider the wave climate of areas, using wave forecasting techniques (Trenhaile, in preparation), wave refraction diagrams (McLean 1967), and fetch distances (Flemming 1965, Swan 1971). Many studies have found that platforms are widest on exposed coasts, which is consistent with mathematical modelling (Trenhaile and Layzell 1980, 1981, Trenhaile 1983b). In southern England and on the Singapore Islands, for example, the widest platforms are in areas where wave action is most vigorous, and where erosion is most rapid (Everard *et al.* 1964, So 1965, Swan 1971); whereas other workers have noted that platforms are often widest along more sheltered coasts, which are protected from waves of more than moderate force (Johnson 1933, Bartrum 1935, Edwards 1941, Hills 1949, 1971, Bird and Dent 1966). This could be because of the slow retreat of the low tide cliffs in mesotidal areas, although Abrahams and Oak (1975) found that there is no significant relationship between platform width and exposure on the south coast of New South Wales.

In Japan, average platform widths of about 60, 50, and 40 m on the exposed Pacific, Japan Sea, and sheltered Inland Sea coasts, respectively, broadly reflect variations in wave energy (Takahashi 1977). Analysis of the raw data on platform width and mean swell and wind wave heights provided by Takahashi for a number of areas in southern Japan, however, failed to identify any significant correlations (Trenhaile, in preparation). In the Vale of Glamorgan in south Wales, platforms facing the prevailing and dominant southwesterly winds are between 150 and 400 m in width (Trenhaile 1972). Much narrower platforms are found on the sheltered, easterly facing coast, and intermediate widths on the coastal section which faces southwards. Although these differences in width partly reflect differences in rock type, a comparison of the width of platforms in the same rock types on the western and southfacing coasts shows that the former are significantly wider (Trenhaile, in preparation). Variations in wave energy seem to be of less significance in Gaspé, Québec. Using the Raleigh test (Curray 1956), it was found that in most cases, the preferred orientations of the wider platforms are different from those of the narrower platforms, but the differences were slight, and could not be attributed to variations in the wave energy reaching the coast.

Although it might be expected that platform width would be greatest along exposed coasts, the relationship is apparently quite complex. This is probably because of the effects of other factors, such as variations in rock structure and lithology, cliff height and the presence and nature of the superficial deposits on, and particularly at the back, of the shore platforms. Robinson (1977c) found that variations in platform width in northeastern England can be partly attributed to the presence of suitable abrasive materials, and differences in the degree of protection afforded to the cliff by deposits at its base. The width of the platforms decreases in order as sandy beaches, bare rock, boulder beaches, and talus cones occur at the cliff base. The evidence from the Vale of Glamorgan provides some support for Robinson's hypothesis. Wide platforms are especially common where the cliff foot is bare and unprotected from wave quarrying, and to a lesser extent where there are pebble beaches. Narrower platforms tend to occur where there are boulder beaches, although they also occur in Keuper marls and in the shaley *angulata* limestones where there are sandy beaches at the cliff foot. Using Takahashi's (1977) data from the Kii Peninsula in southern Japan, however, it was found that the differences in the width of platforms backed by sandy beaches or bare rock were inconsistent and statistically insignificant (Trenhaile, in preparation). Robinson's hypothesis merits further consideration, but in most areas it is probably necessary to determine the amount and potential mobility, as well as the type of cliff foot deposits, to determine their protective or abrasional contribution to platform development.

Edwards (1941, 1958) suggested that there is an inverse relationship between cliff height and platform width, because of the amount of debris produced by cliff erosion. Low cliffs have been found to retreat more quickly than high cliffs in some areas (Shepard and Grant 1947, Kawasaki 1954, Williams 1956, Sparks 1972), but many other factors serve to obscure this relationship. Attempts to determine the effect of cliff height on platform width in the Vale of Glamorgan and in Gaspé, Québec were generally inconclusive (Trenhaile 1972, 1978).

Platform Elevation

Efforts to identify palaeosea-levels from platforms within and outside the contemporary intertidal zone, have been confounded by the controversy over the elevation at which shore platforms develop in different environments and rock types. Kinahan (1866) correlated terraces in Galway Bay, Ireland, with the levels of storm wave attack, high spring tides, high neap tides, low neap tides, and low spring tides. Although Kinahan's model represents the first attempt to explain the way in which the tide directs the erosive work of the waves, it is based upon a misunderstanding of the way in which wave energy is distributed within the intertidal zone. Many workers have argued that the height of shore platforms in relation to specific tidal levels is a fundamental genetic criterion, but there is no consensus on the way in which platforms are formed, and consequently on the elevations at which they develop. Bartrum (1916, 1924, 1926, 1935, 1938, 1952) distinguished Old Hat platforms, slightly below the level of normal high tides, from storm wave platforms up to several metres above (see also Jutson 1939, Cotton 1963). Tricart (1959) thought that supratidal benches in Brazil are contemporary features formed by salt spray weathering. A similar explanation has been proposed to account for platforms and benches at or above the high tidal level in sites which receive maximum levels of insolation in humid temperate climates (Johannessen *et al.* 1982). Sanders (1968a) considered that the elevation of most horizontal, high tidal platforms in Tasmania is determined by the factors which control the saturation level in the rock. Gill (1972b) proposed that platforms are ultimately graded to the low water level, but that there has not been enough time to achieve this in resistant rocks, since the sea reached its present level.

The possible effect of high Holocene sea levels on platform development in the southern hemisphere, and possibly elsewhere, has not been adequately considered. A few workers have argued that the present elevation of some shore platforms, relative to specific tidal levels, is indicative of changes in the relative sea level. Platforms near Sydney have been ascribed to recent emergence (Andrews 1916, Hedley 1924, Jardine 1925, Steers 1929, 1937, Voisey 1934), but this has not been generally supported by other evidence (see Chapter 7). Fairbridge (1952) thought that shore platforms develop at the low tidal level, and that platforms which are now close to the high tidal level must therefore have formed when relative sea level was higher; a similar view has been expressed by Teichert (1947), and earlier by Buddington (1927).

In Australasia, and elsewhere in warm, meso- and macrotidal environments, contemporary shore platforms have been reported at all elevations, ranging from the low tidal level to well above the high tidal level. Few of these platforms, however, have been precisely surveyed, or their elevations referred to reliable benchmarks or tidal data. With regard to the fairly slight differences between tidal levels in meso- and microtidal environments, the lack of precise survey and tidal reference data, the effect of lithological and structural variations and exposure to wave attack (Duckmanton 1974, Kirk 1977, Reffell 1978), as well as the secondary lowering of the platforms by weathering processes, it is surprising that platform elevation has been accorded such genetic significance. The great interest in subhorizontal platforms which are close to, or above, the high tidal level, has tended to obscure the fact that intertidal platforms close to the midtidal level are probably more typical of much of Australia and New Zealand. Most of the subhorizontal platforms in Gaspé, Québec (Trenhaile 1978) and western Newfoundland are at or very close to the mid-tidal level. Dana (1880) thought that the level of greatest wear is just above the half-tide level. Hills (1972) also believed that the platforms of southern England, western Europe, and southern Australia are ultimately reduced to the half-tide level, but this could only occur in areas with a small tidal range where the platforms are almost horizontal. In the macrotidal environments of western Europe, platforms extend continuously from about the high to the low tidal level, although there is some evidence of slight flattening about the mid-tidal level (So 1965, Trenhaile 1972).

The Cliff–Platform Junction

In macrotidal environments, the junction of the cliff with the sloping shore platform can be several metres above mean platform elevation. In areas with small tidal ranges, unless there is a well defined ramp at the cliff base, junction elevation can be almost the same as the mean elevation of the subhorizontal platform. Cliff–platform junctions tend to be close to the high tidal level, although there is usually considerable variation as a result of geological influences and variations in exposure to wave attack (Zeuner 1958, Wright 1970, Trenhaile 1972, 1978). In the Vale of Glamorgan, for example, the junction is usually closer to the spring than to the neap high tidal level, but variations in the height of the spring high tidal level only account for 22% of the variation in junction height (Trenhaile 1972). The height of the junction, and of the platform itself, also vary along indented coasts, although they are higher on the headlands on some coasts and higher in the embayments on others (Bartrum 1935, So 1965, Wood 1968,

Wright 1970, Hills 1971, 1972, Abrahams 1975, Takahashi 1977, Sunamura 1978a).

The Ramp

Ramps are fairly steep slope elements (usually 2.5 to 10°) situated between the more gently sloping platform and the cliff base. They have been most frequently reported from the swell wave environments of Australasia and elsewhere around the Pacific (Wentworth 1938, Hills 1949, 1971, 1972, Fairbridge 1950, Healy 1968a), and occasionally from the storm wave environments of the north Atlantic (Trenhaile 1969, 1974a, Robinson 1977b,c). It has been suggested that the occurrence and morphology of ramps are partly related to the strength and frequency of the swash of storm waves, to the waves of translation which sweep across the platforms, and to the presence of abrasive material at the cliff foot (Hills 1949, 1971, Ollier 1969). Theory suggests that the prominence of concave ramps in Australasia is a natural consequence of the generally low tidal range and the dominance of swell waves in this area (Trenhaile and Layzell 1981). Field observation in Gaspé, and the apparent lack of any relationship between ramp gradient and elevation, however, suggest that the presence and form of the ramps are determined by rock structure and lithology at the cliff foot. Thick shale beds and other weak materials near the high tidal level seem to be particularly suitable for the development of prominent ramps in this area (Trenhaile 1978).

Low-Tide Cliffs

Subhorizontal shore platforms in meso- or microtidal environments often terminate abruptly seawards in low tide cliffs or ramps (Figs. 9.2 and 9.5), which are undercut in the form of subtidal notches in some types of rock (see, for example, Hills 1971). They have been attributed to the former edge of a drowned hill (Hills 1971), and to marine erosion cutting into the stepped littoral platform, as it attempts to produce the classical marine profile (Bartrum 1926). Low-tide cliffs rarely occur in Britain, but there are low-tide ramps or elements of steeper gradient at the seaward margins of the Chalk platforms of the Isle of Thanet in southeastern England, where the tidal range is only about 4.5 m (So 1965, Trenhaile 1974a). Many of the subhorizontal platforms in mesotidal Gaspé (Trenhaile 1978) and western Newfoundland terminate abruptly in low-tide cliffs or rugged, structurally controlled ramps. Other platforms terminate in a series of low, structural scarps, each of which functions as a low-tide cliff at different stages of the tide. Many low-tide cliffs are structurally controlled, but their association with subhorizontal platforms in areas with a small tidal range suggests that geology is not the fundamental reason for their occurrence.

Little work has been done on the morphology of low tide cliffs. Reffell (1978) proposed that near Sydney, Australia, there is a positive relationship between platform elevation and the steepness of the low-tide cliff, and between steepness and the exposure to wave attack. Geometrically, the height of the low-tide cliff should increase as the tidal range decreases (Trenhaile 1978). Theory also implies that low-tide cliffs are a natural consequence of a low tidal range, and possibly a fairly weak wave environment (Trenhaile and Layzell 1981).

Other Features

Hills (1971) has described several other features of the aeolianite platforms of Victoria, which have also been reported from similar platforms in Western Australia and elsewhere (Teichert 1947, 1950, Fairbridge 1948, 1950, Bedi and Rao 1984). The notch at the cliff base, which has been attributed to solution, sometimes has a visor above it and a plinth below. The visor is a ledge consisting of a band of indurated rock, much harder than the rock above. According to Hills, this induration results from the deposition of calcium carbonate when fresh rain water moving down through the cliff comes into contact with rock saturated with sea water. A similar explanation has been provided for aeolianite visors above the notches on the southern coast of Bermuda (Taillefer 1957). This may explain why the height of the visor declines as it is traced into sheltered areas (Stearns 1935, Wentworth 1939, Hills 1971). The plinth is a slight prominence projecting from the outer edge of the base of the notch. Hills suggested that it could be related to the height to which water is drawn by capillary action above the platform surface. The gutter (or moat of Wentworth 1939) is a channel occasionally found at the base of the ramp, which has been eroded by sand, pebbles, and small boulders. It has also been found on a supratidal greywacke platform in western Victoria (Jutson 1949a, Gill and Lang 1983).

Platform Morphology and the Effect of Geological Factors

Geological factors exert an important influence on the development and form of shore platforms in several ways (Trenhaile 1980):

1. The structure, lithology, and mineralogy of the rocks govern the efficacy of the erosional processes. Wave quarrying usually dominates when the rocks are thinly bedded and well jointed, even in fairly sheltered environments. Frost and chemical weathering are important in rocks which are able to absorb large quantities of water. Geological influences can control the form of platform profiles: Gill (1972b) for example, found that platforms in southeastern Australia are subhorizontal on highly soluble rocks, but gently sloping on soft, non-soluble rocks. Hills (1971) made a

detailed comparison of the processes and morphology of the aeolianite and arkosic sandstone platforms of Victoria. Geological factors also play an important role in determining the amount and the type of deposits which accumulate at the cliff base, which in turn, appear to affect the width of shore platforms (Robinson 1977c).

2. Rock dip affects the surface roughness of shore platforms formed in alternations of weak and resistant beds. Smooth platform surfaces can be coincident with the bedding planes of the more resistant members of gently dipping sedimentary alternations. Corrugated or washboard relief usually develops in steeply dipping rocks, unless dense jointing, thin beds, or other factors allow the lithologically more resistant beds to be effectively planated.

3. The degree to which shore platforms are partially inherited from periods when sea level was similar to today's depends upon the susceptibility of the rocks to the erosive processes; this will be discussed later.

The greatest obstacle to assessing the effect of geological factors on platform development is in determining the strength of the rock in relation to the processes acting on it. It has been suggested that there is a relationship in southern Japan between the rate of cliff recession and the compressive strength of the rocks (Horikawa and Sunamura 1967, Sunamura 1973). The distribution of the data is extremely skewed, however, and the relationship as presented is almost certainly statistically insignificant. Nevertheless, in the Vale of Glamorgan, an average compressive strength of 123 MNm^{-2} in the Liassic limestones and 12 MNm^{-2} in the shales (Williams and Davies 1980), appears to be broadly consistent with their relative susceptibility to erosion. This is not always the case. In southern Japan, mudstones on a shore platform are eroded more quickly than interbedded tuffs, even though the abrasion hardness, and the compressive and impact strength of the mudstones are greater than the tuffs. This may be because of mudstone flaking induced by the absorption or adsorption of water (Suzuki *et al*. 1970).

Geological factors influence the gross form of individual platform profiles, although over large areas, their influence is often restricted to providing local variations about regional means which are determined by tidal range, and other aspects of the morphogenic environment (Trenhaile 1974a, 1978, Trenhaile and Layzell 1981). This could be because well-developed shore platforms show a marked preference for a rather limited range of rock types which erode fairly easily. They include weathered granites, some metamorphic rocks, chalk, aeolianites, and limestones, and particularly alternations of sandstones, siltstones, limestones, and other sedimentary rocks, with shales or weak mudstones. Most igneous rocks are usually too resistant, even in vigorous wave environments, unless they have been intensely weathered by physical or chemical processes.

Platform Gradient

The slope of shore platforms tends to increase when traced into rocks which are more resistant to erosion. In Glamorgan, residuals from the regression between platform gradient and tidal range tend to be positive in the resistant Carboniferous Limestones, and negative in the weaker Littoral Trias, and in the shaley Liassic upper *bucklandi* and *angulata* (Trenhaile 1969, 1972). In northeastern and southeastern England, variations in rock structure and lithology account for differences in platform gradient of 0.5° or more, within short distances along the coast (Trenhaile 1974a). Mean platform gradients are also significantly greater on the more resistant outcrops in Gaspé, Québec (Trenhaile 1978), and on the Kaikoura Peninsula in New Zealand (Kirk 1977). Most of the platforms in Tasmania are subhorizontal, but sloping ramps do occur on the igneous rocks of the eastern coast (Sanders 1968a). The increase in platform gradient with rock hardness is also consistent with the results of mathematical modelling (Trenhaile and Layzell 1980, 1981, Trenhaile 1983b). Variation in platform gradients between headlands and embayments also reflects changes in rock hardness, although variations in the exposure to wave attack must also be considered. Chalk platforms on the Isle of Thanet in southeastern England are steeper in the embayments (So 1965, Wood 1968), presumably because weaker wave action in the sheltered bays more than compensates for the presence of less resistant rocks than on the adjacent headlands. In the Vale of Glamorgan, platforms are steeper on the headlands, probably because differences in rock resistance are more significant in this area than differences in wave intensity (Trenhaile 1972). Whether platforms are steeper in the embayments or on the headlands therefore depends upon the degree of variation in rock hardness and wave intensity along the coast.

The relationship between platform gradient and rock hardness is complex. Gill (1972b) proposed that the ultimate profile of equilibrium in most soft rocks in Victoria is a gently sloping ramp, but in aeolianite and other soluble rocks it is subhorizontal. On the Otway coast of Victoria, Gill and Lang (1983) distinguished between supratidal, horizontal platforms in greywackes, and sloping, intertidal ramps graded to the low tidal level in siltstones. They suggested that these platforms represent different stages in the same evolutionary process, but that sufficient time has only been available since the sea reached its present level for the weaker siltstones to attain the ultimate equilibrium form, graded to the subaqueous profile. Gill and Lang provided no information on the gradient of the siltstone platforms on the Otway coast, although a siltstone platform in south Gippsland slopes seawards at only about 0.6° (Gill 1967). Their data show that the greywacke platforms usually slope seawards at between 0.17 and 0.8°, but structural control caused three surveyed greywacke profiles to slope landwards at up to 1.8°. The relative width of the platforms (Gill 1973a) and several microerosion meter (MEM) stations, which were unfortunately mainly on the greywackes, suggested that the siltstones are eroded more quickly than the greywackes. Their data, however, show that two of the three MEM sites on the siltstones recorded downwearing rates which were less than the mean for the greywackes. The proposition that platform gradients steepen through time, as well as the observation that gradients are gentler on the more resistant rocks in Victoria, is contrary to the observations and models of other workers (Flemming 1965, Bradley and Griggs 1976, Kirk 1977, Sanders 1968a, Trenhaile and Layzell 1981, Trenhaile 1983b). Nevertheless, it should be noted that although platform gradients in Victoria vary according to rock type, the slope of the platforms is quite low in both lithologies. Variations take place within a fairly narrow range, with a mean which is essentially consistent with the relationship between gradient and tidal range.

Platform Width

Wave-cut shore platforms require a fairly vigorous wave environment, and rocks which are moderately resistant to wave action (Dana 1849, Edwards 1941). The range of suitable conditions for shore platforms appears to be quite narrow (Trenhaile and Layzell 1981, Trenhaile 1983b). The presence of fairly wide shore platforms on the sheltered sides of islands, and in other sheltered areas, such as along the eastern coast of the Vale of Glamorgan (Trenhaile 1972), suggests that the occurrence of rocks which are susceptible to mechanical wave erosion, and in some cases to weathering, is the single most important criterion for their development.

It would seem logical to assume that the widest wave-cut shore platforms develop in the weakest rocks. This is the case in parts of southwestern and northeastern England (Agar 1960, Everard *et al*. 1964), and in southern Japan, where they are usually absent along the igneous rock coasts (Takahashi 1977). Platforms also tend to be widest in the weaker rocks in central California, and in particularly well-joined rocks in Alaska (Bradley and Griggs 1976, Chastain 1976). The relationship between platform width and rock hardness, however, is probably quite complex. In Australia, rapid erosion of the low-tide cliff restricts the width of shore platforms formed in weak rocks, and the broadest surfaces are therefore associated with rocks which are moderately resistant to erosion (Edwards 1941). On the Isle of Thanet in southeastern England, and on the northeastern coast of Yorkshire, quite narrow platforms are also found in weak rocks (So 1965, Robinson 1977c). In the Vale of Glamorgan, the widest platforms tend to be in the fairly resistant Littoral Triassic conglomerates and limestones in the southeastern part of the region. This may be partly the result of cliffs being low to absent in this area. Probably of greater significance is the fact that the platforms in the shaley Liassic *angulata* are significantly wider than those of the more resistant *bucklandi* limestones in the same area.

The dip of the rock determines the degree of protection afforded to the weaker beds by the stronger members of an alternating sequence. Washboard relief develops when differential erosion causes the more resistant beds to stand up as ridges, which then protect the weaker beds intervening in the troughs from the full force of the waves. Resistant dipping beds in the cliff face can also provide protection to the weaker beds behind, slowing cliff recession. Rock dip and platform width are significantly correlated in alternations of sandstones, siltstones, and mudstones ($r = 0.48$) in the southern Kii Peninsula of Japan, and in the shales, greywackes, and argillites of Gaspé, Québec ($r = -24$). The correlations increased to -0.7 and -0.35 in Japan and Québec, respectively, when the dip was considered in a direction perpendicular to the cliff face, which more closely corresponds to the dip encountered by the short, refracted storm waves (Trenhaile, in preparation).

It has been proposed (Everard *et al.* 1964) that in steeply dipping rocks, the shore platforms are widest when the beds strike perpendicularly to the cliff face. When the dips are low, the platforms are widest when the strike is parallel to the cliff face. To investigate this possibility further, eight combinations of rock dip and strike were classified. Ranking these classes in terms of their relative susceptibility to erosion is difficult because other factors, including bed thickness, joint density, and variations in the direction and force of the waves, must also exert an important influence. Nevertheless, assuming that the rocks consist of alternations of beds of variable resistance to wave erosion, and that the waves reach the coast in a direction normal to the cliff face, it is possible to recognize a hierarchy which seems to roughly describe the effect of rock dip and strike on rates of cliff recession, and correspondingly, on platform width and surface roughness (Fig. 9.9):

A) When the rock is horizontal, as in the Liassic limestones of the Vale of Glamorgan (Trenhaile 1972), the erosion of the weaker beds, which are exposed to wave attack, leads to the progressive undermining of the more resistant beds in the cliff and platform (Trenhaile 1971). The platform generally consists of a flight of low terraces, formed along the bedding planes of the resistant strata. The surfaces are usually quite smooth, and large waves are able to reach the cliff base at the high tidal level. Subhorizontal strata, therefore, are probably most susceptible to rapid cliff erosion and the formation of very wide platforms.

B) Rapid rates of cliff erosion and very wide platforms can also be associated with beds which dip parallel to the cliff face. The weaker strata in the cliff are exposed to the waves, but their discontinuity in the

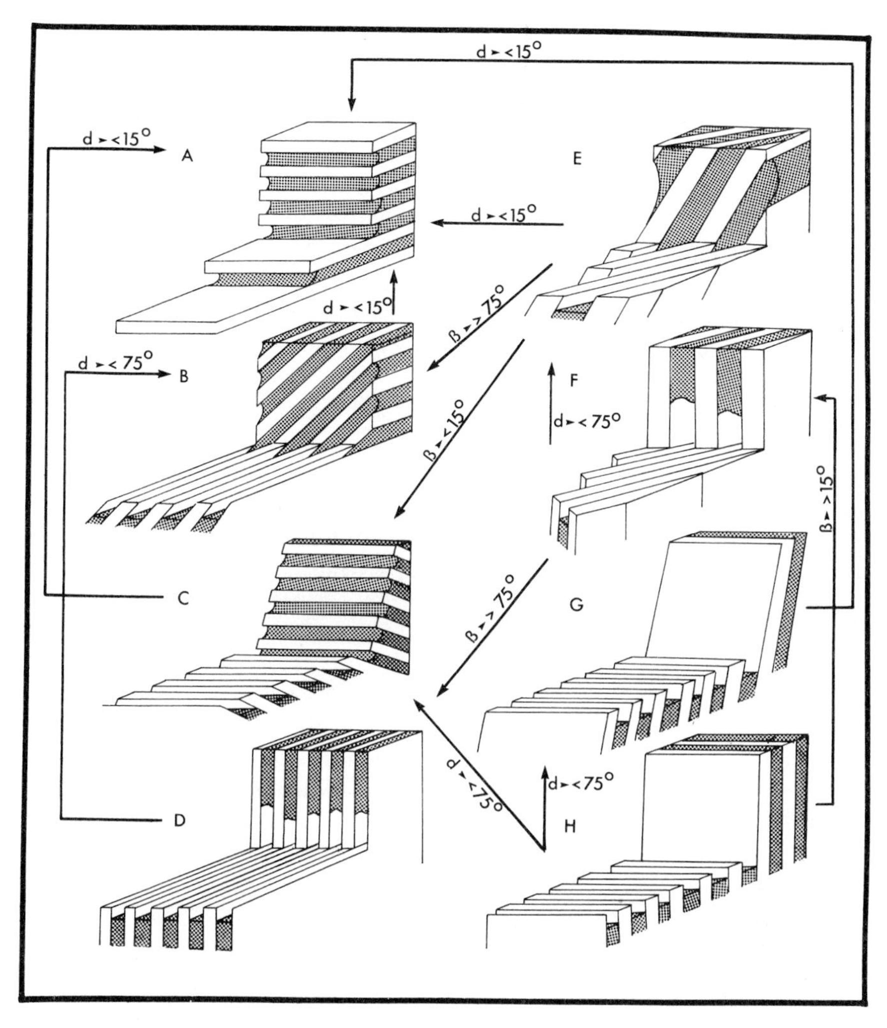

Figure 9.9. Structural classes based upon variations in rock strike and dip. Arrows represent promotions within the class hierarchy according to changes in strike and dip.

horizontal plane at the cliff base retards cliff recession to some extent. Ridges of resistant strata project above the general platform surface affording some protection to the intervening troughs of weaker rock, particularly when the waves approach the coast obliquely.

C) The third most rapid cliff recession may be in strata which dip landwards, striking parallel to the cliff face. As the weaker beds in the cliff and platform are eroded, they become increasingly inaccessible to the waves, unless the projecting beds of more resistant rock are removed. In the cliff, rock slides are inhibited by the landward dip, although rock falls can occur where the beds are undermined.

D) Intermediate rates of erosion and platform widths may occur in vertically dipping rocks which strike perpendicularly to the cliff face. Cliff recession is probably quite rapid initially in the weaker strata, but continued erosion depends upon removal of the more resistant material on either side of the recesses in the cliff. The weaker beds in the platform form troughs, which run towards the cliff between ridges of more resistant rock. The development of this washboard relief eventually reduces the wave energy reaching the cliff base, particularly when the waves approach the coast obliquely.

E) When the strike of dipping rocks is at an angle to the cliff, the weaker rocks are protected by upstanding resistant beds in the platform, and by resistant rocks in the cliff. Classes D and E were considered to be about equally susceptible to wave attack.

F) Vertically dipping strata which strike obliquely to the cliff initially provide easy access to wave attack. As erosion precedes, however, the weaker strata are increasingly protected by the ridges of more resistant rock on either side. Washboard relief on the platform also interferes with wave action at the cliff base, and the rate of cliff recession may therefore be quite slow.

G) Cliff recession in seaward-dipping beds striking parallel to the coast eventually presents a wall of resistant rock to the incoming waves. Platform and cliff erosion is therefore probably quite slow, and the platforms are narrow.

H) When vertical beds strike parallel to the coast, the cliff face eventually becomes essentially stationary for long periods along particularly resistant strata. Continued cliff recession depends upon the waves breaching these resistant beds, which is made more difficult by the development of washboard relief on the platform. This class appears to be the least suitable for cliff recession and the formation of wide platforms.

This model essentially extends the hypothesis of Everard *et al*. (1964). Classes A (those with very low landward dips) and C (low dips only) produce platforms that are wide relative to the regional mean, where dips are low and the strike is parallel to the cliff face. Classes B (steep dips only) and D (vertical dips) also provide fairly wide platforms in steeply dipping rocks striking perpendicularly to the cliff.

The model was tested using data from Japan, Glamorgan, south Wales, and Gaspé, Québec. Rock dips less than 15° were classified as horizontal, and those greater than 75° as vertical. Strikes less than 15° to the cliff were considered parallel to the cliff, and those more than 75° were classified as perpendicular. The use of classes defined by arbitrary parameters to describe a continuous spectrum of forms inevitably causes the classification to be sensitive to slight changes in rock dip and strike about the 15 and 75° boundaries. The greatest rank promotion in the width hierarchy occurs when classes E and G change to class A, as dips become less than 15°. Despite these and some other extreme cases, the classification is quite robust, in that most changes in dip and strike are responsible for changes in rank of only one or two places (Fig. 9.3).

The Liassic rocks in the Vale of Glamorgan are subhorizontal, falling almost exclusively within class A. Very wide platforms occur in this area, as predicted by the model, although this must also reflect the vigorous wave environment. Comparison of the effect of dip and strike within areas was made for Gaspé, where classes C, H, E, and F are well represented; and on the Kii Peninsula of Japan, where classes A and E are most common. Using the Mann–Whitney U-test to compare the mean widths of platforms, it was found that there is general agreement between model predictions and recorded variations in platform width within these two areas. The widest platforms are associated with classes A and B, and the narrowest with F and H. The hierarchical position of some of the intermediate classes was generally similar to model predictions, although there were a few slight variations, possibly as a result of the effect of a number of other factors on platform width (Trenhaile, in preparation).

The width of platforms can increase or decrease between headlands and the adjacent embayments (Edwards 1941, So 1965). This is partly a result of the conflicting influences of variations in rock resistance and exposure to wave attack. It should be noted that changes in coastal configuration also induce changes in the rock structure, relative to the orientation of the cliff. Refracted waves therefore encounter different dips and strikes along an indented coast, which can contribute to variations in the rate of cliff recession and in platform width and surface roughness (Fig. 9.10).

Platform Elevation

Mean platform elevation is very sensitive to geological influences, particularly in areas with a small tidal range, where only a few metres distinguishes a supratidal from a mid-tidal surface. Unfortunately, there has not been much discussion of the effect of rock structure and lithology on the elevation of platforms (an exception is Bird and Dent 1966). The mean elevation of shore platforms usually increases with the hardness of the rocks (Gill 1967, 1972b, Trenhaile 1969, Hills 1971, Kirk 1977, Takahashi 1977, Reffell 1978, Gill and Lang 1983), and high platforms are therefore usually narrow (Duckmanton 1974). Theory also suggests that mean platform elevation increases with rock hardness (Trenhaile and Layzell 1981), and this could explain why headland platforms are higher than those in the adjacent embayments (Bartrum 1935, So 1965, Wright

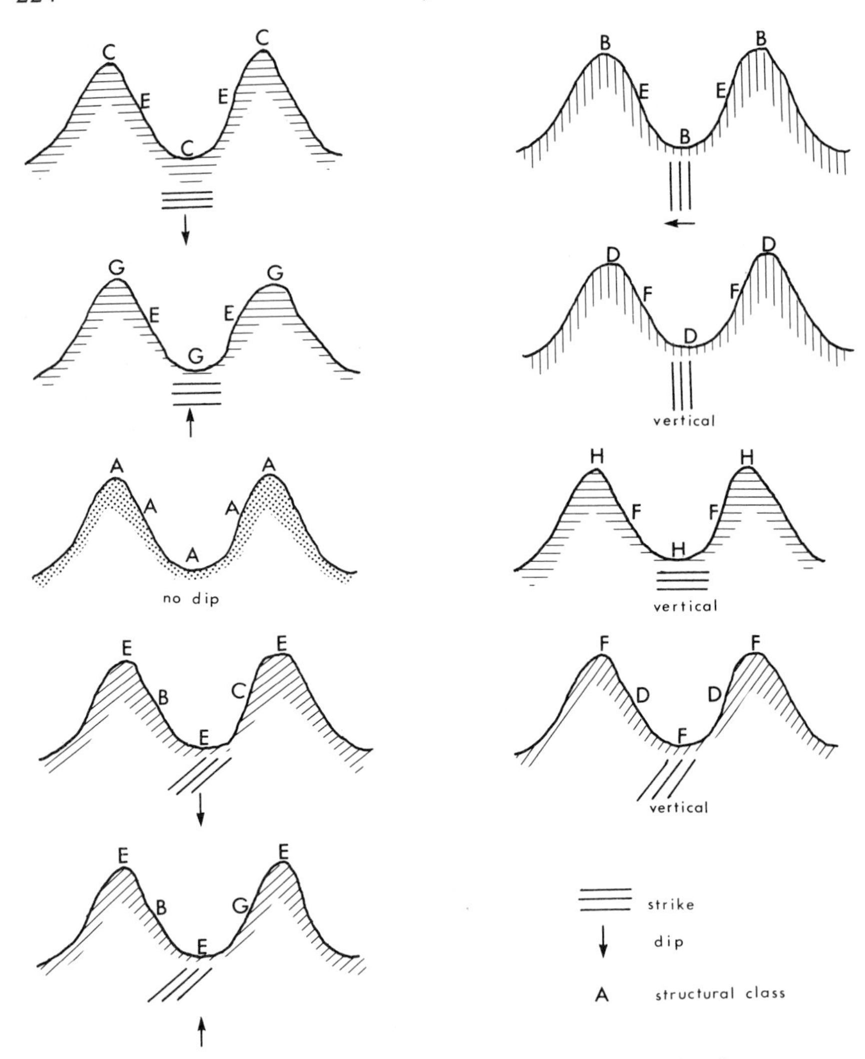

Figure 9.10. Variations in structural class along indented shorelines.

1970, Hills 1972), although this is not the case in all areas (Wood 1968, Sunamura 1978a).

Differential erosion of weak, horizontal, or gently dipping strata etches out ledges at various elevations on cliff faces, well above the level of the highest tides (Jutson 1939, Edwards 1941, Baker 1943, 1958). Ferrar (1925) thought that Bartrum's Old Hat platforms in northern New Zealand are geologically controlled, the result of the exposure of joint surfaces, and differences in the cohesion along bedding planes. Ongley (1940) described ledges in New Zealand formed by contemporary spray and splash, which are between 17 and 24 m above sea level. In the Vale of Glamorgan, relic and modern ledges are found within and above the intertidal zone, where weak shale and marl beds overlie more resistant strata. In the resistant Triassic and Carboniferous rocks, the ledges were formed during a period of higher sea level, but in the weaker Liassic limestones and shales, they are the result of contemporary storm waves, spray, and splash (Trenhaile 1971). None of these geologically controlled ledges can be used to identify the elevation of former sea levels with any degree of precision (Trenhaile 1974b).

The Cliff—Platform Junction

The junction between the shore platform and the cliff is usually close to the high tidal level, but there can be considerable variation according to local structural and lithological conditions (Everard *et al*. 1964). Particularly resistant strata tend to raise the junction above the regional average in southern England (Wright 1967). In the Liassic rocks of Glamorgan, Wales, junction elevation varies by several metres, according to whether thick limestones or friable shales occupy the cliff base. Residuals from the regression between junction elevation and the height of the mean high water spring tides showed that high, positive residuals are typical in the resistant Carboniferous Limestones in the west, and in the Littoral Trias in the east. Low negative residuals occur in the shaley upper *bucklandi* and *angulata*, where the junctions in places are only slightly above, or even in some cases below, the level of the mean high water neap tides (Trenhaile 1969, 1972).

Other Features

It has previously been noted that ramps are often related to situations where fairly weak rocks outcrop at the cliff base, in the presence of suitable abrasive materials (Hills 1949, Trenhaile 1978). Ramparts can be formed in resistant rocks which hold up the retreat of the seaward terminus of shore platforms (Johnson 1938, Edwards 1951, Gill 1972a), and low-tide cliffs and ramps are often associated with particularly resistant strata, although in most cases they cannot be the basic reason for their occurrence.

Models of Platform Development

Early attempts to model stages in platform development were qualitative, and usually structured within a cycle of erosion. A diagram accompanying Davis's (1896) paper inferred that parallel slope retreat takes place through time, with a progressive increase in platform width but without a concomitant change in platform gradient. It was Johnson (1919), however, who first applied the Davisian cyclical concept to the development of shore platforms, and his model, occasionally slightly modified, appeared in many geomorphological textbooks (Wooldridge and Morgan 1937, Von Engeln 1942, Thornbury 1954). In Johnson's model, slope progressively declines as the platform broadens, although the change in gradient seems to be quite slight during the early stages of development. Challinor (1949) challenged the classical view. He proposed that platforms undergo parallel slope retreat, without any change in their width or gradient. His hypothesis requires that the downcutting of the platform proceeds rapidly enough to allow continued wave attack at the cliff base. Other workers have suggested that there can be a balance between the rates of erosion at the high and the low tidal levels, and that platform width can therefore remain constant through time (Edwards 1941, Bird 1968, Trenhaile 1972). This may be because of the feedback mechanism associated with the effect of platform morphology on the wave energy reaching the cliff base. When the erosion rate is greater at the high than at the low tidal level, an increase in platform width and a reduction in gradient cause the waves to become weaker at the cliff base, until a balance is attained with the energy received at the low tidal level. A similar argument will pertain if wave energy and the rate of erosion are greater at the low than at the high tidal level.

Two attempts have been made to study the formation of shore platforms in the laboratory, using plaster and cement blocks in wave tanks and flumes to simulate wave attack on rock coasts. The necessity of using materials which erode more quickly than natural rock, however, introduces considerable uncertainty over the applicability of the experimental results to natural environments. One of the main conclusions of Sanders's (1968b) study was that there was no evidence to suggest that horizontal platforms could be produced by hydraulic action alone, in homogeneous, unbedded and unjointed material. Gill (1972c) argued that the results are inconclusive because insufficient time was allowed for a true platform to have been cut. The wave tank–flume experiments of Sunamura (1973, 1975, 1976, 1977, 1982b) also formed a part of a much larger study dealing with the erosion of rock coasts. He found (1975) that the rate of development of the platform as well as its elevation are determined by the type of wave which reaches the base of the cliff (see Chapter 1).

There have been several recent attempts to mathematically model platform development. Most have either been concerned with tideless seas, or have ignored the intertidal zones of coastal regions (Flemming 1965, Horikawa and Sunamura 1967, Scheidegger 1970, Sunamura 1977). Flemming found that the rate of cliff recession slows through time, but never completely ceases. Japanese attempts to develop mathematical models have been facilitated by the collection of field data on wave height at the cliff base, and by measurements of the compressive and impact strength of the rock. Horikawa and Sunamura (1967) found that:

$$dx/dt = C_R \times f,$$

where dx is the eroded cliff distance in time dt, C_R is a coefficient representing the erodibility of the rocks, and f is the erosive force of the waves. This equation, and an earlier one devised by Horikawa and Sunamura (1966), were later modified to incorporate field data from laboratory experiments and from field measurements in southern Japan (Sunamura 1977). In its modified form, the erosional equation was:

$$dx/dt \propto \ln (fw/fr),$$

where dx/dt is the mean cliff erosion rate, fw is the assailing force of the waves, and fr is the resisting force of the cliff materials.

Sunamura (1978b) modelled the development of submarine platforms by considering wave-induced cliff erosion and wave dynamics in shallow water. This model suggested that the rate of platform formation declines exponentially through time. A state of equilibrium is finally reached as a result of the effect of increasing platform width on the attenuation of wave energy at the cliff base. The ultimate platform profile is wider and flatter where the rocks are weak. Sunamura (1978c) has also modelled the development of continental shelves during the Holocene marine transgression.

Work by Trenhaile and co-workers has been concerned with the form of intertidal shore platforms. One model attempted to determine the way in which the tidal distribution of wave energy sculptures the profiles of platforms (Trenhaile and Layzell 1980, 1981, Trenhaile 1983b). It was based upon the equation:

$$R_{n,t} = tWF_n \tan \alpha_{n-1}/V,$$

where $R_{n,t}$ is the erosion (in centimetres) occurring in t years at intertidal level n; W is the deep water wave energy delivered per hour; F_n is the tidal duration value or the accumulated time that the still water level has been at level n (hrs yr^{-1}) (see Chapter 1); and α is the submarine slope. If W/V is constant, then it can be substituted by A (cm hr^{-1}), which is the erodibility constant.

The equation, which is quite similar to Horikawa and Sunamura's (1966) and Sunamura's (1977) expressions for erosion at the cliff foot, was used to simulate the development of shore platforms in Britain, eastern Canada, and Australia. In response to the feedback mechanism (Sunamura 1976, 1978b), the gradient of each portion of the intertidal profile varied until the rates of erosion became equal at all points along the platform. This state of dynamic equilibrium was achieved when the differences in the gradient of each segment compensated for variations in the tidal duration values within the intertidal zone. The equilibrium width increased with wave intensity and with tidal range, and inversely with rock hardness; whereas the gradient at a particular intertidal level increased with the strength of the rock, and decreased with wave energy and with the tidal duration value at that elevation. The model suggests that the general form of the tidal duration curves produces platforms which are concave at the cliff base and convex at the seaward margins, particularly where the rocks are resistant, waves are weak, and the tidal range is low. This may provide a partial explanation for the prominence of ramps and low tide cliffs in meso- and microtidal environments. The model was quite successful in simulating the morphology of shore platforms in the three study areas, and it provided further support for the contention that subhorizontal platforms can be cut by mechanical wave action in areas with a small tidal range.

Consideration of the shape and symmetry of tidal duration curves (Chapter 1) suggests that most of the platform surface should fall within the neap tidal range, and that mean platform elevation would tend to be fairly close to the mid-tidal level. This provides support for Dana's contention that the level of greatest wear is a little above the half-tide mark. It is also consistent with slight flattening of the platforms about the mid-tidal level in Britain (So 1965, Trenhaile 1972), which So considered was a reflection of the greater frequency of storm wave action at that level. Hills (1972) suggested that shore platforms in southern England, western Europe and southern Australia are ultimately reduced to the half-tide level. As noted previously, this is a gross exaggeration in areas with a high tidal range where the platforms are sloping, but it may be true in areas such as Newfoundland and Gaspé, where the tidal range is low.

In meso- and microtidal environments, structural and lithological factors, as well as the rise in the water level during storms, can cause platforms to develop well above the high tidal level. The model considered tidal variations in the elevation of the still water level, but not variations in the height at which waves actually operate. Most of the wave pressure models discussed in Chapter 1 suggest that maximum pressures exerted by standing, breaking, and broken waves are at or only slightly above the still water level. Nevertheless, the water level can be raised by the piling up of coastal waters associated with low pressure systems. Furthermore, wave action is probably most effective when water levels are particularly high, because of strong winds and the greater water depths near the shore. These factors, together with the effects of local structure and lithology, may account for the occurrence of many of the supratidal and high tidal platforms in Australasia.

The elevation of shore platforms is influenced by the resistance of the rock. In weak rocks, even quite weak waves are capable of accomplishing some erosion, so that the entire spectrum of wave sizes reaching the coast plays a role in determining the elevation of the resulting platforms. The mean elevation of these platforms would be fairly close to the mid-tidal level. In resistant rocks, only large waves would achieve much erosion (see Sunamura 1973), and these usually occur when water levels are high. This concept is consistent with observations that platform elevation increases with rock hardness (Gill 1967, 1972b, Takahashi 1977). The gradient of wave-cut shore platforms is determined by the deviation about the modal value of the wave energy–elevation distribution curve (Fig. 9.11). The deviation declines with the tidal range, but it must also be particularly low in swell wave environments, where wave energy varies comparatively little from the mean (Davies 1972). The deviation must also decrease with platform elevation, because high tidal platforms in resistant rocks are the product of a narrow range of larger waves. This relationship provides an alternate explanation which could, in some cases, account for the presence in Australasia of horizontal high tidal platforms in resistant rocks, and lower, gently sloping platforms in weaker rocks (Gill and Lang 1983). Geological factors often play an important role in the formation of high tidal surfaces, however, and they are sloping where the jointing or bedding planes are strongly inclined (Sanders 1968a).

Trenhaile (1983b) used a second mathematical model, which was concerned with the relative rates of erosion at the high and the low tidal levels rather than with the distribution of wave energy within the tidal range. The model is based upon the assumption that cliff recession takes place in two stages: undercutting to the point of collapse; and removal of the debris. The total time required for these stages is given by:

$$T_{n,x} = (T_r \cot \beta + nx - E \sum_{T_0}^{Tn-1} T)(C_1 + C_2 nx)x,$$

where platform width is represented by the first set of parentheses; T_r is the tidal range; β is the gradient of the inherited slope; n is the number of times the cliff has been undercut to the point of collapse and the debris removed; x is the maximum depth of the undercut before collapse; and $E \sum_{T_0}^{Tn-1} T$ is the total amount of erosion which has occurred at the low tidal level. C_1 and C_2 are constants equal to tan α/UT_r and tan α/ST_r, respectively where U is the amount of cliff undercutting per year, α is the present gradient of the platform, and S is the amount of debris removed each year. The model uses morphological data which can be obtained in the field.

Model runs using data representing Glamorgan, Wales, and Gaspé, Canada showed that platform development is initially rapid. Platform extension becomes progressively slower, until an equilibrium state, defined by equal rates of erosion at the high and low tidal levels, is attained. The platforms were close to their equilibrium states after the equivalent of 2 to

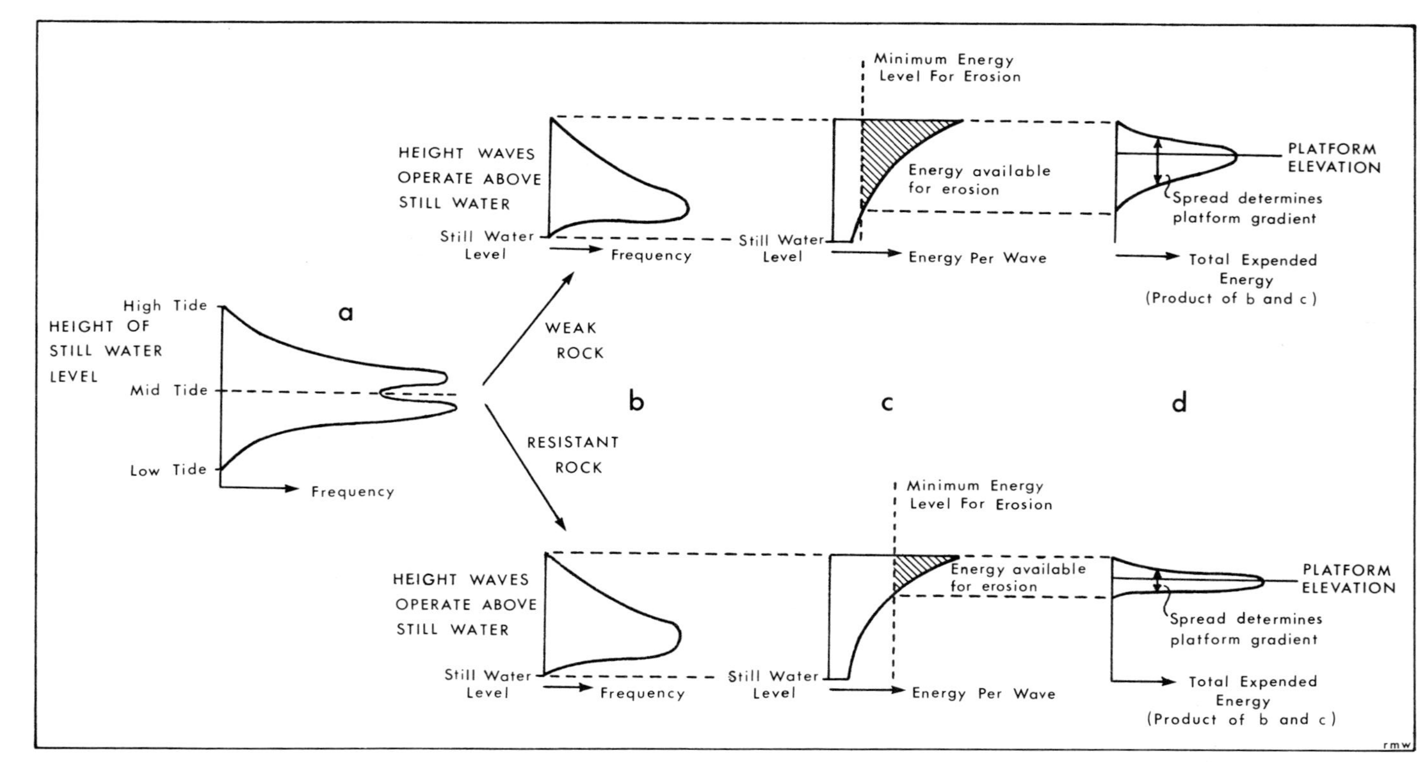

Figure 9.11. The effect of rock resistance on the elevation and gradient of wave-cut platforms.

2.5 ka BP, although only a few runs attained absolute equilibrium after 5 ka. The width of the simulated profiles after about 3 ka was similar to the width of the platforms in the field. Equilibrium states in both models depend upon a constant rate of erosion immediately below the low tidal level. If this rate progressively declines because of a significant reduction in the submarine gradient as a result of erosion at the low tidal level, then the intertidal platforms would eventually attain a state of static equilibrium, as the high tidal erosion rate approached zero. Rates of subtidal erosion off most rocky coasts are probably quite slow, however, so that it is unlikely that there could have been much reduction in offshore slopes since the sea has been at its present level.

The erosional model has been slightly modified to study the Holocene development of rock coasts, based upon Clark and Lingle's (1979) palaeosea-level model (see Chapter 7). Their model considers the effect of Holocene sea levels which were higher than at present, as well as levels which rose asymptotically to the present position. The results suggest that on a uniform Earth, shore platforms would be wide and gently sloping in areas which were once ice-covered; narrow and steep in areas peripheral to the ice, and in the collapsed forebulge regions; and moderately gentle and wide to very gentle and wide in areas which were much further from the ice. The effects of sea level change in the field, however, are obscured by variations in climate, tidal range, wave intensity, and rock resistance (Trenhaile and Byrne 1986). The model has also been used to study the effect of middle and upper Pleistocene changes in sea level on the development of eroded continental shelves and coastal terraces (Trenhaile, in press).

Inheritance

It has been proposed that wave action is presently modifying ancient erosional surfaces inherited from a period when the relative sea level was similar to today's. This hypothesis is consistent with the paleaosea-level record of the middle and late Pleistocene, when the sea was close to its present level on several occasions (see Chapter 7). The presence of raised beaches on exposed coasts has frequently been cited as evidence of very slow rates of contemporary marine erosion. As Sparks (1972) has noted, it may be that a considerable portion of the raised beach has been removed by contemporary wave erosion. Nevertheless, in many igneous and metamorphic rocks, and in some massive carbonates, the present rate of erosion seems far too low to account for the formation of wide shore platforms since the sea reached its present level. The sea is exhuming an old abrasion surface on the north and northeastern coast of France and in the Channel Islands, for example, where the lower Normannian shoreline

is found in the intertidal zone (Guilcher 1969, Ters and Peulvas
Inherited coastal features have also been identified in western a
ern Britain and in Ireland (Stephens 1957, Orme 1962, Hopley 1963, Synge 1964, Everard *et al*. 1964, Whittow 1965, McCann and Richards 1969, Phillips 1970a,b, Dawson 1980, Sissons 1981), from the northern shore of Lake Superior (Phillips 1977, 1978, 1980), and on the Singapore Islands (Swan 1971). In south Australia, shore platforms in granite and gneiss have been exhumed from a late Pliocene/early Pleistocene etch surface, fortuitously situated within the present spray and intertidal zones (Twidale *et al*. 1977). Near Melbourne, a sandstone platform is being exhumed and slightly lowered from beneath an overlying sandy bluff (Bird *et al*. 1973).

Most clear cases of inheritance have been reported from resistant rock coasts, but there has also been some reference to partial inheritance where the rocks are more susceptible to contemporary erosional agents (Dionne 1972, Gill 1972b), Hansom 1983). On the south coast of New South Wales, several workers have proposed that the platforms have been lowered and otherwise modified from surfaces originally formed during Pleistocene interglacials (Bird and Dent 1966, Abrahams 1975, Abrahams and Oak 1975). Agar (1960) also suggested that the Liassic shale platforms of northeastern Yorkshire are largely inherited. He believed that patches of talus and cemented beach deposits on the platforms formed in the period preceding the fall in sea level at the end of the Eemian interglacial. This interpretation was questioned by a number of workers (see discussion following Agar's presentation), and more recently by Robinson (1977a), who showed that the deposits are postglacial in age, possibly as little as two hundred years old. There is a general lack of evidence of inheritance on weak, sedimentary coasts, where rapid rates of erosion may have removed till covers, raised beaches, structural remnants, and other evidence of ancient surfaces in or slightly above the present intertidal zone.

Many workers have argued that shore platforms in fairly weak rocks are contemporary features. In Alaska, Chastain (1976) thought that the breccia, sandstone, and argillite platforms are only a few thousand years old. On Aoshima Island in Japan, a 200 m wide shore platform in alternating sandstones and mudstones is thought to have been cut in the last 6 ka (Kino 1958, Sunamura 1973), and those on the Kii Peninsula in the last 3 ka (Takahashi 1977). In southern Victoria, Hills (1971) attributed the morphology of the platforms in aeolianites, sandstones, and mudstones to the processes presently operating on them while the sea has been at its present level; this view has been supported by Gill (1972b). In South Australia, the aeolianite platforms, and probably some of those in crystalline and Precambrian sedimentary rocks, testify to the erosional efficacy of contemporary wave action (Twidale *et al*. 1977). The limestone and mudstone platforms on the Kaikoura Peninsula in southern New Zealand have also developed in the last of 6 ka, although inheritance, related to fluctuations in erosive activity, has occurred during that time (Kirk 1977). Postglacial shore platforms on Spitsbergen have been formed by a combination of wave and frost action (Moign 1974a,b). Rudberg (1967) thought that cliff retreat in the marls of Gottland is usually too slow for the platforms to have formed since the sea reached its present level. Because the erosion rates would have been greater when the platform was narrow, he did not exclude the possibility that the platforms are essentially contemporary. Initial rates of erosion are usually very rapid on new volcanic shores (see Chapter 8). On Surtsey, for example, rapid retreat of the poorly consolidated lava cliff caused the formation of a shore platform in tephra (Norrman 1980). This may be because of steep offshore slopes which permit very effective wave erosion, although the special geological and hydrological characteristics of new volcanic coasts make it difficult to apply these results to other areas.

Fairbridge (1971, 1977) is correct in asserting that contemporary rates of erosion in hard crystalline rocks are extremely slow, and that wide shore platforms in areas such as Devon and Brittany are largely inherited. There is far less justification for arguing that the presence of inherited features on resistant coasts necessarily implies that shore platforms in weaker rocks are also inherited. Fairbridge proposed that the wide shore platforms of southern England, northern France, southern Australia, and elsewhere were cut by frost and floating ice in the early Pleistocene, when low glacial sea levels were thought to have been similar to today's. This hypothesis is not consistent with modern views on Pleistocene sea levels or with the results of more than a century's work on shore platforms in these areas, and it places an emphasis on the erosive efficacy of floating ice which cannot be reconciled with the field evidence.

Relationships between platform morphology in fairly unresistant rocks, and various aspects of the morphogenic environment (Wright 1970, Trenhaile 1972, 1978, 1980, Takahashi 1977, Brodeur and Allard 1983), emphasize the degree to which shore platforms have been able to adjust to the present level of the sea. These relationships suggest that contemporary erosional processes acting on unresistant rocks have been able to accomplish much more than the slight modification of inherited surfaces. Shore platforms in the St Lawrence Estuary near Québec City, for example, may have been partly inherited from the Micmac surface, but they have subsequently become completely adjusted to the contemporary environment (Brodeur and Allard 1983).

The lack or weakness of relationships between platform morphology and wave and tidal conditions might be used to provide support for inheritance in more resistant rocks (Phillips 1970a,b, Everard *et al.* 1964), and a measure of the degree to which platforms have adjusted to the present sea level. In Favorite Channel, Alaska, for example, there are some very steep platforms with gradients of 7 to 8°, although the tidal range is only about 4 m. This may be because the platforms have been slow to attain their equilibrium state, a result of the small wave fetch and several isostatically induced fluctuations of 3 to 5 m in relative sea level within the last few thousand years (Chastain 1976).

Inheritance may have contributed to the development of shore platforms in some fairly weak rocks, but it is not a necessary condition for their formation. The mathematical models discussed suggest that wide platforms could have developed since the sea reached its present level, partly because of very rapid initial rates of erosion. Consequently, slow contemporary cliff and platform erosion are probably unrepresentative of the average rate for the last few thousand years. Even where the formation of shore platforms in resistant rocks has been facilitated by inheritance, the elevation of the ancient relative sea level would rarely be exactly coincident with its present level. In most cases, some modification of the inherited platform would be necessary to produce a form which is adjusted to the contemporary morphogenic environment.

If the inherited surfaces were slightly higher than the modern platforms, it might be argued, with equal validity, that the contemporary platforms were inherited, or that a modern platform has been cut into a low cliff representing the difference in elevation between the two surfaces. This situation occurs for example, on Majorca on gently sloping coasts (Butzer 1962), and on the Otway Coast of Victoria, where the contemporary platforms are cut into the last interglacial surface a few metres above (Gill and Lang 1983). What degree of modification should be designated to distinguish an inherited from a non-inherited or contemporary platform? Very little modification may have taken place in very resistant rocks since the sea reached its present level, but in weaker rocks the intertidal zone may be completely modern. The term 'inheritance' means different things to different people, and much of the debate over the question of its significance originates from a lack of precise definition.

Unglaciated areas with particularly resistant rocks are generally fairly stable tectonically, and relative sea levels in the interglacials may have been similar to today's. In less stable regions, which are often composed of weaker rocks, relative sea levels may have varied considerably throughout the Pleistocene, so that conditions would have been generally unsuitable for inheritance of ancient surfaces without substantial modification. True inheritance, therefore, may be more likely to have occurred in unglaciated areas on very resistant rock coasts, rather than on less stable, and more easily eroded, soft rock coasts.

Rates of Erosion

Although they may be much less than in the past, contemporary rates of erosion have often been cited as evidence for the inheritance of shore platforms. There are many estimates of the rates of platform development, but few reliable measurements. Reported backwasting rates for cliffs (see Chapter 8) and platforms range from the negligible up to 50 to 70 m yr^{-1}. Downwasting rates on shore platforms range between 0.1 and 35 mm yr^{-1}, being greatest in vigorous, wave-dominated environments (Sunamura 1973, Kirk 1977). Unfortunately, much of the data are based upon evidence of questionable reliability, including sequential photography (terrestrial and aerial), and old maps. Data on downwasting rates often refer to the work of single erosional agents, such as various fauna and flora (Healy 1968b, Evans 1968) (Chapter 4), and chemical processes (Revelle and Emery 1957), rather than to the total rate of platform lowering. The most reliable information on absolute rates of platform downwasting has recently been provided by the use of the microerosion meter (MEM). This instrument is capable of measuring very slight changes in the elevation of rock surfaces, but it cannot measure the effects of quarrying large rock fragments or joint blocks. Where this erosive process is dominant, measurements obtained by the MEM method are not representative of the total rate of erosion. Furthermore, the processes responsible for platform erosion must still be inferred from the data. Improvements in the design and use of the MEM allow checks to be made on the accuracy of the instrument (see, for example Gill and Lang 1983), but they cannot eliminate the inherent limitations of this technique.

The MEM was first used on coastal rocks by Trudgill (1976b) on Aldabra Atoll, Kirk (1977) on the Kaikoura Peninsula in southern New Zealand, and Robinson (1977a,b,c) in northeastern England. On the Kaikoura Peninsula, mean platform lowering was found to be 1.53 mm yr^{-1}, although it was generally higher on mudstones than on limestones, and where abrasional material is available. Granular disintegration occurred on both rock types, as well as flaking and chipping in the mudstones. The upper portion of the platform is largely eroded by subaerial processes, the lower by marine. The platforms were thought to be postglacial and subject to quite rapid modification. According to Robinson, erosion rates in the Liassic shales of northeastern Yorkshire are between 0.1 and 2 mm yr^{-1} on the gently sloping parts of the platform, and 1 to 30 mm yr^{-1} on the steeper, debris-covered ramp at the cliff base. This compares with Agar's (1960) estimate of a rate of downcutting of about 0.25 mm yr^{-1} on the lower shale foreshore, and greater rates on the gravel-covered upper foreshore. Robinson thought that corrasion is the main erosive process operating on the ramp, particularly in winter, when the waves are most vigorous. On the plane, however, the main process is associated with the alternate wetting and drying of the shales, which causes expansion and contraction of the clay lattices and the breakup of the rock into fragments, which are then washed away (Robinson 1977c). Erosion rates on the gently sloping planes and at higher elevations are greatest in summer, when the drying period is longest, thereby effectively smoothing and lowering the platform surface. At the cliff foot, rates of erosion are 15 to 18.5 times greater when there is a beach than when the rock surface is bare (Robinson 1977b). Where there is a beach, effective erosion is restricted to a narrow zone extending from about 14.5 cm below the beach surface, where erosion is at a maximum, up to about 10 cm above. The main processes are corrasion, which eroded at about 5.8×10^{-3} cm tide^{-1}, and wedging, a form of quarrying associated with the presence of fine-grained beach material, with a mean erosion rate of 11.05×10^{-3} cm tide^{-1}. Quarrying dominates at the cliff base (about 20 cm above the beach surface) and wherever beach deposits are absent.

The microerosional data suggested that the ramp and the plane in Yorkshire are becoming progressively gentler. This is contrary, however, to the morphological evidence, which suggests that the platforms have retreated in an approximately parallel fashion. Robinson (1977a) has proposed that this may be the result of morphological changes induced by variations in the amount and size of the deposits on the platform, and in the structure and lithology of the cliff, as it retreats landwards. On Aldabra Atoll, backwearing rates in coral limestone are 3 to 4 mm yr^{-1} in exposed areas, and 1 to 2 mm yr^{-1} in more sheltered areas (Trudgill 1976b) (Chapter 10). Biological activity is secondary in importance to abrasion and wave action in the exposed areas, but is prominent in the sheltered sites. According to Kirk (1977), Trudgill (unpublished) has also measured downwasting rates of 0.3 to 0.4 mm yr^{-1} on the Carboniferous Limestones of County Clare in Ireland.

In southern Devon, England, greenschists in the spray zone just above the high water mark are being lowered at between 0.55 and 0.64 mm yr^{-1} (Mottershead 1981, 1982a). Other microerosional data have been obtained from the northern Adriatic (Torunski 1979) and from the Otway Coast of Victoria (Gill and Lang 1983). On the horizontal, supratidal greywacke platforms of Victoria, the rate of backwearing is about 9 mm yr^{-1}, and downwearing 0.35 mm yr^{-1}. Erosion may be faster on the sloping siltstone ramps, where backwearing has been estimated at 18 mm yr^{-1}, and downwearing between 0.5 and 1 mm yr^{-1}, although two of the three sites provided rates which were less than the mean for the greywackes (backwearing rates were estimated by Gill 1973a).

Other techniques, less precise, have been used to estimate rates of cliff and platform erosion. Off eastern and western Australia, Hodgkin (1964) used steel rods and plaster casts to obtain downwasting rates of 0.6 to 1 mm yr^{-1} in the limestones of the upper intertidal zone. Emery (1941) used dated inscriptions cut into the sandstones of La Jolla, southern California. The general rate of downwasting updated from the earlier study, appears to be between 0.3 and 0.55 mm yr^{-1} (Emery and Kuhn 1980). Nails driven into the sandstone cliff about 1 m above its base, suggested that it has eroded at about 30 mm yr^{-1} (Pinckney and Lee 1973), although an extension of this study (Lee *et al*. 1976) found that the median value is only about 6 mm yr^{-1}. Using old photographs, Emery and Kuhn (1980) estimated that the sandstone benches have been lowered by 1 to 20 mm in the last 35 years, compared with backwearing of 350 mm in the same period on the steep edges of the platform. Block removal has caused the cliff to retreat by up to 7 m in the last 40 years. In the western Vale of Glamorgan, south Wales, consideration of the spacing, height, and rate of retreat of the Liassic limestone scarps, partly based upon the evidence of old photographs, suggested that the rate of platform downwasting is between 0.36 and 0.53 mm yr^{-1} (Trenhaile 1974a, Trenhaile and Layzell 1981). The rate of downcutting in this area is consistent with the results from other regions, considering the very exposed wave environment and the steep platform gradient in Glamorgan, but is much less than the downwasting rate of 24.5 mm yr^{-1} estimated for the Chalk platforms of the Isle of Thanet in southeastern England (So 1965).

The amount, and particularly the quality, of the data on platform and cliff erosion are rapidly increasing. It must, however, be emphasized that these data refer only to processes operating today. They are not necessarily representative of the processes and rates at an earlier period when the platforms were narrower and steeper. Wave quarrying is still very significant in many areas, but it would probably have been even more important in the earlier stages of platform development, when the surfaces were steep and irregular. Nevertheless, contemporary backwasting and downwasting rates in many areas do appear to be generally within the range of values necessary to explain the formation of shore platforms in the weaker sedimentary rocks since the sea reached its present level.

Submarine Erosion

Very little is known about the nature or efficacy of the erosional processes operating on the submarine slopes off rock coasts. These data are needed to model the long-term development of marine erosional surfaces. Erosion of the bottom changes the water depth and the bed slope, and causes concurrent changes in the size and energy of the waves approaching the intertidal zone, and consequently in the rate of intertidal erosion. If submarine erosion is very slow, then despite some climatically induced oscillations, there would have been only a very slight decrease in the average wave energy reaching the nearshore zone in the few thousand years since the sea reached its present level. Rapid submarine erosion on the other hand, would have caused considerable reduction in the offshore slope and in the wave energy reaching the coast. If average wave energy has been approximately constant, then the intertidal platform could attain a state of dynamic equilibrium; but if it has progressively decreased because of a reduction in the offshore slope, then it would tend to approach a state of static equilibrium (Trenhaile and Layzell 1981, Trenhaile 1983b).

There has been considerable debate over the maximum depth to which sand can abrade a submarine rock surface. Gulliver (1899) introduced the term 'wave base' to describe the ultimate depth to which a marine abrasion surface can develop. He did not give an estimate of its depth, but Johnson (1919) placed the limit of erosion at 183 m, and Barrell (1920) at about 90 m. Rode (1930) considered that abrasion could only take place down to between about 46 and 92 m, but recent workers (Dietz and Menard 1951, Longwell and Flint 1955) have been even more conservative, placing the limit at about 9–10 m. Dietz (1963) thought that wave-cut shore platforms form at sea level, although they can extend a little lower if the sea is stable. Bradley (1958) found that pyroxene grains off Santa Cruz, California, become significantly less abraded below a depth of about 9 m. On the southern Crimean coast, erosion has taken place at the base of limestone blocks at depths down to 10 m, but abrasion forms are absent on the bare rock surfaces. On the Caucasian coast of the Black Sea, flysch is abraded down to at least 20 m, and between 5 and 12 m in Lake Baikal, where the waves are smaller (Zenkovitch 1967). Erosion of the overconsolidated clay tills in Lake Ontario seems to slow down considerably below a depth of 4 m (Davidson-Arnott and Askin 1980). Unconsolidated clay particles in exposed environments can be moved by waves under heavy swell conditions down to almost 200 m, but they are incapable of significant abrasion of the bottom. Several workers have considered the minimum size of sand grains which can act as an effective abrasive. Rode (1930) suggested that the minimum diameter is 1 mm, whereas Twenhofel (1945) thought that little abrasion can be accomplished by grains less than 0.5 mm in diameter.

With regard to the fairly short time in which the sea has been at its present level, it is the rate of submarine erosion, rather than its maximum depth, which is of greater significance for the development of intertidal shore platforms. It is generally believed that submarine erosion is much slower than erosion in the intertidal zone, where water hammer, hydraulic quarrying, and abrasion operate either exclusively or most efficiently. Near Sochi on the Black Sea, reflected waves from a sea wall move pebbles over the underlying flysch surface. At the water's edge, the surface has been lowered by about 15 cm yr^{-1}, at a depth of 2 m by 6 cm yr^{-1}, and at 3 m by no more than 1 cm yr^{-1}. At greater depths, a lowering of 4 mm yr^{-1} was almost entirely the result of *Pholas* and *Barnea* boring (Zenkovitch 1967). Rock borers, including echinoids, sponges, and pholad pelecypods are active at considerable depths, but they can only lower the regional submarine slope very slowly. Zenkovitch (1967) derived simple equations to provide a rough estimate of the rate of submarine lowering, using data on cliff erosion rates and the shape and slope of the submarine bottom. They suggest that calcareous rocks in the Black Sea are being worn down at an annual rate of only tenths of a millimetre. In the Vale of Glamorgan, the rate of downwearing calculated by this method, is only about 0.06 mm yr^{-1}, and is less than 1 mm yr^{-1} in other parts of southern Britain. Zenkovitch considered that mechanical erosion usually ceases when the gradient of the bottom is from 0.01 to 0.05. Horikawa and Sunamura (1970) used Zenkovich's equations to calculate rates of submarine erosion on mudstones and tuffaceous sandstones in southern Japan. The rates are quite rapid, but the technique is based upon several debatable assumptions. These include the assumption that cliff erosion rates and the submarine gradient have been constant in the last 2 ka. Actual observations of the submarine extensions of shore platforms are very few. Sanders (1968a) found little evidence of subaqueous abrasion in Tasmania, but Reffell (1978) thought that it was significant near Sydney.

10

Limestone Coasts

Many workers have tried to define the elements of limestone coasts which could distinguish them from other coasts within the same climatic environments. In some parts of the world, and in some calcareous rocks, the processes and landforms are similar to those of other rock types. In the vigorous storm wave environments of the mid-latitudes of the northern hemisphere, for example, mechanical wave action plays an important role in the erosion of coasts consisting of the geologically younger, and usually physically weaker, limestones. The older, more resistant limestones, such as the British Carboniferous, however, are usually sufficiently resistant to wave action to allow other mechanisms to operate efficiently. The most characteristic features of limestone coasts are generally associated with warm climates and young calcareous rocks. Coral reef limestones occur between latitudes 30°N and S, and aeolianites (calcarenites), which are former Quaternary dune sands, between 15 and 45°N and S. Generally, weaker waves, higher temperatures, and an enormously varied marine biota favour chemical and biological activity on the calcareous rocks of the lower latitudes. Furthermore, the prevalence of magnesium-rich coral and aragonite, rather than pure calcite, may facilitate chemical action in tropical waters.

The term 'corrosion' is generally used in this discussion in preference to 'dissolution' or 'solution', to refer to the development of a variety of features which are characteristic of coastal limestone. This avoids the implication that the process involved in the formation of these features is necessarily chemical solution. As has been noted in previous chapters, there is much recent evidence to suggest that the causative mechanisms are, at least in part, salt fretting, and biochemical and bioerosional grazing and boring.

Corrosional Zones

Guilcher (1953, 1958a) proposed that the form of the littoral in limestone regions can be classified according to the temperature and tidal regime. He distinguished and described four main sequences of landforms (Fig. 10.1):

(*a*) In cool temperate regions and in limestones resistant to mechanical wave action, as on the Carboniferous Limestones of south Wales, southern Ireland, and Aran (Guilcher 1952a), the spray zone is pitted by small

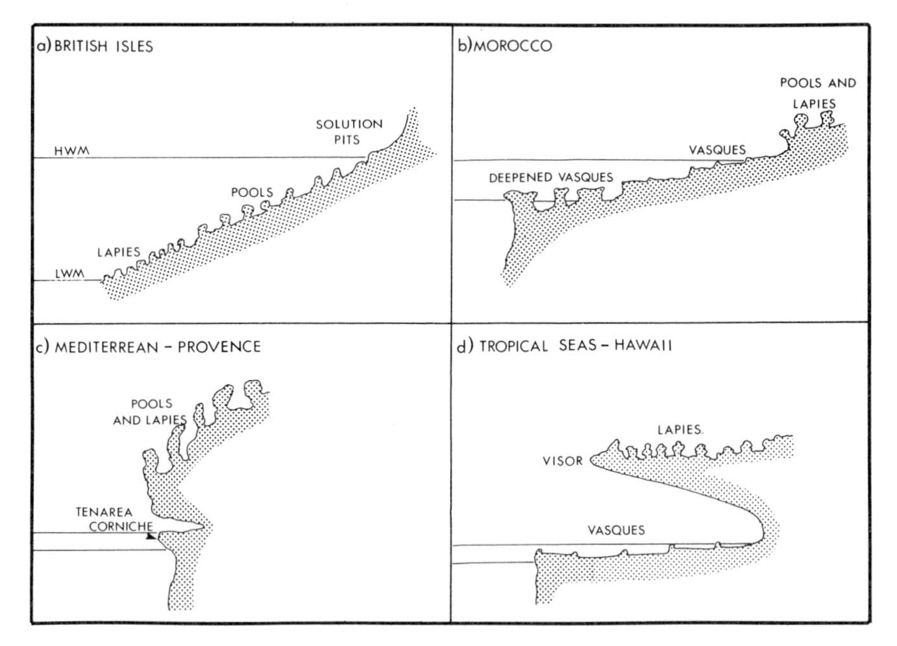

Figure 10.1. Littoral limestone zonations (according to Guilcher 1953).

corrosion hollows a few millimetres in depth and less than five millimetres in width (microaveoles, vermiculations). These features are restricted to the wetter parts of the rock, and they disappear above the level which is frequently attained by spray. Lower down the platform, in the upper parts of the intertidal zone, the main feature is the presence of pools with fairly flat bottoms and overhanging edges. Sharp pinnacles or lapies (marine *karren*) coexist with these pools in the lower portions of the intertidal zone. They occur on subhorizontal as well as subvertical surfaces, but their development is best in exposed regions. Their average depth is between about 10 and 30 cm. The 'British' type of zonation has also been recognized in parts of northern Spain, northern Portugal, and western France (Guilcher 1958b), where the limestones are not too weak and wave action is strong. In parts of the northern Spanish coast, however, a platform a few metres in width slopes steeply seawards (mean slope 20°) beneath a visor and a notch at about mean sea level. The platform ends abruptly seawards in a low tidal cliff. The zonation is somewhat similar to that in tropical and subtropical waters, but although the platform has a number of deep pools with overhanging edges on it, wide shallow pools or vasques are absent. Guilcher suggested that the occurrence of this tropical-like zonation in the southeastern Bay of Biscay is a result of the higher temperatures in this area than elsewhere around the Bay. The zonation in the western Crimea is similar to that in Britain, although many of the solutional hollows in this area have been modified by pothole abrasion (Givago 1950). Mechanical wave erosion is also dominant in the upper portions of the Carboniferous Limestone platforms of northern Britain, although corrosion is of more importance at lower elevations (Common 1955).

(*b*) In warm, tidal seas, jagged lapies are found further up the platform in the zone of spray and splash, as along the Atlantic coast of Morocco, southern Portugal (Guilcher 1957), and Barbados (Tricart 1972). These lapies, which are more dissected and pronounced than in Britain, are separated by small, flat-bottomed pools with overhanging rims. Vasques break the upper portions of the intertidal zone into a series of steps or terraces, each slightly lower than the preceding one. In the lower parts of the intertidal zone, the pools are much deeper (up to several decimetres), and they have overhanging sides. The intertidal zone terminates abruptly seawards in a small vertical or overhanging low tidal cliff. The upper pitted lapies zone may be separated from the vasques zone by an overhanging visor, but it is smaller than in tropical regions. A similar series of forms in aeolianite has been described by Fairbridge (1950) on Point Peron in Western Australia, and by Hills (1971, 1972) and Bird (1974) in Victoria.

(*c*) In the Mediterranean—a fairly warm sea without notable tides—lapies and pools with overhanging rims have formed in the spray and splash zones in Provence near Marseilles, and in Catalonia. There may be a notch at about the high tidal level, together with a lip or visor which can overhang by between 0.5 and 2 m. A constructional corniche of *Tenarea tortuosa* can form a second lip at lower elevations in the tidal zone (Guilcher 1953, Froget 1963, Nicod 1972). The lip-and-notch profile, however, is not always present in Mediterranean France. Near Nice, lapies and dish-shaped pools with overhanging sides dominate the spray and splash zones in dolomitic limestones. There is no corniche in this area (Debrat 1974). In Lebanon, corrosion features have been described by Dalongeville (1977) on a trottoir a few metres in width and about a quarter of a metre above sea level. The supralittoral zone is a fossil trottoir characterized by pools which are generally crater shaped, with concave bottoms and pinched-in sides. The fossil surface is separated from the contemporary trottoir by a low cliff with vermiculation-like depressions. Alveoles with overhanging rims, 5 to 8 cm in diameter and about twice as deep, occur at the back of the modern trottoir and in the lower supralittoral. Lower down the platform, the alveoles are smaller and shallower and there are some vasques, although they are not common in most parts of the Lebanon. Basins about a metre in diameter and almost as deep are found in the lower portions of the platform, where they are supplied by swash and always flooded.

(*d*) In very warm seas, as in Hawaii (Wentworth 1939), the Red Sea (Guilcher 1955), Madagascar (Battistini 1977, 1981), the Caribbean (Focke 1978), and India (Bedi and Rao 1984), the zonation in coral limestone consists of: lapies in the spray and splash zone; a lip and associated notch 1 or 2 m in depth at about the high tidal level; a platform with vasques occupying the intertidal zone; and a low tidal cliff. The lip and notch are conspicuous elements of warm seas, possibly because of high temperatures and small tidal ranges in many areas; this will be discussed later. Corrosional basins and deep notches and undercuts are found on the upper parts of the reef flats on Bikini and other nearby atolls (Revelle and Emery 1957). Residual or detrital rocks on the flats are often undercut ('negroheads'), providing convincing evidence that corrosion is effected by sea water, rather than by fresh water as was suggested by Wentworth (1939).

The degree to which variations in coastal corrosional forms in limestones can be attributed to differences in climate is still the subject of much debate. Some broad relationships between climate and landforms can be identified, such as the presence of vasques and deep corrosional notches and protruding visors in warm climates, and the generally greater efficacy of corrosional processes in these areas. Other factors, such as tidal range, the degree of exposure to wave action, and rock structure account for considerable differences in the character of limestone coasts between and within climatic regions. It is likely, for example, that the presence of deep notches and protruding visors in warm seas is at least partly a reflection of the generally low tidal range in tropical and Mediterranean areas. Furthermore, the usually more vigorous wave environments in the mid-latitudes of the northern hemisphere must exert an influence on the operation of erosional processes and on the form of the resulting landforms to a much greater degree than in tropical and Mediterranean environments.

The effect of variations in tidal range and exposure to wave action is manifested on the limestone coasts of Bermuda (Taillefer 1957, Neumann 1966). Two profiles can be distinguished. On the sheltered Harrington Sound coast, where tidal range is only 15 to 20 cm, the profile consists of a single notch, and a narrow and deep visor in the spray zone, finely sculptured with alveoles and lapies. On the more exposed northern and southern coasts, where the tidal range is 1 to 2 m, notches are found at a variety of elevations. On the southern coasts, notches have formed at the high and the low tidal levels, where they define the upper and lower limits of a platform of lapies and pools. The intertidal lapies become deeper and more evolved towards the sea, but they are less angular and the ridges are less sharp than those in the spray zone. There are also two types of profile on Oahu, Hawaii (Wentworth 1939). The bench profile consists of a level corrosional bench, up to 1 m above mean sea level, terminating abruptly seawards in a low tidal cliff. The bench is between 1.5 and 15 m wide, and its passes inland into a gradually rising pitted zone. In some places an abrasion ramp, a moat, or a quarried surface is found at the landward margin of corrosional benches. The second major type of profile, also described by Guilcher (1953), consists of a notch and an accompanying visor protruding outwards by up to 3 m. The top surface of the visor is pitted. A seaward-sloping platform or corrosional bench can extend for a metre or so from the base of the notch.

A particularly interesting study has been made of the effects of wave action on coastal lapies in the Bristol Channel in southwestern Britain (Ley 1977, 1979). Solutional relief is greatest in this area just below the mid-tidal level. Near the high tidal level the rock is pitted, joints have been partially enlarged, and there are some small, shallow pools with flat bottoms and extensive divides. Near to the low tidal level, the platform surface is flat and quite smooth, apart from the overdeepening and widening of joint and bedding planes. Ley (1979) proposed that the relief or surface area of the solutional features is proportional to the wave energy expended within the intertidal zone. Relief is greatest between the neap tidal levels, where the greatest amounts of energy are expended (see Chapter 1), and it declines with the reduction in wave energy towards the spring tidal extremes. As the platform is lowered by erosion, the surface area above the mid-tidal level increases as it becomes exposed to a zone of higher wave energy. Below the mid-tidal level, platform lowering causes a decline in wave energy, progressive removal of the lapies, and a reduction in surface area. Ley considered, therefore, that a state of dynamic equilibrium exists, in which a constant amount of erosive energy is expended per unit area. As the tidal range increases, the degree of concentration of wave energy between the neap tidal levels decreases, so that the degree of microrelief about the mid-tidal level must also decline; this occurs towards the eastern parts of the Bristol Channel. Alternatively, increasing wave fetch increases the energy expended between the neap tidal levels, and therefore the degree of microrelief. Ley found that the degree of development of the lapies increases with the purity of the limestones.

Some typical corrosional features of limestone regions are also well developed in calcareous aeolianites, but in the calcareous sandstones of southern California, the dominant elements are shallow tidal pools, many of which have flat bottoms and raised rims (Emery 1946). Similar forms have been reported from other areas (see Chapter 2). It should also be noted that features such as aveoles, lapies, and shallow pools are produced in non-calcareous rocks (such as basalts, granites, and other igneous lithologies) by salt fretting, particularly but not exclusively in the low latitudes (Guilcher *et al.* 1962, Tricart 1972, Consentius 1975, Guilcher and Bodere 1975).

Plates-formes à Vasques

The term *plates-formes à vasques* originates from the work of Guilcher (1953) and Guilcher and Joly (1954) in Morocco, although it describes a feature which is essentially analogous to the solution benches of Hawaii, previously described by Wentworth (1939). Vasques are wide (up to several decimetres), shallow pools with flat bottoms, which form a network consisting of a tiered, terrace-like series of steps on limestone, and particularly aeolianite, platforms. The pools are separated from each other by sinuous, narrow, lobed ridges, between 1 and 20 cm in height, running continuously for dozens of metres (Fig. 10.2). Vasques can develop on trottoirs which are fronted by vermetids. The *plates-formes à vasques* develop between the high and the low tidal levels. They are submerged at high tide, but fed by breaking waves at the low tidal level, with the return flow cascading down into the successively lower pools.

The rims surrounding each pool can be residual corrosion features marked by the pinnacles of lapies; built by organisms such as calcareous algae, vermetids, or even serpulids; or a combination of the two (Guilcher 1958a). Guilcher (1953) suggested that the pools are more corroded than the rims, which are partly protected by organic growths; although in southern and southeastern Madagascar, some vasques have developed on the lower portions of trottoirs as a result of the enlargement of boreholes (Battistini 1980). True *plates-formes à vasques* have only been recorded from intertropical and Mediterranean climatic regions. Battistini and Guilcher (1982) and Dalongeville and Guilcher (1982) made a survey of

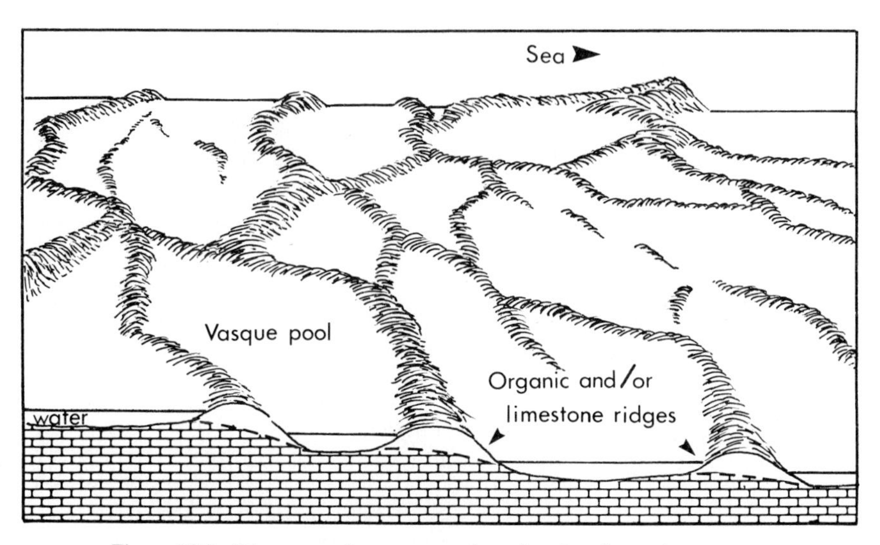

Figure 10.2. Diagrammatic representation of a *plate-forme à vasques*.

the literature to determine their occurrence, and the nature of the intervening rims. In the Mediterranean, the platforms occur as far as 40°N. in Italy and in Sardinia, which is further north than in the Atlantic. They have been reported from Syria, Lebanon (Dalongeville 1977), Israel (Safriel 1966), Cyprus, Crete, Turkey, Morocco (Guilcher 1953), Malta (Paskoff and Sanlaville 1978), Tunisia, Algeria (Guilcher 1954), Spain, Italy, and Sardinia. As elsewhere, there are rims which are constructed and others which are non-constructed, as well as ridges consisting of country rock with a cover of organic material. These various forms can coexist in the same areas. Exposed environments appear to facilitate the development of these platforms (Dalongeville and Guilcher 1982). Outside the Mediterranean (Battistini and Guilcher 1982), vasques have been described in Hawaii (Wentworth 1939), Guam (Emery 1962, Tracey *et al.* 1964), New Caledonia, the New Hebrides (Guilcher 1974b), Tonga, western and southeastern Australia (Fairbridge 1950, Hills 1971), Puerto Rico (Kaye 1959), Desirade in the Lesser Antilles, Barbados (Tricart 1972), the Netherlands Antilles (Focke 1978), Costa Rica, northeastern Brazil, Madagascar (Battistini 1980, 1981), Inhaca off Mozambique, Kenya (Bird and Guilcher 1982), the Red Sea (Guilcher 1952b, 1955), and in southern Portugal (Guilcher 1957).

Corniches and Trottoirs

Organogenic formations assume an important role in the development of calcareous coasts in warm climates. The terms 'trottoir' and 'corniche' have been used to describe organic protrusions which grow out from steep rock surfaces at about sea level, as well as rock ledges which are cut into the littoral rock and coated with a thin crust of organic material (Molinier 1955a, Pérès 1968). Although both phenomena provide narrow pavement or sidewalk-like paths at the foot of marine cliffs, in this chapter, 'corniche' will be used to describe the former situation, and, in accordance with the original definition of Quatrafages (1854), 'trottoir' (literally, footpath or pavement) will be used to refer to the erosional rock ledges and their organic veneers (Fig. 10.3).

Three major types of organic reef have been distinguished in the western Mediterranean (Pérès and Picard 1952, 1964, Molinier and Picard 1953, 1954). They are:

1. reefs formed of calcareous red algae at about the mean sea level;
2. corniche composed of polychaete worms (Serpulidae), which develop below mean sea level and are usually submerged—Folke (1978) has argued that these deposits often turn out to be misidentified Vermetid tubes; and

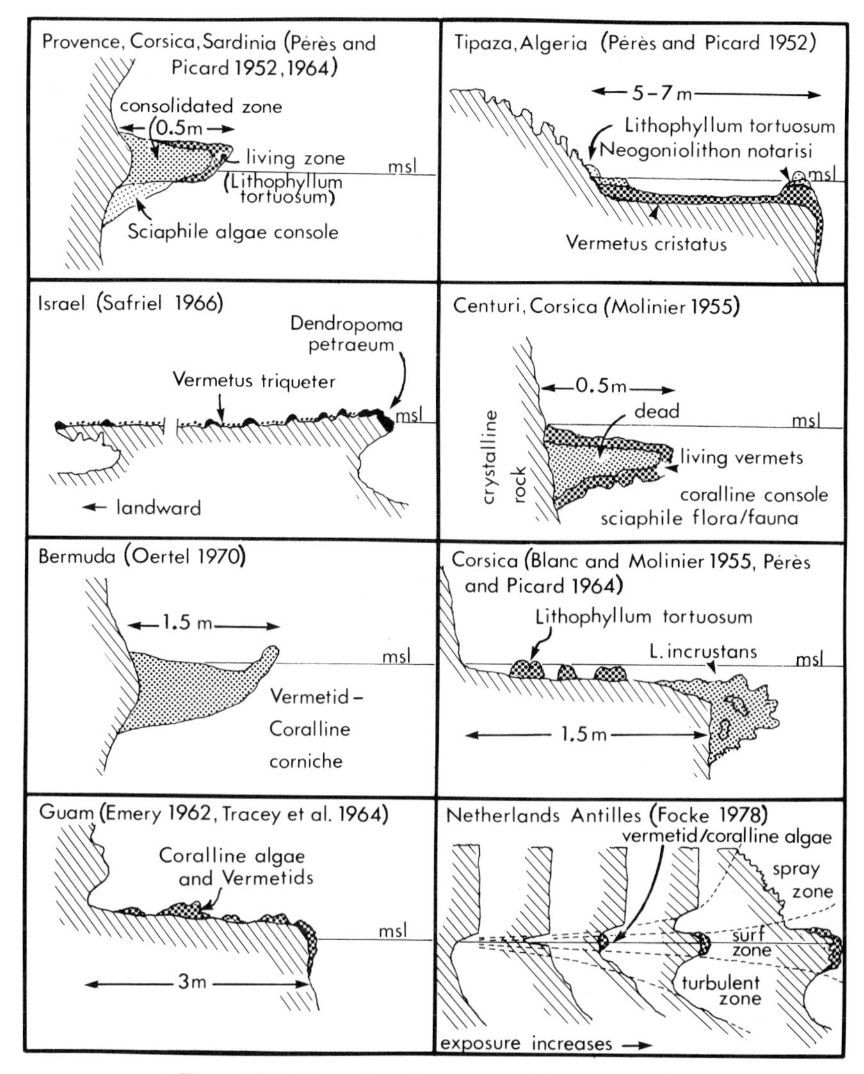

Figure 10.3. Examples of coastal corniche and trottoir.

3. erosional ledges, slightly below mean sea level, with a veneer of sessile Vermetid gastropod tubes.

In the Mediterranean, corniches are usually composed of the calcareous alga *Tenarea tortuosa* (= *Lithophyllum tortuosum*), and other diverse Melobesieae algae, such as *Neogoniolithon notarisi*. They are oranic accumulations, which protrude out about 0.5 to 2 m from steep coastal slopes, at about the mean sea level (Guilcher 1953, 1958a, Froget 1963, Nicod

1972). Corniches are essentially intertidal structures which are absent in calm, sheltered areas. They cannot withstand the impact of the strongest waves, however, and are therefore best developed in the inlets of exposed coasts (Pérès and Picard 1952, 1964). Although *T. tortuosa* can survive in areas where it is subject to direct sunshine, it seems to prefer less insolation. This could explain why corniches develop better on very steep slopes than on gentle gradients, and particularly why they favour north-facing surfaces (Pérès and Picard 1952). The outer portion of corniches consist of living algae, but diagenesis of the interior makes the formation very resistant to wave attack.

The growth of algal corniches also provides suitable environments for a variety of fauna and flora. On the surface of well-developed corniches are populations which are tolerant of strong insolation and prolonged periods of emersion. On the lower face, there is usually a population of overhanging sciaphiles (Blanc and Molinier 1955). Cavities within the formation are filled by a variety of bivalves, gastropods, spiders, mites, crustaceans, and worms (Pérès and Picard 1952, 1964, Guilcher 1953, Pérès 1968). Molinier (1955b) and Blanc and Molinier (1955) have described massive corniches 20 to 30 cm in width in Corsica composed of the Melobesieae *Lithophyllum incrustans* mixed with some *Corallina mediterranea*, and on the lower face *Lithothamnium lenormandi*. These accumulations are largely infralittoral, although the upper portions can extend into the lower mesolittoral. They form projecting corniches bordering a small erosive terrace about 1.5 m in width, cut in sandstones. The terrace itself has pads of *T. tortuosa* in semi-shaded hollows. Where the rock is subjected to strong insolation, however, the battered edge of the terrace is densely populated by species of the brown algae *Cystoseira*, and the surface by isolated, thin crusts of *Vermetus cristatus*. Molinier (1955b) and Blanc and Molinier (1955) have also discussed the occurrence, in Corsica and near Marseilles, of fist-sized *bourrelets* or pads of *C. mediterranea* and *L. lenormandi*. They develop in a shaded environment in the infralittoral zone, below a *T. tortuosa* corniche. The coralline algae trap detrital material carried by the sea, which is then cemented by the encrusting *L. lenormandi*. It therefore grows in thickness, always covered by the coralline algae. Sipunculids and foraminifera also contribute to their development.

Corniches of *Tenarea* are not restricted to calcareous substrates. In Catalonia, for example, they have developed on crystalline rocks (Barbaza 1970), but they do not form on friable rocks, presumably because the *Tenarea* thalli need rocks of a certain hardness to become fixed (Pérès and Picard 1952). They are nevertheless usually larger and more regular on limestones and calcareous sandstones, where a notch can develop in the supralittoral zone which is exposed to the physical and chemical effects of spray.

In several parts of the western Mediterranean, as in Corsica and the south of France, there are corniches or 'balconies' composed of Serpulid worms and Melobesieae algae (Pérès and Picard 1952, 1964). They are found on resistant rocks in sheltered environments, below mean sea level in the upper part of the infralittoral. In Corsica, a reef on granite is composed of Melobesieae algae, and several Serpulid species, *Protula* sp., *Serpula vermicularis, S. concharum, Vermiliopsis multicristata*, and particularly, *Pomatostegus polytrema*.

Vermetids do not generally flourish in the northern parts of the western Mediterranean, where most organic reefs consist of calcareous algae and Serpulid worms. Temperatures are more favourable for large Vermetid populations in the southern parts of the Mediterranean, where they grow vigorously on narrow ledges cut into the littoral rock (Safriel 1966). Quatrefages (1854) first mentioned the occurrence, in Sicily, of a trottoir just below mean sea level; although he failed to realize that the Vermetid tubes only provide a thin veneer to an eroded rock surface below. Purely constructional Vermetid and coralline algae corniches can form on fairly resistant, even non-calcareous, substrates. Trottoir platforms, however, develop where the rock is fairly easily eroded, as in weak limestones and sandstones. Erosion occurs in the spray zone, forming an eroded, subhorizontal platform which is quickly covered by Vermetid tubes, especially on the wave-battered fringes of the platform. Continued erosion of the cliff slowly widens the platform, which is protected by the Vermetid encrustations. These platforms are up to 7 to 8 m in width in Sicily and Algeria (Pérès and Picard 1952, Molinier and Picard 1953).

Vermetid encrustations (Keen 1961) have been reported from around the western Mediterranean, in France (Molinier 1955a,b, Debrat 1974), the Balearic Islands (Molinier 1954, Molinier and Picard 1957), Italy (Molinier and Picard 1953), Algeria (Pérès and Picard 1952), and Tunisia (Molinier and Picard 1954). Trottoirs are usually associated with thin crusts (several centimetres thick) of *Vermetus cristatus* Biondi (= *Dendropoma petraeum*) below mid-tidal level in the infralittoral zone. In Algeria and Sicily, the Vermetid crust attains its greatest thickness on the seaward margins of the platform, where the water is most agitated. It is also thicker, however, along the edges of fractures, which may correspond to fissures in the underlying rock. This produces a *plate-forme à vasques* consisting of broad shallow pools separated by elevated Vermetid ridges (Molinier and Picard 1953). At Tipaza in Algeria, *V. cristatus* only occupies the back of the platform, whereas the rest of the platform and the seaward terminus carries *V. triqueter* (Pérès and Picard 1952). If the Vermetid crust is sufficiently built up, it may become covered by *Tenarea* and *Lithothamnium* algae in the lower portions of the mesolittoral zone (Molinier and Picard 1953, Molinier 1954). This generally occurs at the seaward and landward margins of the platform. In Majorca, for example, limestone ledges in the lower mesolittoral zone are partially protected from corrosion by *Tenarea* and *Neogoniolithon notarisi* (Molinier and Picard 1957). In the Balearic Islands, the infralittoral platform is covered by thin veneers of *Vermetus cristatus*, interdispersed with *Laurencia popillosa*, the coralline algae *Jania rubens*, and the calcium-depositing Phaeophyta *Padina* (Molinier 1954). In some semi-shaded places in Corsica, Vermetids have been replaced by *Tenarea* algae on narrow terraces in the infralittoral zone (Molinier 1955b).

Blanc and Molinier (1955) and Molinier (1960) also identified Vermetid *mergelles* or curbs in Corsica. Unlike the true Vermetid trottoir, they are not formed in crusts or pads, but as projecting bulges at the wave-battered fringes of beds of schist which dip slightly seawards. They are cemented on the surface in the lower mesolittoral by the calcareous alga *N. notarisi*, and sciaphile flora develop below the overhang. Vermetid corniches can develop on hard substrates which are resistant to marine erosion. On the island of Centuri, northwest of Cape Corsica, Vermetid tubes have accumulated on the battered fringes of metamorphic rock slabs, forming structures which are similar to projecting *Tenarea* corniches. The presence of Vermetids in the mesolittoral zone on Centuri and in parts of Corsica may be related to particularly intense wave action, which shifts their habitat upwards (Molinier 1955a).

Vermetid trottoirs are also common in the eastern Mediterranean (eg. Safriel 1966, Sanlaville 1972). They are usually quite narrow, but they can be dozens of metres in width in aeolianites (Fevret and Sanlaville 1966, Dalongeville 1977). Safriel (1966) suggested that differences in the form of Vermetid platforms in the western and eastern Mediterranean reflect the slightly higher temperatures in the east, where subtropical conditions may prevail in summer. In northern Israel, horizontal Vermetid-encrusted ledges attached to the land, similar to those in the western Mediterranean and in Lebanon (Fevret and Sanlaville 1966, Dalongeville 1977), were considered to represent immature forms. Mature Israeli platforms on the other hand, are broad, flat, round or elliptical, surf swept and awash at low tide, and possibly separated from the coast by hundreds of metres of water. They have overhanging edges and raised margins. *Dendropoma petraeum* (= *V. cristatus*) occupies the raised platform margins and the edges of the terraced pools, where wave action is strongest. *Vermetus triqueter* (= *V. gregarius*) is found on the platform surface, as it only thrives in areas which are underwater, or sheltered from direct surf action. Safriel (1966) proposed that in Israel, the Vermetid communities develop simultaneously with the formation of the platform, rather than after the platform has been formed. He proposed the following sequence of development:

1. the formation of corrosion basins in limestones or aeolianites;

2. barnacles first colonize the rims of these basins, but *Vermetus* arrives when the basins become deep enough to be permanently filled with water;
3. *Dendropoma* replaces the barnacles on the basin rims when they have been lowered to an appropriate level, and further lowering then almost ceases, because of the protection afforded by this cover; and
4. ledges form as the basins coalesce, as *Dendropoma* fails to protect the increasingly sheltered internal rims.

Trottoirs are also common in tropical seas, in Cape Verde, Senegal, Ghana, and Barbados. The 'boilers', which resemble microatolls, are related forms in Florida and Bermuda (Prat 1935, Guilcher 1953, Laborel 1966).

Safriel (1974) compared the atoll-like reefs of Israel and Bermuda. In both areas, the circular platforms have raised, overhanging rims, and they retard the erosion of exposed promontories. In Israel, however, the Vermetids, cemented by coralline algae, form a thin crust over the underlying limestone, whereas in Bermuda, the reefs are growing, wave-resistant, biogenic structures. The lower interiors of the reefs in Israel are the result of erosion, but in Bermuda they are the result of differential growth. In Bermuda, calcareous algae cement the tubes of *Dendropoma irregulare* and *Petaloconchus nigricans*. On the exposed, tidal southern coast, Vermetids develop between the visor and notch in the mesolittoral and the overhang created by infralittoral erosion below. In sheltered Harrington Sound with its very small tidal range, the Vermetids are less abundant, although they are at the same level as on the southern coast. This area has little biological erosion in the mesolittoral, but much more in the infralittoral (Laborel 1966).

Throughout the tropical Atlantic, a well-developed surf platform occurs within the range of encrusting coralline algae (*Porolithon, Lithophyllum* etc.) and Vermetids. *Millepora*, gargonians, corals, and sponges play a similar but less effective role in some areas (Newell 1961). They have been reported from Brazil (Kempf and Laborel 1968, Delibrias and Laborel 1971), west Africa (Laborel and Delibrias 1976), Puerto Rico (Kaye 1959), Barbados (Newell and Imbrie 1955, Newell 1956, Tricart 1972), the Bahamas (Newell 1961), the Netherlands Antilles (Focke 1978), the Cayman Islands (Woodroffe *et al*. 1983), and Venezuela (Gessner 1970).

Concretionary formations occur in a number of areas in Brazil (Kempf and Laborel 1968). They are composed of calcareous algae (Melobesieae) and *Petalonconchus varians* and *Dendropoma irregulare* Vermetids. They are found on igneous rocks as well as on sandstones and coral limestones, but always in the upper infralittoral, in areas exposed to strong wave action. A recent change in environmental conditions may be responsible for the replacement of *Petalonconchus* by *Dendropoma*, and the Vermetid in general by calcareous algae. On the weaker rocks (aeolianite, silaceous marine sandstones, dead coral), horizontal platforms have formed in exposed environments at the boundary between the meso- and infralittoral zones. These can be covered by Vermetids as in Sicily (Molinier and Picard 1953), but boilers develop if they only form on the outer edge. On hard rocks, *bourrelets* or flanges form a type of corniche on steep rock surfaces in Brazil and Corsica (Molinier 1955b).

Focke (1977, 1978) has described the development of limestone cliffs in Curaçao and elsewhere in the Netherlands Antilles (Figs 10.4 and 10.5). Rapid bioerosion occurs in the intertidal zone. The form of the coastal profile is dependent upon the degree of exposure (Fig. 10.3). In sheltered areas, notch profiles predominate, but without significant organic accumulations. Accretions are common in more exposed areas, but they are restricted to a narrow vertical zone in the middle of the cliff notch, at about the mid-tidal level. In the most exposed areas, the coralline algae and the Vermetid accretions become thicker, better lithified, and higher, and they have given rise to protruding surf benches (trottoirs) up to 10 m wide, and as much as 2 m above mean sea level. These accretions are primarily

Figure 10.4. Notch in coral limestone on the sheltered southern coast of Curaçao, Netherlands Antilles.

Figure 10.5 Surf bench with vasques on the exposed northern coast of Curaçao, Netherlands Antilles. Benches in this area are backed by lapies and pools and occasionally by a notch in the spray zone. The outer portions of these benches are undercut by a subtidal notch.

constructed by the Vermetid gastropod *Spiroglyphus irregularis* (= *Dendropoma irr.*) and the coralline algae *Lithophyllum congestum* and *Porolithon pachydermum*. Sessile foraminifera and internal sediment help to fill the cavities. The level of the platforms is determined by the highest point at which the Vermetids are able to build accretions. Focke noted that there may be just as many frame-building organisms in sheltered as in exposed coastal regions. The occurrence of accretions in exposed areas is therefore apparently related to water turbulence, possibly because of its positive effect upon the supply of nutrients to the organisms, or possibly because lithication is facilitated by the large amounts of water pumped through the accretions. Lithication, by aragonite and magnesian calcite cements, strengthens the accumulations against wave attack and seals the substrates, protecting them from biodegredation. Fevret and Sanlaville (1966) and Focke (1977, 1978) have suggested that as platforms widen they retard erosion in the spray zone and notch, until eventually a state of dynamic equilibrium must be attained. On the Cayman Islands, it has been proposed that the surf benches, which are slightly above mean sea level, are in equilibrium with the present sea level and with wave energy conditions. In one area, for example, the rate of platform lowering is only 0.2 mm yr^{-1} (Woodroffe *et al.* 1983). It has also been argued that other types of shore platforms are in equilibrium with their environments (see Chapter 9).

ne tropical Pacific, the lime-secreting algae, particularly *Porolithon* *les* (Newell and Imbrie 1955), have greater durability and possibly r growth rates. This results in the development of algal ridges near the ward ends of reef flats (Newell 1961). Vermetid-rimmed terraces up to 6 m in width and 3 m above mean sea level have been reported on Guam (Emery 1963, Tracey *et al*. 1964).

Notches

Deeply undercut cliffs, platforms, and reef flat boulders are prominent features of limestone coasts in tropical regions, but notches are found in other climatic environments, and in other types of rock (see for example, Kelletat 1982). Near Tokyo, Japan, the formation of a notch in tuffs has been attributed to solution, hydration, alternate wetting and drying, and organic acids released by attached organisms (Emery and Foster 1956). Wave-cut notches have been produced in homogeneous cement and plaster blocks in wave flumes (Sanders 1968b, Sunamura 1973). In Britain, chalk and limestone notches have been cut by wave action (Common 1955, Wood 1968, Trenhaile 1969). In cool regions, however, notches are generally poorly defined in fairly homogeneous rocks, and locally restricted where they are associated with lithological and structural variations. On Hudson Bay, for example, angular notches have developed along structural planes of weakness, as a result of the effects of waves, the ice foot, and gelifraction (Allard and Tremblay 1983).

Deep, narrow notches in warm seas are usually found in areas with a very low tidal range, probably because the erosive processes are concentrated within a narrow range of elevations. The effects of tidal range and exposure on the development of limestone coasts in Bermuda have been considered by Taillefer (1957) and Neumann (1966). Taillefer compared the sheltered, inland sea coast of Harrington Sound, which has a tidal range of only 15 to 20 cm, with the more exposed southern coast of the island, where the tidal range is 1 to 2 m. Neumann made a similar comparison between the northern coast and the Harrington Sound coast. On the southern coast, Taillefer found that a well-developed notch has formed at the high tidal level, as well as a less continuous one at the low tidal level; these notches are separated by a surf bench. A visor above the high tidal notch was attributed to induration of the aeolianite at the level at which sea water inhibits the downward percolation of fresh water charged with calcium carbonate. Hodgkin (1970), however, has argued that the visor in coral limestones is a residual feature, restricted to areas backed by a low cliff. Neumann found notches above, within, and below the intertidal zone on the northern coast, although there is no continuous intertidal notch in this area. In Harrington Sound, however, the limestone cliffs are usually deeply undercut, overhanging by as much as 3 to 4.6 m. Taillefer reported that this notch is intertidal, but Neumann found that it developed just below the low tidal level, although it is exposed for a few weeks each year during extreme tidal periods. The notch has a flat roof, corresponding to the low tidal level, and is independent of variations in rock structure. Neumann considered the notch to be the result of rock borers, such as sponges, pelecypods, and worms, which are particularly active near the water level, where wave agitation provides a regular supply of nutrients and aids in the removal of the weakened rock particles.

Focke (1977, 1978) showed that the degree of exposure also has a profound effect upon the form of limestone notches in Curaçao in the Netherlands Antilles (Figs. 10.3, 10.4 and 10.5). In sheltered areas, notches have developed about the mid-tidal level, but where the waves are more vigorous the increased turbulence of the water facilitates the accumulation of organic material, largely consisting of Vermetids and calcareous algae. These crusts develop in the centre of the notch and protect the underlying rock, dividing the original notch into two sections, one above, and another below a protruberance. In the most exposed areas, the spray zone reaches up to the top of the cliff, and the upper notch is generally replaced by a corroded slope. Progressing seawards, the coastal profile then consists of this slope, a trottoir, and the lower notch. This relationship between the form of the coastal profile and the exposure to wave action has also been noted on Grand Cayman (Woodroffe *et al.* 1983). A similar organically induced division of a notch into two sections occurs in Puerto Rico (Kaye 1957). These notches are contemporary, but other workers believed that double or multiple notches are evidence of changes in sea level, or intermittent tectonic events.

Double notches are very common in a number of areas, as in the Ryukyu Islands (Takenaga 1968), Borneo and Malaysia (Hodgkin 1970), eastern Indonesia (Verstappen 1960), Guam, Puerto Rico (Kaye 1957, 1959), and Barbados (Tricart 1972). Fairbridge (1948, 1950, 1968b) considered that notches form at about mean sea level, so that those which are now below the low tidal level must reflect former lower sea levels. If Holocene sea levels have been higher than today (Chapter 7), then the contemporary notch may be found a little below a notch or notches formed during recent stillstands of the sea (Tricart 1972). In the Bismarck Archipelago, the presence of a higher notch has been attributed to the sea level responsible for a 1.5 m terrace found throughout the Pacific and Western Australia (Christiansen 1963) (see Chapter 7). MacFadyen (1930) and Guilcher (1952b) have reported the presence of double notches around the Red Sea. Guilcher found that double notches, 1.2 to 1.4 m apart, only exist in sheltered areas, and are replaced by a single notch in exposed sites. The higher notch was attributed to the maximum of the Dunkerkian sea level.

Notch levels are similar to those reported in Western Australia (Teichert 1950).

There are undoubtedly many reasons other than changes in relative sea level, which might account for the formation of multiple notches on limestone coasts—for example, because of variations in rock structure and lithology, and possibly because different notch-forming mechanisms operate most efficiently at different elevations. Some workers have found that the main intertidal notch develops at or close to the high tidal level (Wentworth 1939, Guilcher 1953, 1958a, Newell 1956, Verstappen 1960, Christiansen 1963, Takenaga 1968, Hills 1971, Tricart 1972, Nicod 1972), although notches are frequently found at other elevations. On the northern Adriatic coast of Yugoslavia, for example, the notch is always below the mean high water level (Schneider 1976, Torunski 1979). Several workers in different areas have found that the notch normally develops at or close to the mid-tidal level (Ginsburg 1953b, Guilcher 1958b, Hodgkin 1964, Teichert 1947, 1950, Fairbridge 1948, 1968b, Sweeting 1973, Debrat 1974, Trudgill 1976b, Bird *et al*. 1979, Woodroffe *et al*. 1983). Kaye (1957, 1959) placed the formation of notches in Puerto Rico in the zone extending from the level of the low tidal wave trough up to the level of mean high waves; that is, the upper and lower limits of the notch are determined by mean wave height. On sheltered coasts, he found that the deepest part of the notch is at the mid-tidal level. A different type of notch forms just below the intertidal zone, as in sheltered regions of Bermuda (Neumann 1966, 1968), Curaçao (Focke 1978), and elsewhere in the humid tropics (Tricart 1972).

The debate over the precise level of formation of notches in the mesolittoral zone partly reflects the very small tidal range in nearly all the regions in which the notches are found. Determining the elevation of a notch relative, for example, to the high or mid-tidal level requires decisions involving vertical intervals of fractions of a metre in many areas. Given the general lack of precise measurement and the poor reliability of tidal data in relation to terrestrial bench-mark elevations in many regions, it is difficult at present to confidently determine the levels of notch development in relation to tidal levels. Notch formation seems to be facilitated by a very small tidal range. Distinct notches appear to be associated with warm, microtidal coasts, and they are usually difficult to discern when tidal ranges are greater. As the tidal range increases, notches can develop close to both the high and the low tidal levels (Taillefer 1957, Flemming 1965, Hills 1971, Battistini 1980, 1981).

The tidal range and the degree of exposure affect the morphology of notches. In general, the higher the amplitude of the waves and the higher the tidal range, the greater is the difference in elevation between the notch roof and floor (Newell 1961, Butzer 1962, Christiansen 1963, Neumann

1966, Hodgkin 1970, Focke 1978, Torunski 1979). Takenaga (1968), for example, measured and classified aspects of the morphology of 139 notches on the Ryukyu Islands, Japan. He found that the height of the notch roof corresponds to the upper limit of sea spray, and is therefore greatest on open coasts. Verstappen (1960), and later Russell (1963), concluded that notches on exposed coasts have flat floors and often steeply inclined roofs, caused by the effects of surf and spray, but in sheltered areas they are essentially horizontal incisions with nearly flat roofs. The vertical height range of these notches increases with tidal range (Fig. 10.6*a*).

Notches in the humid tropics typically range from 1 to 5 m in depth (Tricart 1972), but they are considerably deeper under favourable circumstances. In western Barbados, for example, notches in soft coral marl are more than 30 m in depth. Deep notches in Barbados form where the coral lacks joints and other fissures which promote cliff collapse, the debris is quickly removed from the cliff base, and sand is available for abrasion. There is little relationship here between notch depth and wave energy (Bird *et al.* 1979). Notches have formed in sheltered as well as in exposed areas (Wentworth 1939, Guilcher 1953, 1958a, Emery and Foster 1956, Christiansen 1963), but in some places, notch depth does appear to be related in a complex way to the exposure to wave action. In the French Mediterranean, notches have attained their best development on the

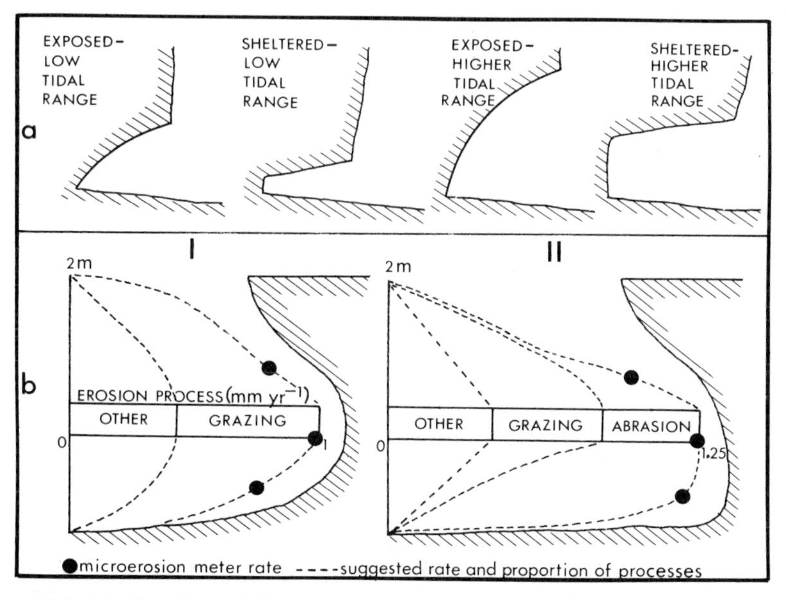

Figure 10.6. (*a*) The effect of tidal range and exposure on the form of the notch in eastern Indonesia (after Verstappen 1960). (*b*) Erosion rates and processes in notches on Aldabra Atoll, where I) sand is absent; and II) sand is present (Trudgill 1976b).

extremities of headlands (Froget 1963, Debrat 1974), and they are absent from the most sheltered areas in Borneo and northwestern Malaysia (Hodgkin 1970). In the western Atlantic (Newell 1961), the Red Sea (MacFadyen 1930), and Oman (Vita-Finzi and Cornelius 1973) the most deeply cut notches are often found in sheltered areas (Fig. 10.4). On the Ryukyu Islands there appears to be a negative relationship between notch depth and the strength of the waves (Takenaga 1968). This is consistent with Trudgill's (1976b) observation that coastal profiles around Aldabra Atoll change from notch to cliff to ramp as the degree of exposure increases. The depth of the notch is also probably related to numerous other aspects of the morphology of the littoral zone. In eastern Indonesia, tidal currents generated in the narrow entrances to bays produce particularly deep notches (Verstappen 1960). In Curaçao (Focke 1978) and elsewhere, it has been suggested that the surf platform is in a state of equilibrium on exposed, windward coasts; this may control the depth of the notch by determining the amount of spray and splash which can reach it, and therefore prevent marked variations in its morphology through time.

The controversy over the origin of notches in warm seas is essentially a specific aspect of the larger debate over the origin of coastal limestone features in general (see Chapters 3 and 4). It was formerly believed that notches are produced by mechanical wave erosion (Agassiz 1895, Prat 1935), but most investigators now consider that chemical or biochemical corrosion, or biological grazing and boring activities, are dominant. This view is partly a response to the observation that notches are often well developed in sheltered locations. MacFadyen (1930) suggested that deep undercuts in Red Sea corals were produced by seawater solution and organic boring, but he was unable to determine which is the more important. Wentworth (1939) thought that notches are formed where fresh groundwater emerges, and below which the rock is saturated with seawater. This theory was devised to circumvent the difficulty of accounting for limestone solution features around the shores of lime-saturated seas. Notches, however, have formed around the base of mushroom-shaped rocks on reef flats ('negroheads') (Umbgrove 1931, Revelle and Emery 1957, Russell 1963), where there can be very little groundwater seepage from the base (Panzer 1949). Furthermore, as Guilcher (1953, 1955) has noted, notches have also developed in arid areas such as the Red Sea, where there is little fresh water. Panzer (1933) considered that notches are formed by seawater solution, and this view has been supported by many other workers (Fairbridge 1948, Kuenen 1950, Guilcher 1953, 1958a, Revelle and Emery 1957, Kaye 1957, Taillefer 1957, Verstappen 1960, Christiansen 1963, Russell 1963, Tricart 1972), although the mechanism is poorly understood. Corbel (1954, 1956) proposed that deep notches in the limestones of Spitsbergen and the Gulf of St Lawrence were produced by solution related to sea water and the presence of snow on the ice foot; it is likely, however, that wave action and frost play an important role in the undercutting of the coastal cliffs in these areas.

Many other workers have emphasized the work of organisms in the development of limestone notches (Newell 1956, 1961, Hodgkin 1964, 1970, Neumann 1966, 1968, Debrat 1974, Focke 1978, Torunski 1979, Schneider and Torunski 1983). Endolithic algae, browsing gastropods, the *Cliona* sponge, and boring pelecypods are often mentioned as being active in the notch zone. Schneider (1976), for example, recognized the polychaete *Polydora ciliata*, the sponge *Cliona vastifica*, and the pelecypods *Lithophaga lithophaga* and *Gastrochaena dubia* in a notch in the northern Adriatic. In this area, the intertidal notch on limestone coasts is thought to be completely biogenic in origin (Schneider and Torunski 1983). In support of the bioerosional origin of notches, it has been observed that they develop in turbulent water, which increases the supply of nutrients to the organisms and helps to remove the weakened rock particles (Neumann 1966). Alternatively, it has been argued that this association is because turbulent water flows quickly, or because it is charged with air bubbles and carbon dioxide, making the water aggressive and facilitating the formation of notches by chemical solution (Guilcher 1953, Kaye 1957, Nicod 1972, Tricart 1972). Some workers have insisted that abrasion and other forms of mechanical wave erosion play a significant role in notch development (Froget 1963, Takenaga 1968). The absence of notches in sheltered areas and the occurrence of smooth forms at exposed sites in the Langkawi Islands in Malaysia, convinced Tjia (1985) that they are the result of abrasion. In Oman, notches are cut by wave action in rocks which have been weakened by the borings of *Lithophaga* (Vita-Finzi and Cornelius 1973). According to Bird *et al*. (1979), solution plays a minor role in western Barbados, where the notch is largely the result of physical processes, increasing in depth where sand is available. Trudgill (1976b) made a detailed investigation of the rates and mechanisms of limestone erosion on Aldabra Atoll in the Indian Ocean (Fig. 10.6*b*). He found that chemical solution by sea water is possible in this area, although it is of minor significance. Where sand is absent, grazing organisms account for between one-third and one-half of the notch erosion on Aldabra. Where there is sand at the foot of the notch, abrasion assumes a major role, accounting for about one-third of the notch erosion. Concentration of abrasion at the base of the cliff forms notches which have much flatter floors and less concave profiles than in areas where sand is absent. Trudgill emphasized that physical processes such as abrasion and other mechanisms of wave action account for a large proportion of the erosion of limestone shores.

Several estimates have been made of the rate of notch erosion and of the contributions of single erosive agencies, but few reliable measurements of the overall rate have been made (Saumell *et al*. 1982). In Oman, the boring rate of *Lithophaga* in notches was judged to be about 9 mm yr^{-1} (Vita-Finzi and Cornelius 1973). Neumann (1966) obtained a figure of 14 mm yr^{-1} for *Cliona lampa*, but this is the maximum obtainable under experimental conditions, and a range of 0.1 to 1 mm yr^{-1} for this species, as measured by Rutzler (1975), seems to be more reasonable. There have been several estimates of the overall erosion rate, which is the result of the contribution of many mechanisms. Verstappen (1960) calculated a rate of 5 mm yr^{-1} on an island in eastern Indonesia, based upon the occurrence of a notch at the foot of a boulder which was presumed to have fallen onto the reef at the time of the Krakatau eruption. Taillefer (1957) considered that notch erosion in sheltered Harrington Sound, Bermuda, is about 10 mm yr^{-1}, but Kaye (1959) thought that in Puerto Rico it is only 1.6 mm yr^{-1}. In Barbados, steel rods placed into the notch recorded rates of erosion of between 0.23 and 2 mm yr^{-1} (Bird *et al.* 1979). In the cool, stormy waters of the St Lawrence estuary, Corbel (1958) estimated that the limestone cliffs are corroded at the rate of 2 to 3 mm yr^{-1}. Hodgkin's (1964) figure of 1 mm yr^{-1}, obtained using steel rods driven into an aeolianite notch in Western Australia, has been widely quoted, and is often used as being representative of limestone erosion rates in warm climates (see Chapter 4). Fairbridge (1968b) considered this figure to be typical of erosion rates in coral. He thought that the harder Mesozoic limestones of the Mediterranean probably erode at about one-tenth that rate, although rates of erosion similar to those in Western Australia have recently been measured in the northern Adriatic (Torunski 1979). Very similar rates of erosion to those in Western Australia, were recorded with a microerosion meter in the notches of Aldabra Atoll (Trudgill 1976b). They suggest that the 2 to 3 m deep notches on Aldabra could have formed within the last 2–3,000 years. On an island in the southern Great Barrier Reef of Australia, the surface in the upper part of the notch, at about the mid-tidal level, has been lowered around pedestals of sedentary rock oysters at the rate of 2.04 mm yr^{-1}. Chiton grazing and excavation of a 'home site' probably accounts for much of this erosion (Trudgill 1983).

Coastal and Submerged Terrestrial Karst

The form of some limestone coasts is determined by the presence of terrestrial karstic features which have been inherited and modified by marine processes, as a result of coastal retreat and changes in the relative level of the sea. In southern Pembrokeshire, Wales, for example, cliff detail in the Carboniferous Limestones has been determined by the sea cutting back into dry karstic water courses, and into caves and grottoes previously formed by terrestrial corrosion (Leach 1933, Steers 1962a).

The interaction between marine and terrestrial processes is exemplified on the Port Campbell coast of Victoria, Australia (Baker 1943). Mechanical wave action in the Miocene limestones, clays, and shales assumes the dominant role in the marine domain in this region, but its efficacy is increased by terrestrial corrosion. The cliff tops on this exposed coast consist of limestone stripped of its vegetation, soil, and clay cover. These bare surfaces, which extend up to 55 m from the cliff edge, have subsequently become zones of small-scale sinkholes and basin-like solutional features (Baker 1958). The area contains a number of streams which flow into sinkholes near to the cliff edge, and then reappear at the cliff base. Caves have been cut by wave action from subterranean, karstic cavities and stalactites and other speleothems give a fluted appearance to the cliff face. Blowholes develop in sinkholes which are connected to the sea through narrow, wave-modified conduits. Particularly spectacular blowholes have developed in similar circumstances in very thinly bedded limestones at Punakaiki, near Greymouth, New Zealand. Narrow geos or long gorges, often associated with joints, are created by the collapse of the roofs of the subterranean stream courses. At Port Campbell, narrow, wave-cut gorges may also be traced back into stream valleys which terminate at sinkholes, or into hanging valleys on the cliff face (Baker 1943). Coastal erosion has also exposed cylindrical hollows, several metres in depth, which were originally solution pipes. Similar features in aeolianite have been reported south of Melbourne (Bird 1970) and in western Australia (Fairbridge 1950).

On the Houtman's Abrolhos Islands off western Australia, sinkholes 1–100 m or more in diameter are found on the reef flats, although some have been filled by sediment. They have been attributed to deep dissection when sea level was low during the last glacial period. These subaerial channels, caves, and potholes were later drowned by the postglacial transgression, and planated by marine processes (Fairbridge 1948). A similar situation occurs in aeolianite on Point Peron, near Fremantle, Western Australia, where numerous sinkholes are connected to the sea through submarine caves and channels, which were formed during a glacial period when sea level was low (Fairbridge 1950).

Marine dolines are common in limestone regions. In Asturias, Spain, Mensching (1965) has described the occurrence of doline valleys, doline bays, and blowholes, as well as marine lapies (see also Guilcher 1958b). Schulke (1968) described five types of marine dolines in Asturias, which are representative of those found in many otner areas (see, for example, Baulig 1930):

(*a*) Submarine dolines are essentially subaerial features which are now continuously under water. Sea water corrosion may have played some role in their formation just before they were submerged by the postglacial transgression, but they are presently inactive unless they contain a submarine resergence.

(*b*) Inundated dolines have floors below sea level, and they are continuously, but only partially, submerged. They are not very common in Asturias, but are encountered more frequently in areas where the tidal range is very small, as in the Dinaric coast in the Mediterranean (Cvijic 1902, Baulig 1930, Milojevic 1952). Similar dolines connected to the sea through conduits occur in coral in the Maltese islands, forming semicircular bays where they have been breached by the sea (Paskoff and Sanlaville 1978).

(*c*) Intertidal/tidal dolines have floors which are within the intertidal zone, and they are therefore periodically partially submerged, according to tidal ebb and flow. They were deepened during glacial periods when sea level was low, and widened by lateral corrosion during transgressions.

(*d*) Other dolines have floors which are above the high tidal level, so that they are never completely submerged. They are, however, episodically and partially or totally washed by spray, and they are connected to the sea through vents. Fountains or blowholes can occur in these dolines when sea water is forced through the conduits during storms.

(*e*) The last category includes dolines which lie above the spray zone, although sea water impregnates the rock below, forming the karst base. Tidal oscillations therefore facilitate the enlargement and extension of conduits in the rock, and place limits on the depth to which they develop.

Schulke (1968) thought that marine dolines develop best where (*a*) coasts are not too high, and are well fissured but resistant to wave attack; (*b*) strong tides and frequent storms displace water in the karst network, and induce regular alternations of wetting and drying; and (*c*) high precipitation aids limestone solution.

Coastal scenery is, in some cases, almost totally dependent upon the character of submerged dolines and other elements of karstic landscapes. In northwestern Yugoslavia, marine invasion has produced a tortuous coastline consisting of rounded bays, peninsulas, coves, and small, elliptical or circular island-hillocks with rounded slopes (George 1948). Cavern collapse and marine invasion has produced most of the harbours, sounds and bays on Bermuda (Bretz 1960), and the drowning of a karst landscape is responsible for much of the form of the sheltered limestone coast of western Florida (Shepard and Wanless 1971). Tower karst has been submerged in parts of Java (Quinif and Dupuis 1985), Vietnam, and Malaysia, possibly as a result of changes in sea level. The coastal landscape in parts of Vietnam consists of a series of enclosed seawater lakes, which may be reached by boat through caves in the surrounding slopes (Jennings 1971).

The presence of partially or totally submerged karst caves influences the development of coastal scenery in many areas. In Malta, semi-circular coves have been formed by the collapse of these caves and their occupation by the sea. Invasion of subterranean caves as a result of wave erosion and cliff retreat was responsible, for example, for the formation of a 'Blue Grotto' in southern Malta (Paskoff and Sanlaville 1978). Submarine caves are particularly numerous in Provence, in the vicinity of Marseilles (Froget 1963). Some of these caves are only partially submerged, containing a lake of sea water on their floors, and communicating with the sea through narrow conduits or siphons. One cave contains cemented, pre-Flandrian sand deposits. Several caves have formed along fault planes, and in calanques. Fragile speleothems have been able to survive in some of the submarine caves because of weak wave action, although it has apparently been strong enough in other cases to clear out loose rubble. Froget also described the network of karstic galleries in the submarine Veyron Bank, an offshore limestone outcrop lying at depths of between −13 and −25 m. Wave action is more important here, and the caves are in a poor state of preservation, often having fractured roofs. Froget attributed the formation of the submarine caves of Provence to a sea level less than 50 m below today's. The pre-Flandrian regression, which was more than 40 m in this area, may explain the formation of most of the caves. In the case of the Veyron Bank, however, a fall of this magnitude would have left it as a small island, with little fresh water for the formation of extensive subterranean karst systems. Froget believed that the Veyron Bank karst was either formed at some time before the pre-Flandrian and followed by land subsidence, or that it is even more ancient. Submerged sinkholes, caves and terraces occur in the Caribbean and on Pacific atolls. Corrosion caves are largely submerged on Bermuda, where they extend down to depths of −24 m (Bretz 1960). The Blue Holes of Belize and the Bahamas are submerged sink holes formed during periods of lower sea level in the Pleistocene. Stalagmites and stalactites are common in the submarine caves of Honduras, but are absent where there are strong currents (Benjamin 1970, Dill 1977).

Submarine springs or resurgences have been described near Dubrovnik, Yugoslavia (Baulig 1930), and on the island of Kefallinia (Cephalonie) in southwestern Greece (Nicod 1963). The karst on this island was forced more deeply during periods of low glacial sea level, and it was invaded by the sea during the Flandrian transgression. Speleothems at a depth of −26 m have been dated to 20 ka BP, and at −3 m they have been dated to 16 ka BP. Submarine springs have also been described on the island of Tavolara, northeast of Sardinia (Siffre 1961). All the littoral limestones of Provence have active and abandoned submarine resurgences, often associated with marine caves, as on Capri (Nicod 1972). Conduits at depths of −30 and −45 m have been found in submarine talus, which suggest that they were emergent during cold periods, when there was active karst circulation Nicod (1967) rejected the alternate hypothesis that attributes the formation of these conduits to the circulation of phreatic water under pressure, during a period of submergence. If the flow of water from the resurgence is more powerful than the hydrostatic pressure of the sea, then the fresh water escapes quite easily. If the pressure of the sea is more powerful, as in deep water, and the flow of fresh water is weak, then the sea invades the main submarine cavities, forcing the fresh water to escape through the narrowest cracks.

Calanques

Calanques are coastal inlets which can be of a gorge-like nature. They have been described in a number of areas in the Mediterranean, including Yugoslavia (George 1948), Malta (Paskoff and Sanlaville 1978), and Majorca, where they are known as *calas* (Butzer 1962), and especially from the limestone coast of Provence. Several types have been identified. According to most investigators, the true calanques or *calanques–rias*, are karstic dry valleys which have been partially drowned. The valleys were deepened during periods of low glacial sea level, and then submerged by the transgression of the Flandrian or Holocene period (De Martonne 1909, Blanchard 1911, Johnson 1919, Butzer 1962, Froget 1963). In Provence, Denizot (1934) distinguished between calanques consisting of short ravines with steep slopes and well defined thalwegs, and less common, wider forms with poorly defined thalwegs. In Malta, the calanques–rias are fault controlled (Paskoff and Sanlaville 1978). Calanques–criques (Berard 1927, Chardonnet 1948, Nicod 1951) are simply an expression of anfractuous coastlines. They seem to be the result of selective wave erosion of fault zones or other areas of weak rock, and are not usually associated with significant terrestrial valleys. Berard (1927) believed that a large proportion of the calanques between Marseilles and Toulon are of this type. Chardonnet also recognized the occurrence in Provence of ancient or fossil calanques consisting of narrow, boxed-in, upstream areas, and wide downstream plains (1 to 3 km) with marine deposits of Pliocene age. He considered that they are former rias which have been fossilized by recent warping. In Malta, a calanque is being created by the marine exhumation of an ancient valley from beneath fossilized materials (Paskoff and Sanlaville 1978).

Chardonnet (1948, 1950) has argued that most calanques are not the result of ria-like invasions of stream valleys. In Provence and in Corsica, he noted that some calanques are not related to a well-developed valley, nor can they be attributed to a recent marine transgression; the evidence, at least between Marseille and Menton, being of recent emergence. Because calanques–rias, unlike some other types, are restricted to limestone regions, he concluded that they must be strongly associated with karst processes. Furthermore, he noted that calanques are best developed in the thickest and most homogeneous limestone formations. He recognized two generations of landforms within the calanques. The first corresponds to an initial period of subaerial erosion, whereas the second commenced when the streams began flowing below the surface. The subterranean systems developed so that parts of the streams flowed in caves which extended below sea level. Eventually, hydrostatic pressure from the marine waters and pressures exerted by the underground streams forced the removal of the rock partitions between the marine and freshwater domains. The rushing of marine water into the subterranean systems then caused the roofs to collapse. Although the sea was thought to have invaded the lower parts of the subterranean systems, this theory does not depend upon the occurrence of a marine transgression. Corbel (1956) proposed that stream waters in a humid, periglacial climate flowed over the frozen ground surface until it came within 500 to 1,000 m of the coast. In this littoral zone the ground was not frozen, and stream flow was underground. Subsequent marine transgression then caused the invasion of the subterranean system, collapsing the roof and forming a ria inlet. Denizot (1934) and Froget (1963) recognized 'false' calanques which are not related to thalweg courses, but appear to be the result of the destruction and retreat of coastal karst. In one case, a submarine cave occurs in the central part of a calanque in Provence. Similar forms can be attributed to the presence of faults.

Calanques–rias and calanques produced by the collapse of subterranean systems have also been identified on the Maltese Islands (Paskoff and Sanlaville 1978). One example consists of a dry valley which has been enlarged downstream by the wave-induced collapse of cave roofs beneath the thalweg. Deep karst has developed to such an extent in Malta, that the collapse of these subterranean systems has formed islands from the side slopes of some calanques. Nicod (1967, 1972) acknowledged the occurrence of calanque-like forms related to the wave excavation of karst cavities and conduits, and also of fault zones and other areas of geologically induced weaknesses, but he did not consider this a satisfactory explanation for calanques–rias. Nicod (1951) believed that the glacial climate of Provence was very damp, which facilitated the cutting of ravines by stream action when sea level was low. The lower portions of these ravines were then invaded as sea level rose in the postglacial period. Further upstream the valleys are dry, and obstructed by material which slid and fell from the sides at the end of the Würm, which was cold and dry in this area.

11

Some Aspects of Coastal Scenery

The Geotectonic Classification

Suess's (1888) geotectonic classification recognises the existence of two basic coastal classes. Atlantic or transverse coasts (Von Richthofen 1886) are cut across the geological trend of the hinterland, whereas Pacific or longitudinal coasts are essentially parallel to the geological grain. A secondary Pacific type includes coasts bordering basins which have recently subsided (Gregory 1912). When they are partially drowned, the Pacific coasts produce the Dalmatian coastal form, consisting of a series of longitudinal islands and intervening sounds. Atlantic coasts are found in parts of Brittany, Ireland, northwestern Spain, Newfoundland, South America, and Africa, although they also occur along parts of the Precambrian coasts of the Indian Ocean. Pacific-type coasts are found along the western side of North and South America, along island arcs extending from the Aleutians to New Guinea and New Zealand, and in parts of the Indian Ocean and the Mediterranean Sea (Fairbridge 1968c).

The geotectonic classification conveniently summarizes and accounts for broad differences in the form and outline of rock coasts. In general, Atlantic-type coasts tend to be indented, partly as a result of the exposure of rocks of variable resistance to wave erosion. Pacific-type coasts are usually fairly straight and regular in form, with only a few indentations where the sea and rivers have broken through resistant strata to expose the weaker materials behind. This has occurred around Lulworth Cove in southern England, for example, where the sea has breached the resistant Upper Jurassic Portland Stone and Purbeck Limestone, and is eroding the Cretaceous Wealden sands and clays, and in places, the Chalk behind.

Suess's classification can now be structured within the framework of plate tectonics (Inman and Nordstrom 1971, Davies 1980). Pacific-type coasts occur along the edges of plate boundaries, whereas Atlantic-type coasts are imbedded in the plates, remote from zones of crustal addition or subtraction. The structural grain of collision, plate-edge, or Pacific-type coasts is parallel to the shore. They are high, straight coasts with narrow continental shelves. There may be a flight of raised terraces on the steep, tectonically mobile hinterland, which is a potentially abundant source of coastal sediment. Most plate-imbedded, or trailing edge coasts, which correspond to Suess's Atlantic-type coasts, have hilly, plateau, or low hinterlands, and wide continental shelves. The structural lineations of these coasts are discordant with the shore.

Whether coasts are formed across or parallel to the geological grain, the scenery which develops is the unique product of a combination of elements, which provides infinite variation to the basic geotectonic form. These elements include:

1. The morphology of the hinterland inherited by the Holocene transgressive sea; this might range, for example, from gently seaward sloping surfaces to rugged upland terrain, with the additional effects of river valleys and marine or subaerial planation surfaces superimposed on them.
2. The climate of the region, past and present, which determines the nature and density of coastal vegetation, and the effect of fluvial, periglacial, chemical, and biological erosional processes.
3. The intensity of the wave environment and the tidal type and range.
4. Changes in relative sea level, including the effect of the uplift or depression of the land.
5. Geological structure and lithology, which are probably the most important factors in determining the general trend and the detailed form of rock coasts.

It is not possible to provide more than a general overview of some of the main elements of coastal scenery. Most geomorphological texts make rather brief comments on the development of rock coasts. More detailed examinations of particular coastlines can be gleaned from geological survey memoirs, and from doctoral and masters' dissertations, but there is still a suprising lack of information on the development, distribution and form of, for example, coves, caves, blowholes, stacks, arches, geos, and other coastal features. Steers's (1964, 1973) survey of the coast of Britain, and Johnson's (1925) and Shepard and Wanless's (1971) work on the coast of the United States provide some useful information on the distribution, and sometimes on the origin, of coastal features, but much work remains to be done.

Bays and Headlands

Most texts make the assumption that headlands generally consist of more resistant rocks than those in the bays. Although this is probably correct, there has been little attempt to substantiate it, either by detailed observation of rock structure or by direct measurement of, for example, the compressive strength of the rock. The presence of headland–embayment sequences along a coast often reflects rather subtle differences in rock structure, which may not be apparent without careful measurement of such factors as joint density and orientation, and the dip, thickness, and lithology of the strata exposed at the base of the cliffs.

The occurrence of small bays and headlands is often the result of variations in rock structure, but larger bays and more prominent headlands often reflect differences in rock lithology. Most igneous rocks, high-grade metamorphics, and some massive carbonates are resistant to wave attack, and they commonly form headlands; whereas friable shales, sandstones, and other rocks with numerous bedding planes and closely spaced joints or faults can be cut back more rapidly to form embayments. Folding and block faulting can bring weaker beds into the zone of wave action between more resistant materials on either side. This situation explains, for example, the formation of the small Stradling Well Bay in southern Glamorgan, Wales (Fig. 11.1), where block faulting has brought down shaley Jurassic upper *bucklandi* beds to the base of the cliff between the more resistant lower *bucklandi* limestones (Trueman 1922). Along this coast, long, shallow bays are associated with gentle folds which bring the weak *angulata* strata into the cliff, between headlands formed in the *bucklandi* beds. Further west in the Gower Peninsula, Oystermouth, Oxwich, and Port Eynon Bays have developed where synclines cause the replacement at the cliff base of the resistant Carboniferous Limestones by the weaker Millstone Grit shales (George 1933). Igneous intrusions account for most of the headlands in western Pembrokeshire (Steers 1964). In part of northern Oregon, almost all the headlands are associated with igneous rocks, and the adjacent bays with weaker sedimentary materials (Byrne 1963, 1964). Within igneous coasts, the occurrence of small headlands and bays reflects variations in jointing, strength, and thickness of the rocks, and the number of flows exposed to wave action and subaerial weathering (Thomson 1979, Bodere 1981).

Many bays have formed where streams enter the sea, particularly where the outlet has been drowned. In some cases, despite a relative rise in sea level, bays have failed to develop, or have subsequently been eliminated.

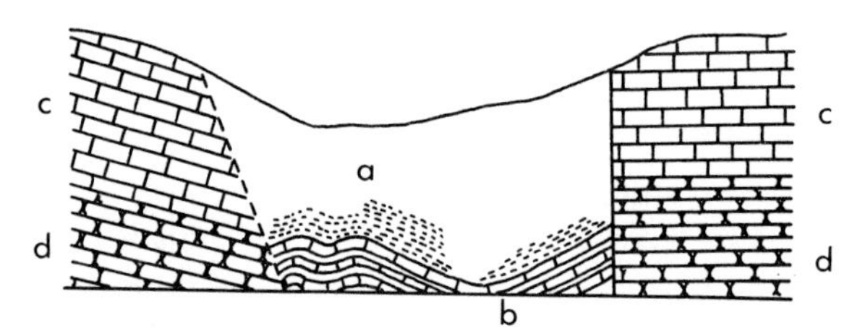

Figure 11.1. Stradling Well, a shallow fault-controlled cove in Jurassic limestones and shales in Glamorgan, Wales. (*a*) is the upper *bucklandi* shales with nodules, (*b*) represents upper *bucklandi*, thin limestones and shales, (*c*) is *bucklandi* limestones, and (*d*) is fairly massive lower *bucklandi* limestones (after Trueman 1922).

This may be because of the large amounts of sediment brought down by large rivers, or by streams with steep gradients flowing from coastal mountains (Shepard and Wanless 1971). The coastline can also be fairly regular when there is little or no stream erosion, and when the rock is fairly homogeneous. At the Seven Sisters of Sussex, England, for example, there are several hanging chalk valleys along a remarkably straight coastal section. Hanging waterfalls or cascades develop in areas where streams have been unable to cut downwards in pace with cliff retreat or with a fall in relative sea level. Arber (1911) has described some fine examples in northern Devon, England. Some of these have resulted from the capture by cliff retreat of the stream flow at some point above its former mouth, leaving the lower section as a dry valley.

When a coast consists of a series of alternating headlands and narrow bays, wave refraction causes energy to be concentrated on the promontories and dissipated in the bays, which can acquire further protection by the deposition of beach material. It has been suggested that if the coast consists of rocks of equal resistance to wave action, differential erosion will gradually cut back the headlands more quickly than the bays, eventually producing a fairly smooth shoreline (Davis 1912, Johnson 1919). The presence of a stream outlet, however, often produces considerable variation in the height of the cliffs in the local area. High cliffs occur at the interfluves, and low or non-existent cliffs at the valley mouth, which may also be in weaker rocks if the stream course is geologically controlled. Faster wave erosion should therefore take place in the valleys than on the higher cliffs on either side. This situation may serve to perpetuate the irregular form of an inherited coastline (Small 1970).

Coasts are usually irregular because they consist of rocks of markedly different resistance to erosion. Many headlands resist wave attack because their structure and lithology are less amenable to breakdown by mechanical wave action than the intervening materials in the bays. This situation can arise because of faulting or folding, which bring rocks of variable resistance to the water level. A number of workers have speculated that in such cases, wave erosion must eventually produce a coast which is in a state of equilibrium, where the recession of the resistant rocks of the exposed headlands is equal to the recession of the weaker rocks in the sheltered bays. Once equilibrium has been attained, the irregular shape of the coast would be maintained through time as it continued to retreat landwards. This possibility was recognized and briefly discussed by Johnson (1919) and others, but was overlooked for example, by Froget (1963), who thought that a crenulated basalt and sandstone coast in the south of France would eventually be straightened and smoothed by wave action. Johnson apparently considered that the topographic irregularities associated with variable resistance to wave attack would be restricted to a series of simple but distinct curves. It seems logical, however, to suppose that the amplitude of the coastal crenulations would increase with the difference in the resistance of the rocks in the headlands and bays, and decrease with the distance between the headlands because of the protection afforded to the bayheads from waves from variable directions. Komar (1976) also noted that wave refraction and other factors determine that there must be a maximum permitted shore relief, and Muir-Wood (1971) used a mathematical model to provide a rudimentary estimate of the depth of a bay at equilibrium.

Although the possibility that crenulated coastlines trend towards an equilibrium form is an interesting concept, the question arises whether there has been enough time for this to develop on rock coasts since the sea reached its present level. Much of the erosive work which would be necessary to cut an equilibrium coast on a tectonically stable landmass would have been accomplished over a number of interglacial periods in the middle and upper Pleistocene, when sea level was similar to today's. Some work, however, would still be required to trim the inherited forms of the last interglacial, when sea level was several metres above its present level.

Unfortunately, there are little reliable data on rates of cliff recession (see Chapter 8), and only a few studies have been made on cliff erosion along crenulated coasts. On the chalk coast near Margate in southeastern England, maps surveyed over a period of 50 years indicate that the shallow bays are still retreating more quickly than the headlands (Wood 1968). In northeastern Yorkshire in northern England, cliff recession in the bays is two to three times greater than on the headlands (Agar 1960). In both areas, the data suggest that equilibrium has not yet been attained, although as with all short-term measurements, we cannot be sure that the data are necessarily representative of longer time periods, particularly in situations where large rock falls and slides make infrequent but major contributions to cliff retreat. A few studies provide some evidence that cliffs are retreating at approximately the same rate along their lengths. In southern England, monthly surveys of the position of the chalk cliff top over a 12-year period, showed that roughly parallel cliff retreat had occurred along a minor headland–embayment sequence (May 1971). In southeastern Japan, air photographs covering a 14-year interval, showed that parallel cliff retreat took place along the irregular Byobugaura coast. A similar situation was noted along the crenulated Taitomisaki coast, by comparing maps representing an 85-year period (Sunamura and Tsujimoto 1982, Sunamura 1973). None the less, the records are presently too scarce, in some cases unreliable, and certainly too short to determine whether crenulated coasts cut in rocks of variable resistance to erosion are in equilibrium with the processes acting on them.

Caves, Arches, Stacks, and Related Features

Mechanical wave erosion is particularly sensitive to variations in rock structure. Small bays, inlets and narrow gorges develop along joint and fault planes, and in the fractured and crushed rock produced by faulting (Fig. 11.2). Stacks, arches, caves and narrow inlets are often found in close association with each other, on coasts which have well-defined planes of

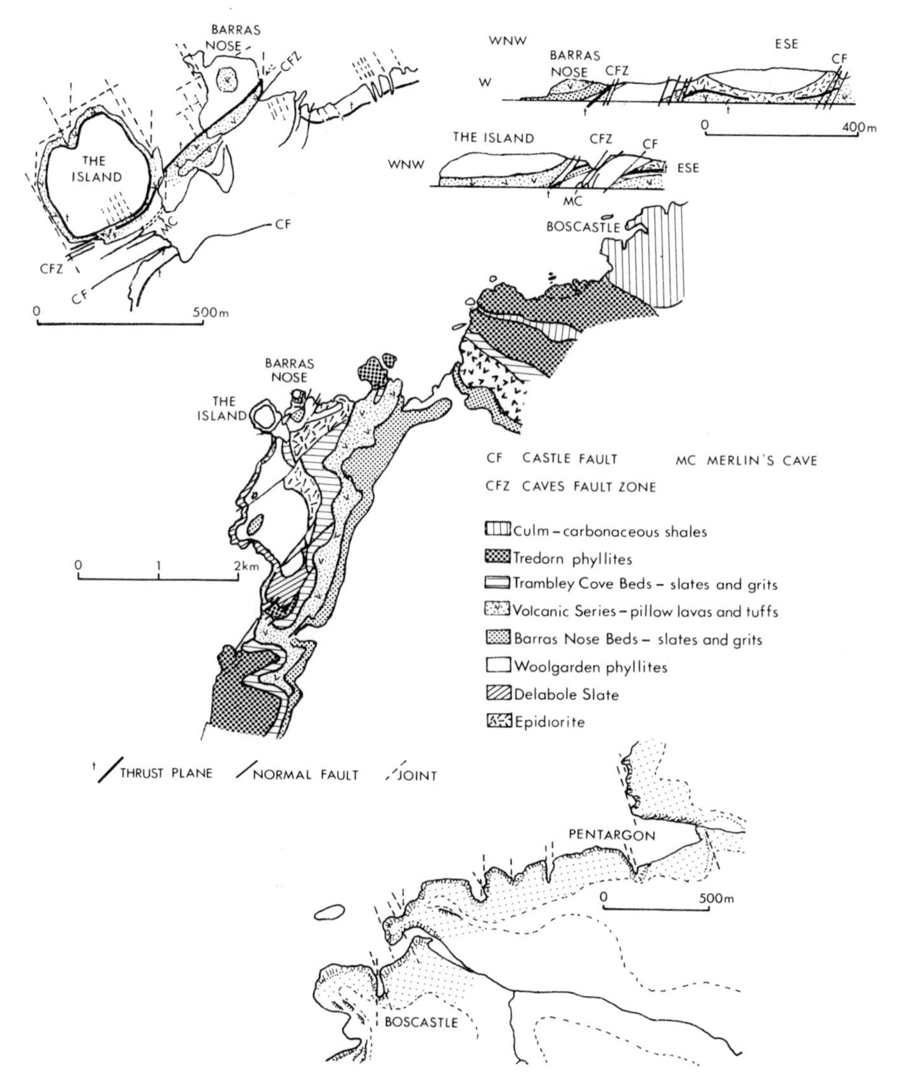

Figure 11.2. Structural and lithological control of the coast around Tintagel, north Cornwall, in southwestern England (after Wilson 1952).

weakness (see, for example, Jutson 1927). The rock itself, however, must also be strong enough to stand as high, near-vertical slopes, and as the roofs of caves, tunnels, and arches. These features are therefore less likely to develop in very weak rock types, or in rocks with a very dense joint pattern. If the joints or other planes of weakness are close together, long, narrow inlets develop in a similar way to the geos of northern Scotland. Small bays can form if the weaknesses are further apart (Baker 1943). The dip of the planes of weakness affect the occurrence and form of erosional features. Geo-like gorges with vertical sides are common in many horizontally bedded rocks, where the joint planes are essentially vertical. In steeply dipping rocks, where the planes of weakness are usually inclined, geos may fail to develop, or will be much more irregular in shape (Steers 1962a). Although joints and faults account for most narrow inlets, some are formed in other ways. On igneous rock coasts, for example, troughs and inlets can be the result of the differential erosion of dykes; and on the basaltic coast of Victoria, Australia, long channels separating islands and reefs from the mainland have been attributed to collapsed lava tunnels (Gill 1973a). Marine gorges may develop through the collapse of the roofs of tunnels. Depending upon the strength of the rock, the inclination of the planes of weakness (including the bedding planes), and the height of the cliff, however, the gorges may have always been open to the sky. This is particularly true of fairly low-cliffed coasts, where the faults or joints are nearly vertical.

Some marine caves are the roofed, landward extensions of gorges which probably developed through collapse (Jutson 1927). These caves appear to have formed by tunnelling below the platform level, whereas others, which probably represent the majority, have developed closer to the limit of wave action at, or somewhat above, the high tidal level (Gill 1973a). Some true marine caves are quite large. Painted Cave on Santa Cruz Island, off the coast of California, has been cut along a fault in volcanic flows. It is up to 24 m high, and boats are able to travel about 152 m into it. The floor of this cave is as much as 12 m below sea level in places (Emery 1954). Wave-cut caves, found at the back of the strandflat in western Norway, are up to 100 m long (Holtedahl 1984). On the shores of the Caspian Sea, one cave is up to 60 m long and 8 m high (Zenkovitch 1967), and two caves on King Island off Tasmania, are 80 and 40 m long (Goede *et al.* 1979). Most marine caves are much smaller, and many, particularly in resistant rocks, are not truly contemporary, although they may have been modified by present wave action (see, for example, Tratman 1971). Amino acid analysis suggests that caves in the Channel Islands and those associated with the Patella Beach on the Gower Peninsula in south Wales are at least as old as the last interglacial period (Keen *et al.* 1981, Davies 1983). Speleothems have developed in caves in folded metamorphic rocks on

King Island, Tasmania, as a result of water seeping down through aeolianite at the cliff top. Uranium series dating suggests that these caves are also at least as old as the last interglacial period, and similar results have been obtained from elsewhere in Tasmania and southern Australia (Kiernan 1979, Goede *et al*. 1979).

The occurrence of marine caves is usually determined by joints, faults, breccias, schistosity planes, unconformities, irregular sedimentation, the internal structures of lava flows or other structural weaknesses; and their form, which can be tunnel or dome-shaped, reflects the number and inclination of these structures (Moore 1954). Caves are particularly common in places where the rock is strongly jointed, as on the island of Staffa off Scotland, where Fingal's Cave has developed in columnar basalts. Very large caves, however, can be associated with fairly insignificant faults (Guilcher 1958a). Panzer (1949) found that very steep to vertically dipping bedding planes or other lines of weakness are the most favourable for cave formation, although flat-roofed forms develop in horizontal strata. Marine caves often develop where two sets of joints cross at right angles.

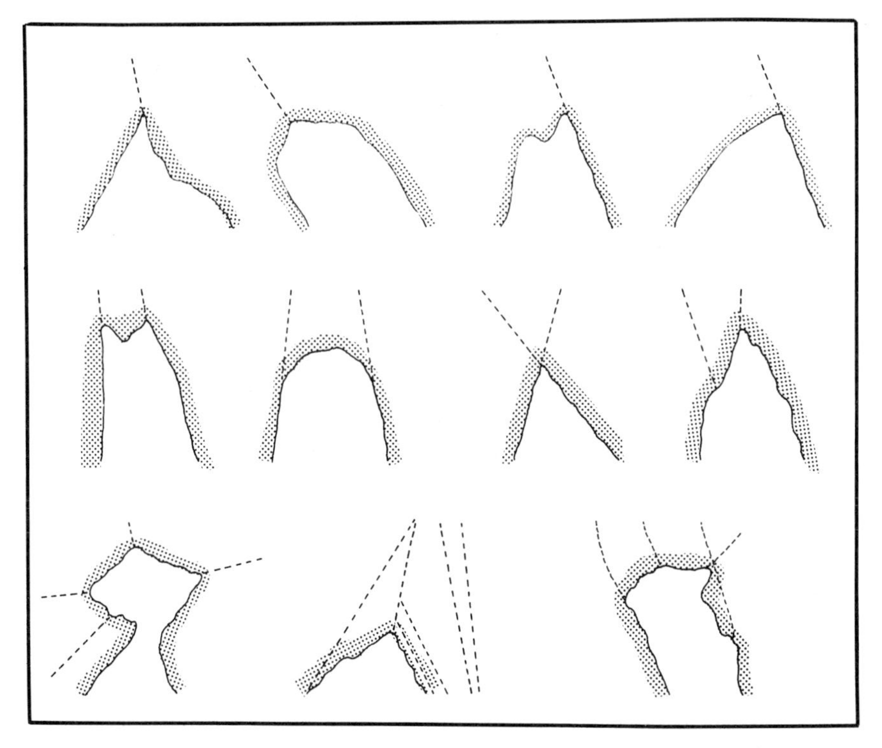

Figure 11.3. Examples of the joint control of the coast of southeastern Italy (after Zezza 1981).

One of the most detailed examinations of the relationship between joint character and cave form and occurrence has been made along the Cape Gargano limestone coast (Fig. 11.3) of southeastern Italy (Zezza 1981). It was found that a single, inclined joint tends to form caves with an inverted V-shape, whereas two parallel joints produce a cave approximating the shape of a rectangle or a parallelogram. Where two joints cross, a triangular cave may form, with its apex at the crossing point and the diverging joint planes forming its walls. More irregular caves are related to the influence of several joints.

Flemming (1965) found that vertical bedding or jointing produces tall, narrow caves. Other caves can be associated with igneous intrusions, where wave action has eroded dykes which are weaker than the country rock, and where basic veins have been etched out in granites. As discussed previously (Chapter 10), some caves in limestone regions have been formed by underground solution, and later inherited and modified by wave action as a result of cliff retreat and changes in relative sea level (eg. Baker 1943). This may be the case on the Port Campbell coast of Victoria, Australia, where wave splash and spray have excavated caves in limestones close to the cliff top, about 30 m above sea level (Hills 1971). Most sea caves are cut by wave action, particularly during storm periods. Pear-shaped caves in Swedish Archaean rocks for example, have been excavated along fissures widened at the base by cobble abrasion (Sjoberg 1981). On the Isle of Thanet in southeastern England, rectangular caves related to faults or joints in the chalk extend up to several metres above the high tidal level. The height of the caves, as measured from floor to roof, increases to the east, and they are deepest on the headlands, possibly reflecting differences in the exposure to storm wave action (So 1965). However, Wood (1968) believed that the occurrence of caves on the headlands of Thanet is the result of the deposition of sand in narrow clefts, which concentrates abrasion along a narrow rock zone. He argued that the presence of sand at the cliff foot along the whole length of the wide bays caused them to retreat more quickly and at roughly the same rate, thereby preventing the formation of caves.

Tunnels and natural arches sometimes develop from marine caves. At Tintagel in northern Cornwall, for example, two through tunnels have developed along a fault zone. One of these, Merlin's Cave (Fig. 11.2), connects the two sides of an island below the castle ruins (Wilson 1952). Many authors have commented that marine arches develop on promontories which can be attacked by waves from two sides. This view envisages arches developing from the coalescence of two caves driven into the headlands from opposing sides (Johnson 1919, 1925, Zenkovitch 1967). Shepard and Kuhn (1983) have noted that there has been little attempt to describe and explain the various types of marine arches found around the world, a comment which might, incidently, be equally valid for caves, stacks, arches, and other related coastal features.

Old photographs have been used by a number of workers to illustrate the development and destruction of arches (Johnson 1925, Byrne 1963, Shepard and Wanless 1971, Gill 1973a). Shepard and Kuhn used a large number of old photographs, covering a period of about a hundred years, to investigate the origin and development of arches around La Jolla, California. Numerous arches in this area developed as a result of the exposure to wave action of an irregular, recently block-faulted and uplifted coast. The only arch now remaining developed as the result of erosion in a bay intersecting the head of a cave. It was found that surge channels, related to an indentation of the coast which piles up the approaching waves on one side of a point, also cut arches, by causing the penetration of the lower, weaker rocks in an otherwise resistant cliff face. Although the photographs showed only one example of an arch developing as a result of erosion on two sides of a point, the literature suggests that this is common in other areas of the world. Magnificent arches have developed at the seaward and landward ends of headlands at Port Campbell, Australia, where joint planes in the limestones can be attacked from two sides. The waves are so vigorous in this area that some arches have been able to form between 6 and 12 m above sea level, and one, formed by the combined action of storm waves and surface drainage down sinkholes, is 27 m from the cliff edge, at an elevation 12 m above the sea (Baker 1943).

A large number of arches appear to have developed from surge channels, and others have formed from the erosion of two sides of corrosional caves (Shepard and Kuhn 1983). Flemming (1965) recognized that arches tend to develop in rocks which dip at less than 45°, although there are examples, on the Gower Peninsula of south Wales for example, in rocks which dip at higher angles. Although massive, nearly horizontal rocks with a few well-defined joints are particularly suitable for the formation of arches, dipping rocks are essential for the development of others. The Green Bridge of Wales, Pembrokeshire, has resulted from the fairly gentle dip which brings thinner, more easily eroded rocks down to the base of a small promontory, below the thicker, more resistant rocks on either side (Steers 1962a).

Marine caves can be enlarged by abrasion and by hydraulic action, involving the compression of air forced into cavities in the rock. Experiments with model caves have shown that high 'shock pressures' are generated against a cave roof when a lense of air is trapped between the wave and the rock surface. These pressures are greatest when the still water level is flush with the cave roof, and when the period of the wave is the same as the resonant period of the water in the cave. These pressures enlarge the cave and can lead to the eventual collapse of its roof (Flemming

1965). If a cave becomes connected to the surface through a joint or fault-controlled shaft, a blowhole or gloup can develop through which fountains of spray are blown out each time a large breaker surges into the tunnel below during storms and high tidal periods (Wilson 1952, Shepard and Wanless 1971). The presence of pebbles around the tops of blowholes in northeastern Scotland testifies to the enormous pressures generated during storms in narrow marine caves (Steers 1962a). Coasts which have experienced a relative rise in sea level, which would aid the erosion of the cave roof, are perhaps more likely to have blowholes than are tideless coasts where the caves are produced at the high sea level (Flemming 1965). Some large blowholes, as at Punakaiki in the northwestern part of South Island, New Zealand, have developed where the sea has invaded solutional tunnels and sinkholes formed along the joint planes in limestone rocks. The floors of these tunnels can be below sea level, as a result of their formation during a glacial low sea level phase (Baker 1943). Larger-scale collapse of part of a cave roof produces pits which are open to the sky. These have been described in vertically jointed quartzite in Brittany, where some occur at the junction of two or more tunnels connected to the sea (Schulke 1968).

Locally more resistant elements of a retreating coast are isolated as rock stacks in bays or off headlands. Stacks usually become separated from the mainland as a result of erosion along joints, faults, or other planes of weakness, in rocks which are otherwise fairly resistant to erosion, although the possible effect of wave refraction, surge channels, and other patterns of wave and current movement have not been adequately considered. The magnificent stacks of Duncansby Head in northeastern Scotland have developed in flagstones which have pores sealed by lime, making them essentially impenetrable to air and water. Water can penetrate easily along the deep, vertical joint planes which cross at high angles, however, cutting the rock into more or less rectangular sections (Steers 1962a). The particular circumstances which allow stacks to develop while the rest of the cliff is retreating more rapidly could explain why stacks often form quite quickly, but are sometimes persistent once they have become isolated from the mainland (see, for example, Gill 1973a). In some cases the lack of sand for abrasion can also contribute to the slow erosion of stacks off headlands (Wood 1968). The presence of a strong local joint pattern often causes arches to be formed in the stack itself (Fig. 11.4). Stacks, such as the Twelve Apostles on the Port Campbell coast of Victoria, Australia (Fig. 11.5) may be the limbs of former arches (Baker 1943), although in other cases they have developed directly from erosion of the cliff face, without any intermediate stage. Strong vertical jointing makes it unlikely, for example, that the isolated pinnacles and stacks which stand out from

Figure 11.4. Arch cut in stack-island in Miocene limestones, clays and shales on the Port Campbell coast of Victoria, Australia.

Figure 11.5. The Twelve Apostles, a series of limestone stacks on the Port Campbell coast of Victoria, Australia.

the columnarly jointed sills of Skye in Scotland, and on Brier's Island in the Bay of Fundy in eastern Canada (Fig. 11.6), were ever the limbs of arches.

Some stacks reflect differences in the resistance of the cliff to wave action, as a result of folding or differences in lithology. Although stacks in sedimentary rocks are probably most common where the strata are essentially horizontal, the Needles of the Isle of Wight in southern England are the result of folding, which has made the chalk in this area more resistant to wave action. On the aeolianite coasts of Victoria, Australia, small stacks have formed where the rock has been highly indurated by sea water (Gill 1973a). The irregular, tubular stacks of Sheringham, in eastern England, are the result of the wave excavation of old solution pipes, which were hardened by secondary lithification by organic acids in percolating

Figure 11.6. Stack in columnarly jointed basalts on Brier's Island, Bay of Fundy, Nova Scotia.

groundwater (Burnaby 1950). Small stacks or 'chimneys' have also formed from solution pipes in the aeolianite of Point Peron in Western Australia (Fairbridge 1950). Dykes and hard masses of tuff, agglomerate, and lava from old volcanic vents form irregular stacks on the sedimentary rock coasts around St Andrews in eastern Scotland (Steers 1962a). On northern Sakhalin Island, large basaltic stacks may have been produced by the removal of tuffs intercalated between the lava flows (Guilcher 1980). Stacks can also be formed from the largest blocks in rock falls (Zenkovitch 1967), and from tropical karst submerged by the postglacial rise in sea level (Guilcher 1980).

Elevated Marine Erosion Surfaces

Gently sloping marine terraces are a conspicuous element of many rock coasts. They often truncate dipping rock strata, and can extend over a variety of rock types. They are usually terminated seawards by the contemporary marine cliff, and landwards by either a gentle break in slope which gradually passes into higher elevations, or by a degraded bluff which represents a former sea cliff. It is possible in many areas to recognize a flight of terraces rising in stages inland, although the older surfaces may have been greatly dissected by streams and other subaerial mechanisms. Sea caves, stacks, and shallow water marine and terrestrial deposits are found on some terraces, often at considerable elevations above the present level of the sea. The presence of marine terraces is responsible for the occurrence of flat-topped cliffs, and they facilitate contemporary marine erosion by reducing regional cliff height, and therefore the amount of debris that must be removed by wave action. The gradient of these terraces is generally similar to contemporary shore platforms, but many are far wider than the platforms being cut at the present level of the sea.

It is neither the intention of this chapter to fully discuss the distribution of elevated coastal plateaux around the world (see Guilcher 1974a, and references in Chapter 7), nor to consider the origin, occurrence, and chronology of narrow benches (often associated with raised beaches) several metres above the present level of the sea (see Chapter 7 and Chapter 9). This chapter will briefly discuss the occurrence of elevated marine planation surfaces in a few well-studied areas, in order to more fully consider the way in which they may have developed. Although many investigators have described the occurrence of emerged coastal plateaux in many parts of the world, few have considered the precise manner in which they may have developed.

Numerous planation levels have been identified in southeastern and western Britain. In a purely statistical study, Hollingworth (1938) recognized principal surfaces in northern and western Britain at 131, 223 to 244,

305 to 326, and 610 m above sea level. Secondary levels were found at 98, 168 to 174, and 274 to 280 m, and in most areas between 344 and 357 m. High level surfaces represented by roughly concordant summit elevations and other erosional remnants (see for example, summaries in Linton 1964, and Steers 1964), may be of subaerial or marine origin, or possibly a combination of the two. There has been considerable debate over the origin and existence of these surfaces, but as they are generally found in upland areas, well removed from the coast, these high upland surfaces need not concern us further here.

At lower elevations, gently seaward-inclined surfaces occupying the cliff tops and coastal hinterlands are generally considered to be the product of marine erosion. They are particularly prominent in southwestern and western Britain. Barrow (1908) identified a terrace on Bodmin Moor, Cornwall, between 229 and 244 m above sea level, and Dewey (1908) recognized it near Tintagel, Cornwall. Green (1936) supported the contention that this surface was formed by wave action. Hollingworth (1938) found that the 131, 229, and 305 m levels are well represented in Devon and Cornwall. Based upon detailed work in the field, Balchin (1937) was able to identify three main surfaces in northern Cornwall, between about 73 and 87 m, 97 and 131 m, and 182 and 250 m. On Exmoor he found marine planation surfaces lying between 61 and 85 m and between 91 and 131 m (Balchin 1952).

Several workers have attempted to divide the coastal Welsh plateau into 61 m and 122 m stages (Goskar and Trueman 1934, Miller 1935). The approximately 61 m platform is prominent in south Wales in the Vale of Glamorgan (Jones 1911, Driscoll 1958), the Gower Peninsula, and southern Pembrokeshire. The 122 m platform is also well developed in Glamorgan, and in southwestern Wales, and a higher surface at about 183 m has been traced over a wide area in southern and western Wales (North 1929, Miller 1937). In north Wales, Greenly (1919) recognized the presence of surfaces between 61 and 91 m (The Menaian), at 123 m, and at 152–183 m. Detailed studies have established the presence of platforms at numerous additional levels along the coastal hinterlands of Wales (Brown 1950, Driscoll 1958). Yates (1965) has shown in graphical form the results of work conducted on the plateau surfaces of Wales. This shows that there is almost no elevation, up to about 762 m, which has not been claimed to be part of a marine erosion surface, although the most frequently reported elevations tend to cluster around 61, 122, and 183 m.

In southwestern and western Scotland, numerous horizontal surfaces, up to elevations of about 700 to 800 m, have been attributed to marine planation (Hollingworth 1938, George 1955, 1960, Jardine 1959, 1966). In the western Southern Highlands, Jardine (1959) recorded nine surfaces, of undetermined origin, at between 60 and 853 m. In northeastern Scotland, Walton (1963) only accorded a marine origin to five surfaces between 60 and 300 m.

In Ireland, the best developed planation surfaces are found along the southern coast in Cork and Waterford counties. Miller (1939a, 1955) considered that there are two main terrace levels, at elevations between 61 and 122 m and 185 and 246 m. There is some morphological evidence of other levels. In the Drum Hills of southwestern Waterford for example, marine surfaces extend up to 213 m. Orme (1964) recognized eight strandlines between 64 m and 183 m, consisting of erosional terraces with gradients of 2° or less and widths up to about 2 km. In Munster, six surfaces have been distinguished on the Clare Plateau between 46 m and 262 m. Those near the coast, and particularly the lowest one between 46 and 61 m, may be marine (Sweeting 1955). In Northern Ireland, Proudfoot (1954) found several surfaces in the Mourne Mountains, some of which might be marine, including one at between 91 and 122 m, which he thought may be the equivalent of the Trevena stage at a similar elevation in northern Cornwall (Balchin 1937).

Emergent marine platforms are prominent features in California and elsewhere along the tectonically mobile Pacific coast of North America (Shepard and Wanless 1971). Most of the lower and younger terraces have a thin covering of beach or shallow marine sediment, covered in turn by much thicker alluvial or talus fan deposits derived from the older surfaces and bluffs above. These thick, non-marine deposits obscure the underlying terrace topography and create problems with terrace stratigraphy (Birkeland 1972, Muhs 1979).

Terraces in Baja California, Mexico, are often more extensive, better preserved, and more continuous than those further north in the less arid areas of California. The early Pleistocene marine limit in northern Baja has been uplifted to more than 300 m in some areas, but depressed close to sea level in others. Multiple marine terraces are found below the marine limit, where there has been appreciable uplift (Orme 1972, 1974). There are three prominent terraces in central Baja. The broadest lies between 54 and 58 m above sea level, and is slightly tilted. It is up to 4.5 km wide, and slopes seawards 0.5 to 2°. A much narrower terrace, up to 300 m wide, lies at elevations between 24 and 31 m. An essentially untilted platform about 7 m above sea level, is up to 1 km wide in some places (Woods 1980).

In the La Jolla area in southern California, terraces are found at elevations of up to 244 m. The prominent, widely distributed Nestor terrace, between 25 and 30 m above sea level, attains widths of 300 to 1,500 m in some areas. The younger Bird Rock terrace is found below the Nestor in some areas, but it is rarely more than 10 m wide. Both terraces have been raised tectonically, although they are still largely horizontal along the coast (Ku and Kern 1974, Kern 1977). Thirteen terraces have been identified up to 451 m above sea level in the Palos Verdes Hills, where some are between 6.4 and 8 km in width (Woodring *et al*. 1946). On San Clemente Island off San Diego, a prominent flight of up to 22 terraces extends all the way up to the summit at 549 m (Lawson 1893, Olmsted 1958, Shepard and Wanless 1971, Muhs and Szabo 1982, Ridlon 1972, Muhs 1983). A series of marine terraces has also been described further north on San Nicolas Island (Vedder and Norris 1963). Five or six terraces have been identified in the San Clemente area on the mainland (Buffington and Moore 1963).

On the Malibu coast in Los Angeles County, Davis (1933) recognized two prominent emerged marine terraces, generally at elevations between about 12 and 40 m, and 61 and 76 m, which he termed the Dume and the Malibu, respectively. Other workers have identified a third intermediate terrace between about 46 and 54 m above sea level, which Birkeland (1972) has termed the Corral. South of Santa Barbara, warped marine terraces extend up to 427 m above sea level. The broadest platform, which is up to 0.5 km wide, is also the lowest, rising from about 64 m in the west to 174 m in the east (Putnam 1937). Upson (1949, 1951) recognized 22 terraces between 1.5 m and 488 m above sea level in the Santa Barbara area. At least five terraces were identified below 61 m, beneath a cover consisting of about 0.6 m of marine sands and gravels, and between 3 and 9 m of alluvial fan deposits. The marine sediments contain shells, and, as elsewhere, the rock platform surface is pitted by pholad borings. The older, dissected terraces above 61 m lack a sedimentary cover, and are distinguished largely on the basis of accordant summits. Further north, at Santa Cruz, south of San Francisco, five prominent, sediment-covered terraces are found between 31 and 259 m above sea level (Bradley 1957, 1958, Sorensen 1968, Bradley and Griggs 1976). The lowest terrace in this area is up to 3.2 km wide.

Origin

Since Ramsey (1846) first enunciated the theory of extensive marine planation, workers have been debating the relative contribution of marine and subaerial agencies to the development of coastal plains and upland erosional surfaces. Few investigators would deny the marine origin of at least some of the gentle, seaward-sloping surfaces which occupy coastal hinterlands, although it is still difficult to account for the formation of surfaces which can be up to 10 kilometres or more in width.

Many early workers believed that very extensive marine surfaces could have been cut over long periods of time while sea level was stationary. Ramsey (1846) considered that with unlimited time, there is essentially no limit to the amount of land which could be reduced to the average height of the sea. Green (1877) concurred with this view, although he thought that planation would be accomplished closer to the low tidal level. A number of American workers (Davis 1896, Gulliver 1899, Fenneman 1902) attributed extensive marine planation during periods of stationary sea level to erosion at considerable depths beneath the water surface. Johnson (1919) believed that marine planation could be effective at depths as great as 183 m (600 feet). In support of this contention that continents could be planated by wave action, he cited as examples of extensive planation the strandflat of Norway, which is up to 64 km wide, and the extensive plain on the east coast of India, which attains a width of 80 km in places. In Britain, Miller (1939b) accepted that marine abrasion surfaces can develop down to depths of 30 m. Balchin (1937) considered that the wide erosion surfaces of northern Cornwall were cut by waves during periodic stillstands of the sea. An explanation similar to Balchin's has been used to account for the *rasas* of eastern Asturias and western Santander in Spain (Hernández-Pacheco 1950).

Other workers have argued that cliff recession will eventually cease when the energy of the incoming waves is completely consumed in crossing a wide shore platform. Von Richthofen (1882) proposed that a rising relative sea level is essential for the formation of extensive marine surfaces, and this view was supported by De Lapparent (1898, 1900), De Martonne (1909), and others (see for example, references cited by Johnson 1919). Davis (1933) recognized that a rising sea is conducive to effective erosion, and most modern workers generally concur with this contention (Zenkovitch 1967, King 1963, 1972). Bradley and Griggs (1976) concluded that the form of the offshore zone in central California is consistent with the effects of a rising sea rather than with the concept of deep marine abrasion.

A rising relative sea level facilitates wave erosion by deepening water depth at and close to the cliff foot, so that wave energy remains sufficient to erode the platform and cliff. Nevertheless, as Sparks (1972) has noted, even if sea level were stationary, a platform could not fully dissipate wave energy indefinitely. Because breaking waves would serve to erode and thereby lower the platform, increased water depth and bottom slope would ultimately permit renewed wave attack on the cliff base. This would, however, be a slow process, and perhaps the greatest objection to the view that extensive marine planation surfaces have been cut during periods of stationary sea level is the lack of evidence of long stillstands during the Pleistocene, and possibly the Tertiary.

Trenhaile (in press) has used a mathematical wave erosion model (equation 4, chapter 9) to study the effect of sea level changes on rock coasts. The model was essentially concerned with wave erosion within the intertidal zone as sea level alternatively rose and fell in response to ice sheet decay and growth. Sea level data from the middle and upper

Pleistocene were used in the model runs (Fig. 11.7). The width of the erosion surface increased with the rate of sea level rise, and decreased with the rate of sea level fall. This result is consistent with the observation that platforms in California are widest where there has been the least tectonic uplift (Bradley and Griggs 1976). As predicted by King (1963) and modelled by Scheidegger (1970), constantly rising sea levels produced a continuously sloping surface. When sea level was steadily falling, a linear slope, parallel to the original surface, was sometimes formed, passing at its base into a wave-cut platform of constant width. This situation was also predicted by King (1963) and Scheidegger (1970), although it only occurred in this investigation when the rate of falling sea level was much greater than the mean for the last glacial period. As noted by Hollingworth (1938), the erosion surface was usually wider when the regional or original slope of the land was low, but the difference was found to be slight. Coastal erosion was particularly rapid when sea level rose at the end of each glacial period. A simulated marine denudation surface between 1 and 3 km in width (depending on rock hardness and wave energy), was produced at the end of one glacial–interglacial cycle. This surface was several times wider at the end of runs representing five glacial–interglacial cycles over the last 600,000 years, and up to 10 km wide when the land subsided at the rate of about 1 m every 10,670 years. The width and gradient (0.5 to 2°) of these erosion surfaces were similar to those of elevated platforms around the world. The concave shape of the simulated profiles was also similar to the offshore zones and emerged marine surfaces in California and Britain (King 1963, Bradley and Griggs 1976, Sorensen 1968, Driscoll 1958). The results generally support the contention that the concave profiles of the marine platform and terraces of California are the result of development during slow submergence.

Most of the simulated erosion surface was cut by waves within the intertidal zone, rather than by deep submarine erosion, even though the greater part of the surface was largely submerged during the interglacial periods. Most erosion occurred when sea level was rapidly rising, at the end of each glacial period. The model therefore provides an explanation for the way in which very wide marine erosion surfaces can be formed.

It is generally assumed that a stepped series of terraces reflect periodic stillstands, or slow changes in the elevation of the sea. This may not be true in all areas. Miller (1939b) suggested that platforms separated by small vertical intervals could be remnants of a single, inclined planation surface. The segments of the platform could have been isolated from each other by fortuitous cliffing, or by subsequent dissection and denudation. Several Soviet workers have proposed that a series of terraces can be formed when there is a uniform change in sea level (Markov 1934, Popov 1957). Zenkovitch (1949) noted that when the land rises steadily, climatically

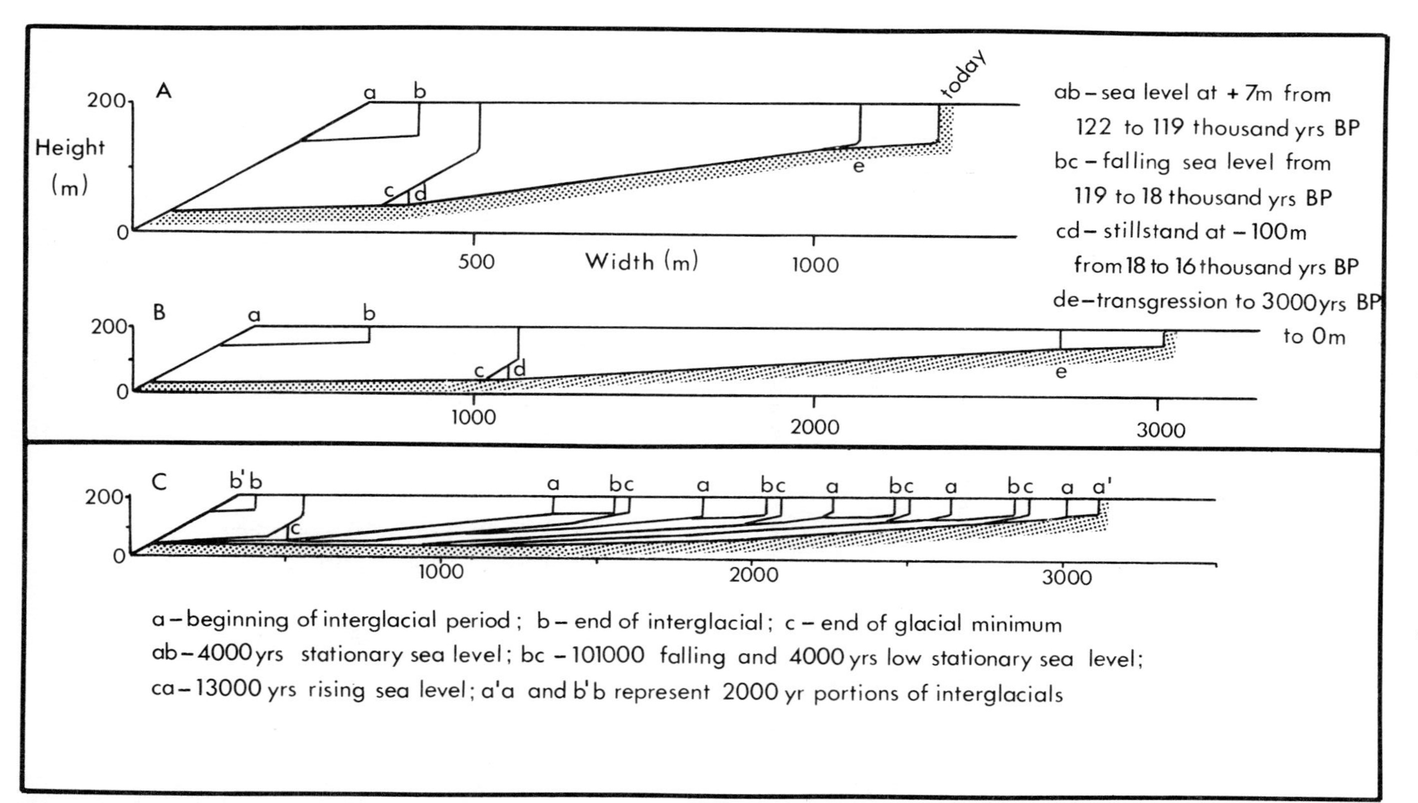

Figure 11.7. Examples of model runs simulating the development of erosional coastal profiles with changing sea level over the last interglacial–glacial cycle (A and B), and over five cycles representing the last 600 ka (*C*). Runs were made for different combinations of wave intensity and rock hardness (after Trenhaile, in press).

induced changes in the amount of sediment brought down to the coast by rivers vary the amount of protection afforded to the cliff and platform. Platforms may only have been cut when the rivers were supplying a small amount of sediment. Variations in wind regime, in wave strength and direction, and in the area and shape of the sea basin, would also affect the formation of marine terraces. Zenkovitch (1967) also argued that when the bottom is free of sediment, the submarine slope is convex; when the land is rising at a uniform rate, wave action will only become sufficient to cut marine cliffs and platforms when the deeper and more steeply sloping portions of the submarine slope are raised into the nearshore zone. Although mathematical modelling failed to produce any evidence that uniformly changing sea level can produce marine terraces (Trenhaile, in press), it suggested that a terrace could under some circumstances develop when sea level falls at a declining rate (Trenhaile and Byrne 1986; see also Scheidegger 1970).

In many areas, marine terraces record the effect of tectonic uplift, warping, and tilting as well as eustatic changes in sea level (Howard 1968). In California, for example, the older, higher terraces are often more tilted than the younger surfaces at lower elevations. The low terraces in some areas are essentially horizontal, suggesting that tilting had ceased before they were formed (Woods 1980).

Flemming (1965) discussed the effects of tectonic uplift and tilting on the form and spacing of marine terraces. If sea level is rising, the effect of steady uplift of the land would either be to reduce the vertical interval between the surfaces, or, if uplift is faster than the rise in sea level, to reverse the elevation-age relationship so that the highest terrace would be the oldest. If the land is subsiding, a rising sea level would cause an increase in the spacing between the terraces. Alternatively, if sea level is falling, uplift of the land would increase terrace spacing. Subsidence of the land while sea level is falling would cause compression of terrace spacing, or, if the rate of subsidence is greater than the fall in sea level, reversal in the elevation–age relationship so that the highest terrace would be the youngest. These conclusions have potentially important implications for variations in terrace form and spacing; and for the elevation–age relationship along tilted coasts, where relative sea level would have changed at different rates, or even in different directions. Flemming recognized that two flights of terraces, related to eustatic sea levels first falling then rising in a glacial–interglacial cycle, could be superimposed on each other on a tilted coast. Where the two sets cross, the terraces would be partly inherited, particularly wide, and apparently younger than they really are. This situation may sometimes explain why there are more terraces along one coastal profile than along another.

Ridlon (1972) thought that because the seaward gradients of the first two terraces on San Clemente Island are significantly greater than the slope of the submarine terrace of probable Wisconsin age, the area may have experienced seaward tilting or unbowing. A similar explanation has been proposed to explain why the higher terraces at Santa Cruz in California, and in New Zealand have a steeper seaward slope than the lower terraces (Bradley and Griggs 1976, Pillans 1983). Cotton (1942), however, suggested that steeply sloping terraces indicate that sea level was rising during their formation. This possibility is supported by mathematical modelling, which suggested that the slope of a marine erosion surface increases with the rate of sea level rise (Trenhaile, in press), although it does not preclude the possibility that the gradients have also been affected by tilting following terrace formation.

Age

It has been suggested that elevated marine terraces were cut in the early Pleistocene, when interglacial sea levels were much higher than today (Depéret 1906, Zeuner 1952a). Fairbridge (1961) considered that the sea was about 100, 50, and 18 m above its present level in the Aftonian, Yarmouthian, and Sangamon interglacials, respectively. This proposal cannot be easily reconciled with modern evidence, which suggests that interglacial temperatures and sea levels were similar to today's, at least in the middle and upper Pleistocene (Chapter 7). Even if all the ice presently on the Earth melted, sea level would only rise by 60 to 80 m; this did not happen in the Brunhes polarity epoch. Many platforms must therefore, have either been tectonically uplifted, or they must be older than about 700,000 years.

Correlation of terrace fragments in California and Mexico on the basis of their elevation is very uncertain because of tilting and deformation by faulting and arching. Furthermore, variations in fossil assemblages are largely attributable in this area to environmental rather than temporal factors. Attempts to derive morphologically significant parameters for terrace correlation have not been convincing (Verma 1973).

Modern dating techniques have recently been used to determine the approximate age, and to assist in the correlation, of elevated marine terraces. This is usually quite difficult because of the general lack of dateable materials on coastal terraces, other than those formed in coral. In New Zealand, terraces have been dated where they are underlain by volcanic ash and tephra (Yoshikawa *et al*. 1980, Pillans 1983), and where speleothems have developed in associated cave systems (Williams 1982). Although corals are quite rare in the marine terrace deposits of western North America, a few fragments have been reliably dated using uranium series methods. These corals have provided important calibration points for the dating of marine shells on other terraces, using the amino acid racemization method.

Rates of tectonic uplift on the west coast of North America, calculated from the age and elevation of marine terraces, generally fall within the range of 0.11 to 0.74 m per 1,000 years in areas characterized by strike–slip rather than by compressional tectonics (Ku and Kern 1974, Muhs and Szabo 1982, Muhs 1983). Assuming that the uplift rate has been essentially constant, terrace cutting would have been most effective when the relative sea level was increasing most rapidly. This would have been some time before the beginning of an interglacial period. Uranium series and amino acid data refer to the age of coral fragments and shells in the marine deposits rather than to the age of the terrace itself. These sediments were usually prograded seawards by the regressive sea, after the terrace had been cut (see, for example, Bradley 1957). The age of the terrace would therefore probably be at least a few thousand years greater than the age of its marine deposits, although the elevation of the cliff-platform junction, or shoreline angle, would correspond approximately to the high tidal level at the interglacial maximum.

The evidence suggests that there was extensive terrace cutting in western America during, or probably just before, the last interglacial period, when sea level was several metres above its present level. Ages of between about 116 and 130 ka BP have been obtained from uranium series analysis of corals on the Eel Point Terrace of San Clemente Island (Muhs and Szabo 1982), on the Nestor terrace around San Diego (Ku and Kern 1974), on a terrace at Cayucos California (Veeh and Valentine 1967), on terrace two on San Nicholas Island (Valentine and Veeh 1969) and on the Magdalena Terrace on Baja California Sur, Mexico (Omura *et al.* 1979). Uranium series dating of shells gave an age of about 130 ka for the Corral Terrace of Point Dume (Szabo and Rosholt 1969); and amino acid analysis provided similar estimates for the age of shells on the Tomatal terrace of central Baja, Mexico (Woods 1980), on terrace four in the Palos Verdes Hills (Wehmiller *et al.* 1977), and on terrace fragments in San Diego County (Karrow and Bada 1980).

One or more terraces are often found below the main interglacial surface. These terraces, which are usually narrower and less pronounced than the interglacial level, can be attributed to the sea level stands of isotopic stages 5c and 5a, at about 103 and 82 ka BP, respectively (see Chapter 7). On San Clemente Island, amino acid analysis has dated fossil shells on terraces lying below the Eel at between 80 and 105 ka BP (Muhs and Szabo 1982, Muhs 1983). Below the Nestor Terrace of San Diego, shells on the Bird Rock terrace have been estimated to be about 80 ka old (Kern 1977). Uranium series dating of shells provided an age of 85 ka for the deposits on the first terrace of San Pedros in the Palos Verdes Hills, and 105 ka for the Dume Terrace of Point Dume (Szabo and Rosholt 1969). The 69 ± 10 ka age of marine shells on the lowest marine terrace between Newport Beach and Laguna Beach, may, on this evidence, be a little low (Szabo and Vedder 1971).

The ages of several older terraces have also been estimated using a variety of techniques. On San Clemente Island, extrapolation of the uplift rate derived from the Eel Point terrace provided ages of up to between 1,250 and 1,500 ka for a flight of terraces, including possible representatives of isotopic stages 7, 9, 11, 15, and either 19 or 21 (Muhs and Szabo 1982). Amino acid analysis provided an estimate of between 415 and 575 ka for the age of the deposits on the fifth terrace on the west side of the island (Muhs 1983). On San Nicholas Island, uranium series dating of coral gave an age of 200 ka at one site on a high terrace (Valentine and Veeh 1969). In San Diego County, amino acid methods provided an estimate of 200 ka for the age of one terrace deposit, and somewhat more than 300 ka for another. The Linda Vista Terrace, which is now more than 100 m above sea level, was too old to be dated in this way, but was estimated to be of the order of one million years old (Karrow and Bada 1980).

The lack of dated marine deposits makes it extremely difficult to determine the age of the coastal platforms of Britain and Ireland. Those workers who consider them to be marine rather than subaerial surfaces have generally attributed them to periods of high relative sea level in late Tertiary, and particularly Pliocene, times (see discussion in Linton 1964, for example). Hollingworth (1938) believed that remnants of marine platforms can be identified at similar elevations throughout western and northern Britain. On the basis of this questionable assumption, he suggested that these unwarped platforms must have developed in the mid-Miocene or Pliocene, after the period of differential earth movement in the late Oligocene/early Miocene. Hollingworth considered that the platforms were cut by the sea during major or minor stillstands, which interrupted the progressive uplift of the land. Wood (1974) thought that the elevated platforms of Cornwall, Wales, and Ireland were cut during prolonged stillstands in the Oligocene and Miocene, before the cutting of the presently submerged, 15- to 18-km wide, late Miocene or early Pliocene surface. In the north of England and southern Scotland, it has been proposed that wave-cut surfaces formed on the fringes of an early Tertiary peneplain, which was first warped, faulted, and depressed, and then uplifted in stages (Sissons 1960). King (1963) has also supported the contention that marine erosion surfaces close to the present coast emerged as a result of Tertiary uplift of the land. Other workers have attributed platform formation to periodic halts of an otherwise progressively falling sea level (see, for example, Driscoll 1958), possibly in the period between some part of the Miocene and the Pleistocene (Steers 1964, 1981). Miller (1939a, 1955) considered that the marine flats of Waterford and Cork Counties, Ireland, were formed as a result of a Plio–Pleistocene transgression. It is also possible that some of the particularly wide marine surfaces may have been reoccupied by the sea on several occasions.

The lack of evidence for sea levels higher than a few metres above the present level in the middle and late Pleistocene (Shackleton and Opdyke 1973), suggests that the marine surfaces were either uplifted, or that they are older than about 700,000 years. Very wide marine platforms either require rapid transgressions of the sea, or extremely long periods of stationary sea level. Tentative sea level data for the Tertiary suggests that (Chapter 7):

1. The sea was 225 to 275 m above its present level from the early Palaeocene to the Oligocene (Vail and Hardenbol 1979).
2. It was about 100 m above today's in the middle Miocene (Glenie *et al*. 1968, Vail and Hardenbol 1979), and possibly 61 to 76 m in the late Miocene of the southern United States (Colquhoun and Johnson 1968, Alt 1968).
3. It was from 43 to 100 m above today's level in the early Pliocene (Colquhoun and Johnson 1968, Lietz and Schmincke 1975, Vail and Hardenbol 1979, Blackwelder 1981).

Although Tertiary marine transgressions were about one hundred times slower than the last postglacial rise in sea level, marine erosion could have been prolonged over much greater periods of time. Furthermore, sea level in the Tertiary appears to have been approximately stationary for much longer periods than at any time in the Pleistocene (Vail and Hardenbol 1979). Two Tertiary periods may have been particularly suitable for the formation of very extensive marine surfaces:

1. The nearly 35-Ma period between the beginning of the Palaeocene and the late Oligocene, when sea level was about 220 to 280 m above today's level.
2. The approximately 11-Ma period extending from the late Oligocene to the middle Miocene, when there was an interrupted rise in sea level from more than 100 m below to about 100 m above its present level.

There is also some recent evidence that the landscape of Devon and Cornwall is, in part, a result of a Cretaceous marine inundation. Scanning electron microscopic examination of some Tertiary deposits has identified reworked marine-polished and aeolian-scratched quartz grains from Cretaceous sediments. The evidence suggests that Dartmoor, Bodmin Moor, and Exmoor were emerged while the metasedimentary plateaux were covered by the Albian–Cenomanian sea (Coque-Delhuille and Gosselin-Vuilleumier 1984).

Mathematical modelling implies that very wide erosion surfaces were cut by waves in the intertidal zone during the Pleistocene, largely as a result of rapid transgressions associated with periods of deglaciation. Flights of Pleistocene terraces develop along tectonically mobile collison coasts, but along more stable, unglaciated trailing-edge coasts, high marine terraces must have been cut in the Tertiary, during long periods of stable sea level or a slow transgression. In glaciated regions, however, the effect of glacio-isostatically induced changes in relative sea level must also be considered.

Strandflats

Strandflats are low and often remarkably horizontal coastal platforms. They have been reported from a number of areas in the high latitudes of the northern and southern hemispheres, although the term has become almost synonymous with the extensive coastal plains on the western and northwestern coasts of Norway. The Norwegian strandflat ranges in width from a few kilometres up to as much as 60 km in a few areas. It is not always conspicuous throughout this region, and in areas of high relief it can consist of a partially submerged flight of terraces, rather than one sharply defined planation surface. Steep mountain fronts often rise abruptly from its inner margin, which is usually about 40 to 50 m above the present level of the sea. Large areas of the strandflat are in resistant crystalline and metamorphic rocks, and it extends over numerous low islands, skerries, and peninsulas, dissected in places by deep fiords and sounds (Fig. 11.8).

Nansen (1922) thought that the Norwegian strandflat was largely produced by frost action, the most important role of the waves being to remove the loose, weathered material. Partly based upon the incorrect assumption that saline water is an unsuitable medium for frost action (see Chapter 5), he proposed that the platform was levelled by accelerated disintegration around the margins of the snow and ice patches at the cliff foot, where a supply of fresh water is readily available. Tidally induced frost cycles could also have played a secondary role, and ice transportation would have provided some assistance to the waves in removing the loose debris. Other workers have agreed that frost played an important role in the development of strandflats, especially in fiords and in other sheltered areas where wave action is weak (Rekstad 1915, Vogt 1918, Grønlie 1924, Nordenskjold 1928, Guilcher 1974a, Moign 1974a,b). A novel explanation has been provided by Tietze (1962), who proposed that the strandflat was levelled by frost action operating under shelf ice, as it rose and fell with the tides. The growth of segregation ice in bedrock cracks under seasonal

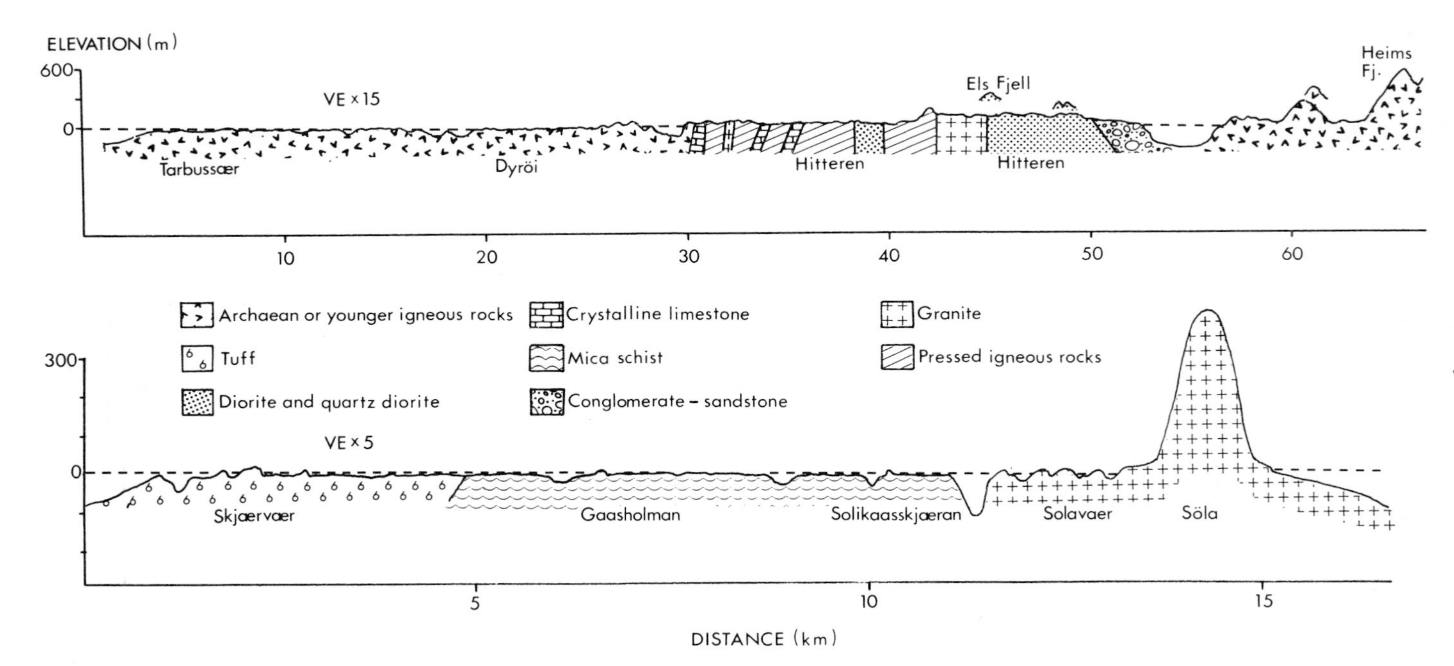

Figure 11.8. Profiles across the strandflat in western and northern Norway (after Nansen 1922).

ice is thought to have been the probable cause for the formation of rock platforms around a former ice-dammed lake in southern Norway. It was suggested that an analogous process occurs on polar coasts, associated with either the ice foot or shore-fast sea ice, although in a coastal environment, the effect of tidal fluctuations in sea level needs to be considered (Matthews *et al*. 1986).

Several workers have promoted shore ice as an effective erosive agent in the coastal zone. According to Nansen (1922), Hansen believed that the Norwegian strandflat was produced by waves and by drifting ice which cut away the sides of the fiords and sounds. The Micmac terrace in the St Lawrence Estuary of eastern Canada has also been partly attributed to the sawing action of floating ice driven back and fore by tidal currents (Goldthwait 1933). More recently, Fairbridge (1977) proposed that the strandflat was cut by sea ice and by frost action, and he argued that the shore platforms of western Europe, the western United States, southern Australia, and Patagonia were formed in a similar way during early Pleistocene glacial periods, when the sea level is thought to have been similar to today. This view has been challenged by workers in the high latitudes. Nansen (1904, 1922) thought that floating ice was an ineffective erosional agent, and Dionne (1978) considered that Fairbridge's theory was based upon an exaggeration of the role of ice in high latitudes. Bird (1967) found that floating ice is rarely important in the Canadian Arctic, and Moign (1974a,b) has rejected its possible role in the formation of platforms in northwestern Spitsbergen.

Nansen (1922) believed that conditions were most suitable for the formation of the Norwegian strandflat before and during each glacial period, when frost and wave action were facilitated by the cold, moist climate. He recognized that glaciation could also have contributed to the formation of the strandflat by eroding and dissecting the coast, thereby increasing the length of the shore which would then be subjected to frost and wave action. Several other workers have proposed that cirque and piedmont glacial erosion played a much more direct role in strandflat formation (O. Holtedahl 1929, Dahl 1946, H. Holtedahl 1960, 1962).

The relative contribution of wave and frost action in the formation of strandflats has yet to be determined (Trenhaile 1983a). Numerous workers have accorded a more important role to wave erosion than was recognized by Nansen (1922). In an earlier work which emphasized mechanical wave erosion, Nansen (1904) noted that the Norwegian strandflat is always narrow in the outer portion of fiords, and absent in the inner portions where wave action is weakest. Nansen (1922) and Moign (1974a,b) believed that wide strandflats only form where waves and frost can operate together. It has been argued that the Norwegian strandflat is primarily a marine abrasion surface (Reusch 1894, Richter 1896, Vogt 1907, Rekstad

1915, Johnson 1919, Strøm 1948, Grønlie 1951), and that the Antarctic and Icelandic strandflats and platforms are essentially the products of marine processes (John and Sugden 1971, 1975, Thorarinsson *et al.* 1959). Wave erosion and transportation were also considered to have been essential for the formation of the strandflats of Spitsbergen (Hoel 1909, Peach 1916, Werenskiold 1953, Moign 1974a,b, Guilcher 1974a).

It is presently possible only to speculate on the processes responsible for the formation of the strandflat. Not only is there little morphological data on the strandflats and shore platforms of high latitude regions, but we know little about the processes associated with frost, ice, and waves in cool climates. Experimental work suggests that saline sea water is a particularly effective medium for frost weathering, although high, critical levels of saturation must be attained in the rocks during freezing, if damaging internal pressures are to be generated. Recent laboratory and field work suggests that when rocks are first exposed to the air by a falling tide, saturation levels can be high enough for effective frost action in the lower portions of the intertidal zone. An additional supply of fresh water from the ice foot or from melting snow may be necessary to attain critical saturation near the high tidal level (Trenhaile and Mercan 1984). This work, however, did not consider the effect of frost operating in water-filled joints and bedding planes. It should also be emphasized that experimental work suggests that frost action is only effective in certain types of rock, particularly in those which are fine-grained, of low tensile strength, and with many prominent lines of weakness. Rocks which have usually been found to be durable to frost include unfoliated metamorphics, and igneous rocks such as basalts, coarse-grained granites, peridotites, and pyroxenites.

The morphology, and especially the gradient, of the shore platforms of eastern Canada vary according to the tidal range. In Gaspé, Québec, where the tidal range is 2.25 to 3.5 m, platform gradient in shales, argillites, and greywackes ranges from 0.1 to 0.6°. In western Newfoundland the tidal range is about 2 m, and the platforms are essentially horizontal in limestones, sandstones, siltstones, mudstones, and shales. In Fundy, Nova Scotia, however, the tidal range is from 13.5 to more than 15.5 m, and the platforms, formed in conglomerates, sandstones, shales and basalts, slope seawards at between 2 and 4.5° (Trenhaile 1980). Also within the Bay of Fundy, platform gradients increase to the northeast as the tidal range increases (Thomson 1979). This relationship between platform gradient and tidal range is consistent with the results from southern Britain, even though frost and ice are much more important in eastern Canada. Even if, as Nansen (1922) suggested, frost action around the fringes of the ice foot is the dominant mechanism for cliff retreat in cool regions, it does not explain how it can cut platforms which slope seawards by several degrees in macrotidal areas, yet are essentially horizontal in areas with a microtidal range. The relationship between gradient and tidal range in southern Britain has been attributed to the way in which wave energy is expended within the intertidal zone (Trenhaile and Layzell 1981), and it is therefore difficult to escape the conclusion that the same explanation is valid in eastern Canada, and possibly in other cool regions. This, of course, is not to deny that frost plays an important role in the retreat of cliffs in high latitudes, and possibly in the modification of the rough and essentially horizontal platform which would be produced by ice foot disintegration, but it is difficult to explain how it can produce gently sloping surfaces in areas with high tidal ranges. Wave action which is strong enough to effectively remove the products of frost action at the cliff base is probably strong enough to assume a significant role in eroding the frost weakened rocks on the shore platform. Wide strandflats may largely owe their development, in otherwise resistant material, to frost weakening of the cliff and platform rocks. Guilcher (1974a) for example, while fully recognizing the significance of wave erosion and transportation to the development of strandflats, emphasized that its efficacy is dependent upon the preparatory work of frost and related processes acting on the cliffs (Fig. 11.9). The final sculpturing of the platform surface, however, which largely determines its morphology, must, in part, be attributed to mechanical wave erosion, possibly during periods when sea level was rising, and to variations in the focus of wave attack as a result of isostatic and eustatic events.

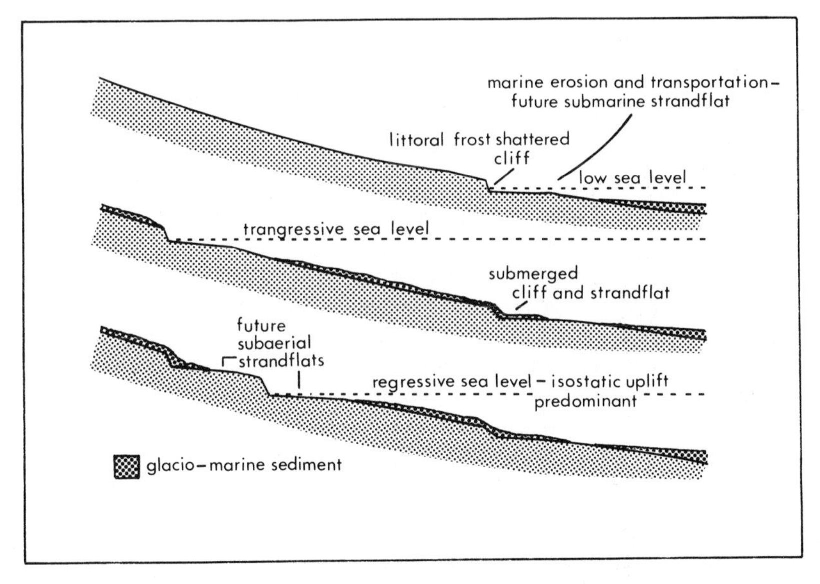

Figure 11.9. The origin of the strandflats of Spitsbergen (according to Guilcher 1974a).

References

Aartolahti, T. 1975: On the weathering cavities in the Rapkivi granite of Vehmaa, SW Finland. *Terra* **87**, pp. 238–44.

Abrahams, A. D. 1975: Some thoughts on the development of shore platforms south of Sydney. *Geogr. Bull.* **7**, pp. 83–4.

—— and Oak, H. L. 1975: Shore platform widths between Port Kembla and Durras Lake, New South Wales. *Austr. Geogr. Stud.* **13**, pp. 190–4.

ACI 1980: *Performance of Concrete in Marine Environments* (Am. Concrete Inst., Detroit), paper SP 65.

Ackermann, N. L. and Chen, P. H. 1974: Impact pressures produced by breaking waves. *Proc. 14th Conf. Coastal Eng.*, pp. 1778–83.

Adams, C. G., Benson, R. H., Kidd, R. B., Ryan, W. B. F. and Wright, R. C. 1977: The Messinian salinity crisis and evidence of late Miocene eustatic changes in the world ocean. *Nature* **269**, pp. 383–6.

Adey, W. H. and MacIntyre, I. G. 1973: Crustose coralline algae: a re-evaluation in the geological sciences. *Geol., Soc. Am. Bull.* **84**, pp. 883–904.

Agar, R. 1960: Postglacial erosion of the north Yorkshire coast from the Tees estuary to Ravenscar. *Proc. Yorks. Geol. Soc.* **32**, pp. 408–25.

Agassiz, A. 1895: A visit to the Bermudas in March 1894. *Bull. Mus. Compar. Zool. Harvard Coll.* **26**, pp. 209–81.

AGI 1957: *Glossary of Geology and Related Sciences* (Am. Geol. Inst., Washington).

Aharon, P., Chappell, J. and Compston, W. 1980: Stable isotope and sea-level data from New Guinea supports Antarctic ice-surge theory of ice ages. *Nature* **283**, pp. 649–51.

Ahr, W. M. and Stanton, R. J. 1973: The sedimentologic and paleoecologic significance of *Lithotrya*, a rock-boring barnacle. *J. Sed. Petrol.* **43**, pp. 20–3.

Alexander, R. S. 1932: Pothole erosion. *J. Geol.* **40**, pp. 305–37.

Alexandersson, T. 1969: Recent littoral and sublittoral high-Mg calcite lithification in the Mediterranean. *Sedimentology* **12**, pp. 47–61.

——. 1972: Intragranular growth of marine aragonite and Mg-Calcite: evidence of precipitation from super-saturated seawater. *J. Sed. Petrol.* **42**, pp. 441–60.

——. 1975: Etch patterns on calcareous sediment grains: petrographic evidence of marine dissolution of carbonate minerals. *Science* **189**, pp. 47–8.

——. 1976: Actual and anticipated petrographic effects of carbonate undersaturation in shallow water. *Nature* **262**, pp. 653–8.

Alexandre, J. 1958: Le Modèle quaternaire de l'Ardenne central. *Ann. Soc. Geol. Belg.* **81**, pp. 213–331.

Allard, M. and Tremblay, G. 1983: Les processus d'érosion littorale périglaciaire de la région de Poste-de-la-Baleine et des îles Manitounuk sur la côte est de la mer d'Hudson, Canada. *Z. Geomorph.* Suppl. Band 47, pp. 27–60.

Alley, N. F. 1979: Middle Wisconsin stratigraphy and climatic reconstruction, southern Vancouver Island, British Columbia. *Quat. Res.* **11**, pp. 213–37.

Alt, D. 1968: Pattern of post-Miocene eustatic fluctuation of sea level. *Palaeogeogr., Palaeoclim., Palaeoecol.* **5**, pp. 87–94.

Ambrosetti, P., Azzaroli, A., Bonadonna, F. P. and Follieri, M. 1972: A scheme of Pleistocene chronology for the Tyrrhenian side of central Italy. *Boll. Soc. Geol. Ital.* **91**, pp. 169–80.

American Society for Testing Materials (ASTM) 1952, 1953, and 1980: *Book of ASTM Standards* (Philadelphia).

Andersen, B. C. 1968: Glacial geology of western Troms, north Norway. *Norges Geol. Unders.* **256**, pp. 1–160.

Anderson, D. M. and Hoekstra, P. 1965: Migration of interlamellar water during freezing and thawing of Wyoming bentonite. *Soil Sci. Soc. Am. Proc.* **29**, pp. 498–504.

—— and Morgenstern, N. R. 1973: Physics, chemistry, and mechanics of frozen ground. In *Permafrost* (North American Contribution to the 2nd Int. Conf., Nat. Acad. Sci. Washington DC), pp. 257–88.

——, Tice, A. R. and Banin, A. 1974: The water–ice phase composition of clay/water systems: 1. The kaolinite/water system. US Army Corps Eng. *CRREL Res. Rept.* **322**, pp. 1–10.

Andreasen, J.-O., Baisgaard, K. and Larsen, P. F. 1978: Saltforvitring. *Georapporter* **5**, pp. 2–26.

Andrews, E. C. 1916: Shoreline studies at Botany Bay. *Proc. J. Roy. Soc. NSW* **50**, pp. 165–76.

Andrews, J. T. 1970: A geomorphological study of post-glacial uplift with particular reference to Arctic Canada. *Inst. Brit. Geogr. Spec. Publ.* **2**, pp. 1–156.

——. 1980: Progress in relative sea level and ice sheet reconstructions Baffin Island NWT for the last 125,000 years. In *Earth Rheology, Isostasy, and Eustasy*, ed. N.-A Mörner (Wiley, Chichester), pp. 175–200.

—— and Dugdale, R. E. 1970: Age prediction of glacio-isostatic strandlines based on their gradients. *Geol. Soc. Am. Bull.* **81**, pp. 3769–71.

—— and Retherford, R. M. 1978: A reconnaissance survey of sea levels, Bella Bella/Bella Coola region, central British Columbia coast. *Can. J. Earth Sci.* **15**, pp. 341–50.

——, Bowen, D. Q. and Kidson, C. 1979: Amino acid ratios and the correlation of raised beach deposits in south-west England and Wales. *Nature* **281**, pp. 556–8.

——, Mears, A., Miller, G. H. and Pheasant, D. R. 1972: Holocene Late Glacial maximum and marine transgression in the eastern Canadian Arctic. *Nature (Phys. Sci.)* **239**, pp. 147–9.

Anon. 1959: Freezing and thawing tests of concrete. *Highway Res. Board Spec. Rept. (Washington, DC)* **47**, pp. 1–67.

Anon. 1966: 'Ordered water' molecular pressures offer new clues to rock failure. *Eng. Mining J.* **167**, pp. 92–4.

Ansell, A. D. and Nair, N. B. 1969: A comparative study of bivalves which bore mainly by mechanical means. *Am. Zool.* **9**, pp. 857–68.

Arber, E. A. N. 1911: *The Coast Scenery of North Devon* (Kingsmead Reprints, Bath).

Arber, M. A. 1940: The coastal landslips of south-east Devon. *Proc. Geol. Assoc.* **51**, pp. 257–71. Reprinted in *Applied Coastal Geomorphology*, ed. J. A. Steers (Macmillan, London), pp. 138–53.

——. 1941: The coastal landslips of west Dorset. *Proc. Geol. Assoc.* **52**, pp. 273–83.

——. 1949: Cliff profiles of Devon and Cornwall. *Geogr. J.* **114**, pp. 191–7.

——. 1962: Coastal cliffs: report of a symposium. *Geogr. J.* **128**, pp. 303–20.

——. 1971: The plane of landslipping on the coast of south-east Devon. In *Applied Coastal Geomorphology*, ed. J. A. Steers (Macmillan, London), pp. 153–4.

——. 1973: Landslips near Lyme Regis. *Proc. Geol. Assoc.* **84**, pp. 121–33.

——. 1974: The cliffs of north Devon. *Proc. Geol. Assoc.* **85**, pp. 147–57.

Arkin, Y. and Michaeli, L. 1985: Shore- and long-term erosional processes affecting the stability of the Mediterranean coastal cliffs of Israel *Eng. Geol.* **21**, pp. 153–74.

Arnfeld, H. 1943: Damage on concrete pavements by wintertime salt treatment. *Statens Vaeginst* (Swed.) 66.

Arni, H. T. 1966: Resistance to weathering. *Am. Soc. Test. Mat. Stand. Spec. Tech. Publ.* **169A**, pp. 261–74.

——, Foster, B. E. and Clevenger, R. A. 1956: Automatic equipment and comparative test results for the four ASTM freezing-and-thawing methods for concretes. *Am. Soc. Test. Mat. Proc.* **56**, pp. 1229–56.

Arthur, M. A. 1979: Paleoceanographic events—recognition, resolution, and reconsideration. *Rev. Geophys. Space Phys.* **17**, pp. 1474–94.

ASTM. *See* American Society for Testing Materials.

Aubry, M. P. and Lautridou, J. P. 1974: Relations entre propriétés physiques, gélivité et caractères microstructuraux dans divers types de roches: craies, calcaires crayeaux, calcaire sublithographique et silex. *Centre de Géomorphologie, CRNS* Bull. **19**, pp. 7–16.

Baggioni, M. 1975: Les côtes du Cilento (Italie du sud): morphogenese littorale actuelle et heritée. *Méditerranée* **20**, pp. 35–52.

Bagnold, R. A. 1939: Interim report on wave pressure research. *J. Inst. Civil Eng.* **12**, pp. 202–26.

Baker, G. 1943: Features of a Victorian limestone coastline. *J. Geol.* **51**, pp. 359–86.

——. 1958: Stripped zones at cliff edges along a high-energy coast, Port Campbell, Victoria. *Proc. Roy. Soc. Vict.* NS. **70**, pp. 175–9.

Bakker, J. P., Kwaad, F. J. P. M. and Muller, H. J. 1968: Einige vorlaufige Bermerkungen uber Salz- und Tonsprengung besonders in Hinblink auf Granite. *Przeglad Geograficzny* **40**, pp. 387–99.

Balchin, W. G. V. 1937: The erosion surfaces of north Cornwall. *Geogr. J.* **90**, pp. 52–63.

——. 1946: The geomorphology of the north Cornish coast. *Trans. Roy. Geol. Soc. Cornwall* **17**, pp. 317–44.

——. 1952: The erosion surfaces of Exmoor and adjacent areas. *Geogr. J.* **118**, pp. 453–76.

Baltzer, F. 1970: Datation absolue de la transgression Holocène sur la côte ouest de Nouvelle-Calédonie sur des échantillons de tourbes à palétuviers: interprétation néotectonique. *Comptes Rendus Acad. Sci.* **271D**, pp. 2251–4.

Bandy, O. L. 1968: Cycles in Neogene paleoceanography and eustatic changes. *Palaeogeogr., Palaeoclim., Palaeoecol.* **5**, pp. 63–75.

Barbaza, Y. 1970: Morphologie des secteurs rocheux du littoral Catalan septentrional. *Mém. Docum. CNRS* NS 11, pp. 1–152.

Barbetti, M. and Flude, K. 1979: Geomagnetic variation during the late Pleistocene period and changes in the production of radioactive nuclides by cosmic rays interacting with the atmosphere. *Nature* **279**, pp. 202–5.

Bardach, J. E. 1961: Transport of calcareous fragments by reef fishes. *Science* **133**, pp. 98–9.

Barnes, H. and Topinka, J. A. 1969: Effect of the nature of the substratum on the force required to detach a common littoral alga. *Am. Zool.* **9**, pp. 753–8.

Barrell, J. 1920: The piedmont terraces of the northern Appalachians. *Am. J. Sci.* 4th Ser. **49**, pp. 227–58, 327–62, 407–28.

Barron, E. J., Thompson, S. L. and Schneider, S. H. 1981: An ice-free Cretaceous? Results from climatic model simulations. *Science* **212**, pp. 501–8.

Barrow, G. 1908: The high level platforms of Bodmin Moor and their relation to the deposits of stream-tin and wolfram. *Quart. J. Geol. Soc. London* **64**, pp. 384–96.

Bartrum, J. A. 1916: High water rock platforms: a phase of shoreline erosion. *Trans. NZ Inst.* **48**, pp. 132–4.

——. 1924: The shore platform of the west coast near Auckland: its storm wave origin. *Rept. Austr. Assoc. Sci.* **16**, pp. 493–5.

——. 1926: Abnormal shore platforms. *J. Geol.* **34**, pp. 793–807.

——. 1935: Shore platforms. *Proc. Austr. NZ Assoc. Adv. Sci.* **22**, pp. 135–43.

——. 1936: Honeycomb weathering of rocks near the shoreline. *NZ J. Sci. Tech.* **18**, pp. 593–600.

——. 1938: Shore platforms. *J. Geomorph.* **1**, pp. 266–8.

——. 1952: Comment on 'Marine erosion' by R. W. Fairbridge. *Proc. 7th Pacific Sci. Congr.* **3**, pp. 358–9.

—— and Turner, F. J. 1928: Pillow lavas, peridotites, and associated rocks from northernmost New Zealand. *Trans. NZ Inst.* **59**, pp. 98–138.

Battisti, C. and Alessio, G. 1957: *Dizionario Etimologico Italiano* (C. Barbèra, Florence).

Battistini, R. 1977: Estrans calcaires et grésocalcaires à Madagascar et dans les îles voisines. *Norois* **95**, pp. 165–72.

——. 1980: Les vasques étagées, formes curieuses des estrans gréso-calcaires au sud de Madagascar. *Mad. Rev. de Géogr.* **37**, pp. 63–86.

——. 1981: La morphogénèse des plateformes de corrosion littorale dans les grès calcaires (plateforme supériere et plateforme à vasques) et le problème des vasques, d'après des observations faites à Madagascar. *Rev. Géomorph. Dyn.* **30**, pp. 81–94.

—— and Guilcher, A. 1982: Les plates-formes littorales à vasques en roches calcaires: répartition dans le monde, mer Méditerranée non comprise. *Karst Littoraux*, Comité National Français de Geographie (Actes du Colloqium de Perpignan, 15–17 Mai, 1982) **1**, pp. 1–11.

Battle, W. R. B. 1960: Temperature observations in bergschrunds and their relationship to frost shattering. In *Norwegian Cirque Glaciers*, ed. W. V. Lewis, Roy. Geogr. Soc. Res. Ser. **4**, pp. 83–95.

Baulig, H. 1930: Le littoral Dalmate. *Ann. Géogr.* **39**, pp. 305–10.
Beach Erosion Board 1961: Shore protection planning and design. *Corps. Eng. Dept. Army (US)*, Tech. Rept. **4**, pp. 1–242.
Beaudoin, J. J. and MacInnes, C. 1974: The mechanism of frost damage in hardened cement paste. *Cement and Concrete Res.* **4**, pp. 139–47.
BEB 1961: *See* Beach Erosion Board.
Bedi, N. and Rao, K. L. V. R. 1984: Nature and evolution of a part of Saurashtra coast, Gujarat, India. *Z. Geomorph.* **28**, pp. 53–69.
Bell, J. M. and Clarke, E. de C. 1909: The geology of the Whangaroa subdivision. *NZ Geol. Surv. Bull.* **8**, pp. 30.
Belperio, A. P. 1979: Negative evidence for a mid-Holocene high sea level along the coastal plain of the Great Barrier Reef Province. *Marine Geol.* **32**, pp. M1–9.
Benayahu, Y. and Loya, Y. 1977: Seasonal occurrence of benthic-algae communities and grazing regulation by sea urchins at the coral reefs of Eilat, Red Sea. *Proc. 3rd Int. Coral Reef Symp. (Miami, Florida)* **1**, pp. 383–9.
Bender, M. L., Fairbanks, R. G., Taylor, F. W., Matthews, R. K., Goddard, J. G. and Broecker, W. S. 1979: Uranium-series dating of the Pleistocene reef tracts of Barbados, West Indies. *Geol. Soc. Am. Bull.* **90** (Part 1), pp. 577–94.
——, Taylor, F. W. and Matthews, R. K. 1973: Helium-uranium dating of corals from middle Pleistocene Barbados reef tracts. *Quat. Res.* **3**, pp. 142–6.
Bénézit, M. 1923: Essai sur les digues maritimes verticales. *Ann. Ponts Chauss.* **93**, pp. 125–59.
Benjamin, G. J. 1970: Diving into the blue holes of the Bahamas. *Nat. Geogr. Mag.* **138**, pp. 347–63.
Berard, L. 1927: La morphologie côtière de Marseille à Toulon et l'origine des calanques. *Ann. Géogr.* **36**, pp. 67–70.
Berger, A. L. 1976: Obliquity and precession for the last 5,000,000 years. *Astron. Astrophys.* **51**, pp. 127–35.
Berger, U. and Kohlhase, S. 1976: Mach-reflection as a diffraction problem. *Proc. 15th Conf. Coastal Eng.*, pp. 796–814.
Berggren, W. A. and Van Couvering, J. A. 1974: The Late Neogene. *Palaeogeogr., Palaeoclim., Palaeoecol.* **16**, pp. 1–216.
Berglund, B. E. 1971: Littorina transgressions in Blekinge, south Sweden: a preliminary survey. *Geol. Fören. Stockh. Förh.* **93**, pp. 623–52.
Berner, R. A. 1967: Comparative dissolution characteristics of carbonate minerals in the presence and absence of aqueous magnesium ion. *Am. J. Sci.* **265**, pp. 45–70.
——. 1975: The role of magnesium in the crystal growth of calcite and aragonite from seawater. *Geochim. Cosmochim. Acta* **39**, pp. 489–504.
Bertram, G. C. L. 1936: Some aspects of the breakdown of coral at Ghardaga, Red Sea. *Proc. Zool. Soc. London* **106**, pp. 1011–26.
Biberson, P. 1970: Index-cards on the marine and continental cycles of the Moroccan Quaternary. *Quaternaria* **13**, pp. 1–76.
Biésel, F. 1952: Equations générales au second ordre de la houle irrégulière. *Houille Blanche* **7**, pp. 372–6.
Bird, E. C. F. 1968: *Coasts* (Australian National University Press, Canberra).
——. 1970a: Shore potholes at Diamond Bay, Victoria. *Vict. Naturalist* **87**, pp. 312–8.

——. 1970b: The steep coast of Macalister Range, north Queensland, Australia. *J. Tropical Res.* **31**, pp. 33–9.
——. 1974: Pitted rocks at Jubilee Point, Victoria. *Vict. Naturalist* **91**, pp. 61–5.
——. 1977: Cliffs and bluffs on the Victoria coast. *Vict. Naturalist* **94**, pp. 4–9.
—— and Dent, O. F. 1966: Shore platforms on the south shore of New South Wales. *Austr. Geogr.* **10**, pp. 71–80.
—— and Guilcher, A. 1982: Observations préliminaires sur les récifs frangeants actuels du Kenya et les formes littorales associées. *Rev. Géomorph. Dyn.* **31**, pp. 113–25.
—— and Hopley, D. 1969: Geomorphological features on a humid tropical sector of the Australian coast. *Austr. Geogr. Studies* **7**, pp. 89–108.
—— and Rosengren, N. J. 1984: The changing coastline of the Krakatau Islands, Indonesia. *Z. Geomorph.* **28**, pp. 347–66.
——, Cullen, P. W. and Rosengren, N. J. 1973: Conservation problems at Black Rock Point. *Vict. Naturalist* **90**, pp. 240–7.
Bird, J. B. 1967: *The Physiography of Arctic Canada* (John Hopkins, Baltimore).
——, Richards, A. and Wong, P. P. 1979: Coastal subsystems of western Barbados, West Indies. *Geogr. Annlr.* **61A**, pp. 221–36.
Birkeland, P. W. 1972: Late Quaternary eustatic sea-level changes along the Malibu Coast, Los Angeles County, California. *J. Geol.* **80**, pp. 432–48.
——. 1974: *Pedology, Weathering, and Geomorphological Research* (Oxford University Press, Oxford).
Birot, P. 1954: Désagrégation des roches cristallines sous l'action des sels. *Comptes Rendus Acad. Sci.* **238**, pp. 1145–6.
——. 1968: *The Cycle of Erosion in Different Climates* (Batsford, London).
Bjerrum, L. and Jorstaad, F. 1968: Stability of rock slopes in Norway. *Norw. Geotech. Publ.* **79**, pp. 1–11.
Black, R. F. 1974: Late Quaternary sea level changes, Umnak Island, Aleutians: their effects on ancient Aleuts and their causes. *Quat. Res.* **4**, pp. 264–81.
——. 1980: Isostatic, tectonic, and eustatic movements of sea level in the Aleutian Islands, Alaska. In *Earth Rheology, Isostasy, and Eustasy*, ed. N. A. Mörner (Wiley, Chichester), pp. 231–48.
Blackwelder, B. W. 1980: Late Wisconsin and Holocene tectonic stability of the United States mid-Atlantic coastal region. *Geology* **8**, pp. 534–7.
——. 1981: Late Cenozoic marine deposition in the United States Atlantic coastal plain related to tectonism and global climate. *Palaeogeogr., Palaeoclim., Palaeoecol.* **34**, pp. 87–114.
——, Pilkey, O. H. and Howard, J. D. 1979: Late Wisconsin sea levels on the southeast US Atlantic Shelf based on in-place shoreline indicators. *Science* **204**, pp. 618–20.
Blake, J. A. 1969: Systematics and ecology of shell boring polychaetes from New England. *Am. Zool.* **9**, pp. 813–20.
—— and Evans, J. W. 1973: *Polydora* and related genera as borers in mollusc shells and other calcareous substrates. *Veliger* **15**, pp. 235–49.
Blake, J. F. 1893: The landslip at Sandgate. *Nature* **47**, pp. 467–9.
Blake, W. 1970: Studies of glacial history in Arctic Canada. I: pumice, radiocarbon dates, and differential postglacial uplift in eastern Queen Elizabeth Islands. *Can. J. Earth Sci.* **7**, pp. 634–64.

——. 1975: Radiocarbon age determinations and postglacial emergence at Cape Storm, southern Ellesmere Island, Arctic Canada. *Geogr. Annlr.* **57A**, pp. 1–71.

——. 1976: Sea and land relations during the last 15000 years in the Queen Elizabeth Islands, Arctic Archipelago. *Geol. Surv. Can. Paper* **76–1B**, pp. 201–7.

Blanc, J. J. 1975: Ecroulements de falaises et chutes de blocs au littoral rocheux de Provence occidentale. *Geol. Méditerranéenne* **11**, pp. 75–90.

—— and Molinier, R. 1955: Les formations organogènes construites superficielles en Méditerranée occidentale. *Bull. Inst. Océanogr. Monaco* **52**, pp. 1–26.

Blanchard, R. 1911: La côte de Provence. *La Géogr.* **24**, pp. 201–24.

Bloch, M. R. 1965: A hypothesis for the change in ocean levels depending on the albedo of the polar ice caps. *Palaeogeogr., Palaeoclim., Palaeoecol.* **1**, pp. 127–42.

Bloom, A. L. 1965: Coastal isostatic downwarping by postglacial rise of sea level. *Geol. Soc. Am. Spec. Paper* **82**, pp. 14.

——. 1967: Pleistocene shorelines: a new test of isostasy. *Geol. Soc. Am. Bull.* **78**, pp. 1477–94.

——. 1970: Paludal stratigraphy of Truk, Ponape, and Kusaie, Eastern Caroline Islands. *Geol. Soc. Am. Bull.* **81**, pp. 1895–1904.

——. 1971: Glacial-eustatic and isostatic controls of sea level since the last glaciation. In *The Late Cenozoic Glacial Ages*, ed. K. K. Turekian (Yale University Press, New Haven, Conn.), pp. 355–79.

——. 1980: Late Quaternary sea level change on south Pacific coasts: a study in tectonic diversity. In *Earth Rheology, Isostasy and Eustasy*, ed. N. -A. Mörner (Wiley, Chichester), pp. 505–16.

——, Broecker, W. S., Chappell, J., Matthews, R. K. and Mesolella, K. J. 1974: Quaternary sea level fluctuations on a tectonic coast: new $^{230}Th/^{234}U$ dates from the Huon Peninsula, New Guinea. *Quat. Res.* **4**, pp. 185–205.

Bodere, J.-C. 1981: Le rôle des influences structurales sur le tracé d'une côte rocheuse volcanique: l'exemple des strandflats du sud-est de L'Islande. *Géogr. Phys. Quat.* **35**, pp. 231–40.

Boekschoten, G. J. 1970: On bryozoan borings from the Danian at Fakse, Denmark. In *Trace Fossils*, ed. T. P. Crimes and J. C. Harper (Seel House Press, Liverpool), pp. 43–8.

Boney, A. D. 1966: *A Biology of Marine Algae* (Hutchinson, London).

Bortolami, G. C., Fontes, J. C., Markgraf, V. and Saliege, J. F. 1977: Land, sea, and climate in the northern Atlantic region during late Pleistocene and Holocene. *Palaeogeogr., Palaeoclim., Palaeoecol.* **21**, pp. 139–56.

Bosence, D. W. J. 1983: Coralline algal reef frameworks. *J. Geol. Soc. London* **140**, pp. 365–76.

Bourcart, J. 1957: *L'Erosion des Continents* (Armand Colin, Paris).

Bourdier, F. 1971: Les anciens rivages quaternaires en France: théories et réalités. *Quaternaria* **15**, pp. 105–10.

Bourman, R. P. and May, R. I. 1984: Coastal rotational landslump (Australian landform example no. 47). *Austr. Geogr.* **16**, pp. 144–6.

Boussinesq, J. 1877: Essai sur la théorie des eaux courantes. *Mém. l'Acad. Sci. Inst. Nat. France* **23**, p. 23.

Bowen, D. Q. 1978: *Quaternary Geology* (Pergamon, Oxford).

——. 1980: Antarctic ice surges and theories of glaciation. *Nature* **283**, pp. 619–20.

Bowman, G. M. 1970: A study of process in water-layer weathering on shore platforms on the south coast of New South Wales. University of Sydney, BA thesis.

Boyer, S. J. 1975: Chemical weathering of rocks on the Lassiter coast, Antarctic Peninsula. *Antarctica* **18**, pp. 623–8.

Bradley, W. C. 1957: Origin of marine-terrace deposits in the Santa Cruz area, California. *Geol. Soc. Am. Bull.* **68**, pp. 421–44.

——. 1958: Submarine abrasion and wave-cut platforms. *Geol. Soc. Am. Bull.* **69**, pp. 967–74.

—— and Griggs, G. B. 1976: Form, genesis, and deformation of central California wave-cut platforms. *Geol. Soc. Am. Bull.* **87**, pp. 433–49.

—— Hutton, J. T. and Twidale, C. R. 1978: Role of salts in development of granitic tafoni, South Australia. *J. Geol.* **86**, pp. 647–54.

——, ——, ——. 1979: Role of salts in development of granitic tafoni, South Australia: a reply. *J. Geol.* **87**, pp. 121–2.

Brancaccio, L., Capaldi, A., Pece, R., and Sgrosso, I. 1978: ^{230}Th- ^{238}U dating of corals from a Tyrrhenian beach in Sorrentine Peninsula (southern Italy). *Quaternaria* **20**, pp. 175–83.

Bray, J. R. 1974: Glacial advance relative to volcanic activity since 1500 AD. *Nature* **248**, pp. 42–3.

Bretz, J. H. 1960: Bermuda: a partially drowned, late mature Pleistocene karst. *Geol. Soc. Am. Bull.* **71**, pp. 1729–54.

Bridgman, P. W. 1912: Water, in the liquid and five solid forms, under pressure. *Proc. Am. Acad. Arts Sci.* **47**, pp. 441–558.

——. 1914: High pressures and five kinds of ice. *J. Franklin Inst.* **177**, pp. 315–32.

Brierly, W. B. 1965: Atmosphere sea-salts design criteria areas. *J. Environ. Sci.* **8**, pp. 15–23.

Brodeur, D. and Allard, M. 1983: Les plates-formes littorales de l'île aux Coudres, moyen estuaire du Saint-Laurent, Québec. *Géogr. Phys. Quat.* **37**, pp. 179–95.

Broecker, W. S. 1965: Isotope geochemistry and the Pleistocene climatic record. In *The Quaternary of the United States*, ed. H. E. Wright and D. G. Frey (Princeton University Press, Princeton, NJ), pp. 737–53.

——. 1966: Absolute dating and the astronomical theory of glaciations. *Science* **151**, pp. 299–304.

—— and Thurber, D. L. 1965: Uranium series dating of corals from Bahaman and Florida Key limestones. *Science* **149**, pp. 58–60.

—— and Van Donk, J. 1970: Insolation changes, ice volumes, and the O^{18} recorded in deep sea cores. *Rev. Geophys. Space Phys.* **8**, pp. 169–98.

——, Thurber, D. L., Goddard, J., Ku, T. L., Matthews, R. K. and Mesolella, K. J. 1968: Milankovitch hypothesis supported by precise dating of coral reefs and deep sea sediments. *Science* **159**, pp. 297–300.

Broikos, A. 1955: Calcul de la stabilité générale des brise-lames à la poussée de la houle rotationnelle. *Génie Civil* **132**, pp. 290–8.

Bromhead, E. N. 1972: Discussion of Wood (1971). *Proc. Inst. Civil Eng.* **53**, pp. 401–16.

Bromley, R. G. 1970: Borings as trace fossils and *Entobia cretacea* Portlock as an example. In *Trace Fossils*, ed. T. P. Crimes and J. C. Harper (Seel House Press, Liverpool), pp. 49–90.

——. 1978: Bioerosion of Bermuda reefs. *Palaeogeogr. Palaeoclimat. Palaeoecol.* **23**, pp. 169–97.

—— and Hanken, N.-M. 1981: Shallow marine bioerosion at Vardø, arctic Norway. *Bull. Geol. Soc. Denmark* **29**, pp. 103–9.

Brotchie, J. F. and Silvester, R. 1969: On crustal flexure. *J. Geophys. Res.* **74**, pp. 5240–52.

Brouwer, D. and Van Woerkom, A. J. J. 1950: The secular variations of the orbital elements of the principal planets. *Astron. Papers Am Ephemeris* part 2, vol. 13, Govt. Printing Off. pp. 83–107.

Brown, E. H. 1950: Erosion surfaces in north Cardiganshire. *Trans. Inst. Brit. Geogr.* **16**, pp. 51–66.

Brown, J. and Sellmann, P. V. 1966: Radiocarbon dating of coastal peat, Barrow, Alaska, *Science* **153**, pp. 299–300.

Browne, F. P. and Cady, P. D. 1975: De-icer scaling mechanisms in concrete. In *Durability of Concrete* (Am. Concrete Inst., Detroit), paper SP–47, pp. 101–19.

Brunauer, S. 1943: *The Adsorption of Gases and Vapors* (Princeton University Press, Princeton, NJ), vol. 1

Bruns, E. 1951: Berechnung des wellenstosses auf molen und wellenbrecher. *Jahrbuch Hafenbautechn. Ges.* **19**, pp. 92–158.

Brunsden, D. 1974: The degradation of a coastal slope, Dorset England. In *Progress in Geomorphology*, ed. E. H. Brown and R. S. Waters, Inst. Brit. Geogr. Spec. Publ. **7**, pp. 79–98.

——. 1979: Weathering. In *Process in Geomorphology*, ed. C. Embleton and J. Thornes (Arnold, London), pp. 73–129.

—— and Jones, D. K. C. 1980: Relative time scales and formative events in coastal landslide systems. *Z. Geomorph.* Suppl. Band **34**, pp. 1–19.

Bryson, R. A. and Goodman, B. M. 1980: Volcanic activity and climatic changes. *Science* **207**, pp. 1041–4.

Buckle, C. 1978: *Landforms in Africa* (Longman, London).

Budd, W. F. 1981: The importance of ice sheets in long-term changes of climate and sea level. In *Sea Level, Ice, and Climatic Change*, Proc. Canberra Symp. Dec. 1979. Int. Assoc. Hydrol Sci. Publ. **131**, pp. 441–71.

—— and Smith, I. N. 1981: The growth and retreat of ice sheets in response to orbital radiation changes. In *Sea Level, Ice, and Climatic Change*, Proc. Canberra Symp. Dec. 1979. Int. Assoc. Hydrol. Sci. Publ. **131**, pp. 369–409.

Buddington, A. F. 1927: Abandoned marine beaches in south-eastern Alaska. *Am. J. Sci.* **13**, pp. 45–52.

Buffington, E. C. and Moore, D. G. 1963: Geophysical evidence of the origin of gullied submarine slopes, San Clemente, California. *J. Geol.* **71**, pp. 356–70.

Burnaby, T. P. 1950: The tubular chalk stacks of Sheringham. *Proc. Geol. Assoc.* **61**, pp. 226–41.

Butawand, F. 1926: Conditions for building jetties with vertical sides. *14th Int. Congr. Navig. Rept.* **38**.

Butzer, K. W. 1962: Coastal geomorphology of Majorca. *Ann. Assoc. Am. Geogr.* **52**, pp. 191–212.

——. 1971: *Environment and Archeology* (Aldine, Chicago), 2nd ed.
——. 1975: Pleistocene littoral-sedimentary cycles of the Mediterranean basin: a Mallorquin view. In *After the Australopithecines*, ed. K. W. Butzer and G. L. Isaac (Mouton Publ., The Hague, Netherlands), pp. 25–71.
——. 1976: *Geomorphology from the Earth* (Harper and Row, New York).
—— and Cuerda, J. 1962: Coastal stratigraphy of southern Mallorca and its implications for the Pleistocene chronology of the Mediterranean Sea. *J. Geol.* **70**, pp. 398–416.
Byrne, J. V. 1963: Coastal erosion, northern Oregon. In *Essays in Marine Geology in Honor of K. O. Emery*, ed. T. Clements (University of Southern California, Los Angeles, Calif.), pp. 11–33.
——, 1964: An erosional classification for the northern Oregon Coast. *Ann. Assoc. Am. Geogr.* **54**, pp. 329–35.
Cadle, R. D. 1966: *Particles in the Atmosphere and Space* (Reinhold, New York).
Cady, P. D. 1969: Mechanisms of frost action in concrete aggregates. *J. Materials* **4**, pp. 294–311.
Cailleux, A. 1953: Taffonis et erosion alveolaire. *Cah. Géol. de Thoiry* 16–17, pp. 130–3.
——. 1962: Etudes de géologie au détroit de McMurdo (Antarctique). *Comm. National Français Rech. Antarctiques* **1**, pp. 1–41.
Calkin, P. E. and Cailleux, A. 1962: A quantitative study of cavernous weathering (taffonis) and its application to glacial chronology in Victoria Valley, Antarctica. *Z. Geomorph.* NF **6**, pp. 217–24.
—— and Nichols, R. L. 1972: Quaternary studies in Antarctica. In *Antarctic Geology and Geophysics*. ed. R. J. Adie (Universitetsforlaget, Oslo), pp. 625–43.
Calman, W. T. 1936: Marine animals injurious to submerged structures. (Revised by G. I. Crawford.) *Brit. Mus. Nat. Hist. Econ. Ser.* **10**, pp. 1–38.
Campbell, J. M. 1917: Laterite. *Min. Mag.* **17**, pp. 67–77, 120–8, 171–9, 220–9.
Cant, R. V. 1973: Jamaica's Pleistocene reef terraces. *Geol. Mijnbouw* **52**, pp. 157–60.
Carey, S. W. 1975: The expanding Earth—an essay review. *Earth Sci. Rev.* **11**, pp. 105–43.
Carr, A. P. and Graff, J. 1982: The tidal immersion factor and shore platform development. *Trans. Inst. Brit. Geogr.* NS **7**, pp. 240–5.
Carr, J. H. 1954: Breaking wave forces on plane barriers. *Calif. Inst. Techn. Hydrodyn. Lab. Rept.* E-11.3.
Carr, P. A. and Van der Kamp, G. S. 1969: Determining aquifer characteristics by the tidal method. *Water Resources Res.* **5**, pp. 1023–31.
Carriker, M. R. 1969: Excavation of boreholes by the gastropod *Urosalpinx*: an analysis by light and electron microscope. *Am. Zool.* **9**, pp. 917–33.
—— and Smith, E. H. 1969: Comparative calcibiocavitogy: summary and conclusions. *Am. Zool.* **9**, pp. 1011–20.
Carroll, D. 1970: *Rock Weathering* (Plenum Press, London).
Carry, C. 1953: Clapotis partiel. *Houille Blanche* **8**, pp. 482–94.
Carson, M. A. 1976: Mass-wasting, slope development and climate. In *Geomorphology and Climate*, ed. E. Derbyshire (Wiley, London), pp. 101–36.

—— and Kirkby, M. J. 1972: *Hillslope Form and Process* (Cambridge University Press, Cambridge).

Carter, R. W. G. 1982: Sea-level changes in Northern Ireland. *Proc. Geol. Assoc.* **93**, pp. 7–23.

Cathles, L. M. 1975: *The Viscosity of the Earth's Mantle* (Princeton University Press, Princeton, NJ).

——. 1980: Interpretation of postglacial isostatic adjustment phenomena in terms of mantle rheology. In *Earth Rheology, Isostasy, and Eustasy*, ed. N. A. Mörner (Wiley, Chichester), pp. 11–43.

CERC. *See* Coastal Engineering Research Center.

Challinor, J. 1931: Some coastal features of north Cardiganshire. *Geol. Mag.* **68**, pp. 111–21.

——. 1948: A note on convex erosion slopes with special reference to north Cardiganshire. *Geography* **33**, 27–31.

——. 1949: A principle in coastal geomorphology. *Geography* **34**, pp. 212–5.

Chapman, R. W. 1980: Salt weathering by sodium chloride in the Saudi Arabian desert. *Am. J. Sci.* **280**, pp. 116–29.

Chapman, V. J. and Chapman, D. J. 1973: *The Algae* (Macmillan, London).

Chappell, J. 1973: Astronomical theory of climatic change: status and problem. *Quat. Res.* **3**, pp. 221–36.

——. 1974a: Relationships between sea levels, O-variations and orbital perturbations during the past 250,000 years. *Nature* **252**, pp. 199–202.

——. 1974b: Geology of coral terraces, Huon Peninsula, New Guinea: a study of Quaternary tectonic movements and sea-level changes. *Geol. Soc. Am. Bull.* **85**, pp. 553–70.

——. 1974c: Late Quaternary glacio- and hydro-isostasy on a layered Earth. *Quat. Res.* **4**, pp. 429–40.

——. 1981: Relative and average sea level changes, and endo-, epi-, and exogenic processes on the Earth. In *Sea Level, Ice, and Climatic Change*, Proc. Canberra Symp. Dec. 1979. Int. Assoc. Hydrol. Sci. Publ. **131**, pp. 411–30.

——. 1983a: A revised sea-level record for the last 300,000 years from Papua New Guinea. *Search* **14**, pp. 99–101.

——. 1983b: Evidence for smoothly falling sea level relative to north Queensland, Australia, during the past 6,000 years. *Nature* **302**, pp. 406–8.

—— and Polach, H. A. 1976: Holocene sea-level change and coral-reef growth at Huon Peninsula, Papua New Guinea. *Geol. Soc. Am. Bull.* **87**, pp. 235–40.

—— and Thom, B. G. 1978: Termination of the last interglacial episode and the Wilson Antarctic surge hypothesis. *Nature* **272**, pp. 809–10.

—— and Veeh, H. H. 1978a: Late Quaternary tectonic movements and sea-level changes at Timor and Atauro Island. *Geol. Soc. Am. Bull.* **89**, pp. 356–68.

——. 1978b: ^{230}Th/^{234}U age support of an interstadial sea level of −40 m at 30,000 yr BP. *Nature* **276**, pp. 602–3.

——, Broecker, W. S., Polach, H. A. and Thom, B. G. 1974: Problem of dating upper Pleistocene sea levels from coral reef areas. *Proc. 2nd Coral Reef Symp. (Queensland, Austr.)* **2**, pp. 563–71.

——, Rhodes, E. G., Thom, B. G. and Wallensky, E. 1982: Hydro-isostasy and the sea-level isobase of 5,500 BP. in north Queensland, Australia. *Marine Geol.* **49**, pp. 81–90.

Chardonnet, J. 1948: Les calanques provençales, origin et divers types. *Ann. Géogr.* **57**, no. 308, pp. 289–97.

——. 1950: La côte française de Marseille à Menton. *Bull. Soc. Roy. Géogr. Egypte* **23**, pp. 185–261.

Charlesworth, J. K. 1957: *The Quaternary Era with Special Reference to its Glaciation* (Edward Arnold, London).

Charre, J.-P. and Lautridou, J.-P. 1975: Expériences de cryoclastie sur des grès et roches vertes. *Revue Géogr. Alpine* **63**, pp. 253–61.

Chastain, C. 1976: Hydrodynamic processes and sea cliff/platform erosion, Favorite Channel, Alaska. Univ Calif. Santa Cruz, Ph.D. thesis.

Chatterjee, S. P. 1961: Fluctuations of sea level around the coasts of India during the Quaternary Era. *Z. Geomorph.* Suppl. Band **3**, pp. 48–56.

Chave, K. E. and Schmalz, R. F. 1966: Carbonate–seawater interactions. *Geochim. Cosmochim. Acta* **30**, pp. 1037–48.

—— and Suess, E. 1970: Calcium carbonate saturation in seawater: effects of dissolved organic matter. *Limnol. Oceanogr.* **15**, pp. 633–7.

——, Deffeyes, K. E., Weyl, P. K., Garrels, R. M. and Thompson, M. E. 1962: Observations on the solubility of skeletal carbonates in aqueous solutions. *Science* **137**, pp. 33–4.

Chesher, R. H. 1969: Destruction of Pacific corals by the sea star *Acanthaster planci. Science* **165**, pp. 280–3.

Chesselet, R., Morelli, J. and Buat-Menard, P. 1972: Variations in ionic ratios between reference sea water and marine aerosols. *J. Geophys. Res.* **77**, pp. 5116–31.

Chorley, R. J. 1969: The role of water in rock disintegration. In *Water, Earth, and Man*, ed. R. J. Chorley (Methuen, London), pp. 135–55.

Christiansen, S. 1963: Morphology of some coral cliffs, Bismarck Archipelago. *Geogr. Tidsskr* **62**, pp. 1–23.

Clague, J. J. 1983: Glacio-isostatic effects of the Cordilleran ice sheet, British Columbia, Canada. In *Shorelines and Isostasy* , ed. D. E. Smith, and A. G. Dawson (Academic Press, London) Inst. Brit. Geogr. Spec. Publ. **16**, pp. 321–43.

—— and Bornhold, B. D. 1980: Morphology and littoral processes of the Pacific coast of Canada. In *The Coastline of Canada*, ed. S. B. McCann, Geol. Surv. Can. Paper 80–10, pp. 339–80.

Clapp, W. F. and Kenk, R. 1963: Marine borers: an annotated bibliography. *Off. Naval Res. Dept. Navy*, ACR-74.

Clark, J. A. 1976: Greenland's rapid postglacial emergence: a result of ice–water gravitational attraction. *Geology* **4**, pp. 310–2.

——. 1980: A numerical model of world-wide sea level changes on a viscoelastic Earth. In *Earth Rheology, Isostasy, and Eustasy,* ed. N. -A. Mörner (Wiley, Chichester), pp. 525–34.

—— and Lingle, C. S. 1977: Future sea level changes due to west Antarctic ice sheet fluctuations. *Nature* **269**, pp. 206–9.

——.——. 1979: Predicted relative sea level changes (18,000 years BP to Present) caused by late-glacial retreat of the Antarctic ice sheet. *Quat. Res.* **11**, pp. 279–98.

——, Farrell, W. E. and Peltier, W. R. 1978: Global changes in post-glacial sea level: a numerical calculation. *Quat. Res.* **9**, pp. 265–87.

Cloud, P. E. 1952: Preliminary report on geology and marine environments of Onotoa atoll, Gilbert Islands. *Atoll Res. Bull.* **12**, pp. 1–73.

——. 1959: Geology of Saipan, Mariana Islands 4: submarine topography and shoal-water ecology. *US Geol. Surv. Prof. Paper* 280-K, pp. 361–445.

——. 1965: Carbonate precipitation and dissolution in the marine environment. In *Chemical Oceanography*, ed. J. P. Riley, and G. Skirrow (Academic Press, London), pp. 127–58.

Coastal Engineering Research Center 1977: *Shore Protection Manual* (US Army, Washington DC), 3 vols.

Cobb, W. R. 1969: Penetration of calcium carbonate substrates by the boring sponge *Cliona*. *Am. Zool.* **9**, pp. 783–90.

Coen-Cagli, E. 1935: Design of breakwaters with vertical sides. *16th Int. Congr. Navig. Rept. (Brussels)* **80**.

——. 1936: L'action des lames de tempête sur les digues maritimes à paroi verticale. *Génie Civil* **109**, pp. 177–82.

Coleman, J. M. and Smith, W. G. 1964: Late Recent rise of sea level. *Geol. Soc. Am. Bull.* **75**, pp. 833–40.

——, Gagliano, S. M. and Smith, W. G. 1966: Chemical and physical weathering on saline high tidal flats, N. Queensland, Australia. *Geol. Soc. Am. Bull.* **77**, pp. 205–6.

Coles, B. P. L. and Funnell, B. M. 1981: Holocene palaeo-environments of Broadland. *Spec. Publs. Int. Assoc. Sedimentologists* **5**, pp. 123–31.

Collins, A. R. 1944: The destruction of concrete by frost. *Inst. Civil Eng. J.* **23**, pp. 29–41.

Collins, I. A. 1970: Probability of breaking wave characteristics. *Proc. 12th Conf. Coastal Eng.*, pp. 399–414.

Colquhoun, D. J. and Johnson, H. S., Jr. 1968: Tertiary sea-level fluctuation in South Carolina. *Palaeogeogr., Palaeoclim., Palaeoecol.* **5**, pp. 105–26.

Common, R. 1955: Les formes littorales dans les calcaires en Northumberland septentrional. *Ann. Géogr.* **64**, pp. 126–8.

Conca, J. L. and Rossman, G. R. 1982: Case hardening of sandstone. *Geology* **10**, pp. 520–3.

——,——. 1985: Core softening in cavernously weathered tonalite. *J. Geol.* **93**, pp. 59–73.

Consentius, W.-U. 1975: Untersuchungen zur morphodynamik tropisch-subtropischer Kusten. III. Verwitterungs- and abtragungserscheinungen an felskusten Ost-Thailands. *Wurzburger Geographische Arbeiten* **43**, pp. 36–43.

Cook, H. K. 1952: Experimental exposure of concrete to natural weathering in marine locations. *Am. Soc. Test. Mat. Proc.* **52**, pp. 1169–80.

Cook, P. J. and Polach, H. A. 1973: A chernier sequence at Broad Sound, Queensland, and evidence against a Holocene high sea level. *Marine Geol.* **14**, pp. 253–68.

Cooke, C. W. 1971: American emerged shorelines compared with levels of Australian marine terraces. *Geol. Soc. Am. Bull.* **82**, pp. 3231–4.
Cooke, R. C. 1977: Factors regulating the composition, change, and stability of phases in the calcite–seawater system. *Marine Chem.* **5**, pp. 75–92.
Cooke, R. U. 1979: Laboratory simulation of salt weathering processes in arid environments. *Earth Surface Processes* **4**, pp. 347–59.
—— and Doornkamp. J. C. 1974: *Geomorphology in Environmental Management: An Introduction* (Clarendon Press, Oxford).
—— and Smalley, I. J. 1968: Salt weathering in deserts. *Nature* **220**, pp. 1226–7.
—— and Warren, A. 1973: *Geomorphology in Deserts* (University of California Press, Berkeley, Calif.).
Coope, G. R. 1977: Quaternary Coleoptera as aids in the interpretation of environmental history. In *British Quaternary Studies: Recent Advances*, ed. F. W. Shotton (Clarendon Press, Oxford), pp. 55–68.
Cooper, M. R. 1977: Eustasy during the Cretaceous: its implications and importance. *Palaeogeogr., Palaeoclim., Palaeoecol.* **22**, pp. 1–60.
Cooper, R. I. B. and Longuet-Higgins, M. S. 1951: An experimental study of the pressure variations in standing waves. *Proc. Roy. Soc. (London)* **206A**, pp. 424–35.
Coque-Delhuille, B. and Gosselin-Vuilleumier, J. 1984: L'Évolution géomorphologique du massif de Devon–Cornwall (GB) au Crétacé et au Tertiaire: apport de la microscopie électronique à balayage. *Rev. Géol. Dyn. Géogr. Phys.* **25**, pp. 245–56.
Corbel, J. 1954: Sols polygonaux et 'terrasses marines' du Spitsberg. *Rev. Géogr. Lyon* **29**, pp. 1–28.
——. 1956: Un karst Méditerranéen de basse altitude: le massif des calanques et la formation de son relief. *Rev. Géogr. Lyon* **31**, pp. 129–36.
——. 1958: Les karsts de l'est Canadien. *Cah. Géogr. Québec* **2**, pp. 193–216.
Cordon, W. A. 1966: Freezing and thawing of concrete—mechanics and control. *Am. Concrete Inst. Monogr.* **3** (Detroit, Michigan), pp. 1–93.
Correns, C. W. 1949: Growth and dissolution of crystals under linear pressure. In *Crystal Growth* (Butterworths, London), *Discuss. Faraday Soc.* **5**, pp. 267–71.
Cotton, C. A. 1916: Fault coasts in New Zealand. *Geogr. Rev.* **1**, pp. 20–47.
——. 1922: *Geomorphology of New Zealand, Part 1: Systematic* (Dominion Museum, Wellington).
——. 1942: *Geomorphology* (Whitcombe and Tombs, Christchurch, NZ).
——. 1951a: Atlantic gulfs, estuaries, and cliffs. *Geol. Mag.* **88**, pp. 113–28.
——. 1951b: Sea cliffs of Banks Peninsula and Wellington: some criteria for coastal classification. *NZ Geogr.* **7**, pp. 103–20.
——. 1952: Cyclic resection of headlands by marine erosion. *Geol. Mag.* **89**, pp. 221–5.
——. 1963: Levels of planation of marine benches. *Z. Geomorph.* **7**, pp. 97–110.
——. 1967: Plunging cliffs and Pleistocene coastal cliffing in the southern hemisphere. In *Mélanges de géographie physique, humaine, economique, appliquée offerts à M. Omar Tulippe*, ed. J. Sporcy, Duclot, Gembloux, **1**, pp. 37–59. [This and other papers by Cotton were reprinted in 1974 in *Bold Coasts* (Reed, Wellington, New Zealand).]

——. 1968: Relation of the continental shelf to rising coasts. *Geogr. Rev.* **134**, pp. 382–9.
——. 1969: Marine cliffing according to Darwin's theory. *Trans. Roy. Soc. NZ Geol.* **6**, pp. 187–208.
—— and Wilson, A. T. 1971: Ramp forms that result from weathering and retreat of precipitous slopes. *Z. Geomorph.* NF. **15**, pp. 199–211.
Cousteau, J. -Y. 1952: Fish men explore a new world undersea. *National Geogr. Mag.* **102**, pp. 431–72.
Cowell, P. J. 1980: Breaker type and phase shifts on natural beaches. *Proc. 17th Conf. Coastal Eng.*, pp. 977–1015.
Cronblad, H. G. and Malmgren, B. A. 1981: Climatically controlled variation of Sr and Mg in Quaternary planktonic foraminifera. *Nature* **291**, pp. 61–4.
Cronin, T. M. 1980: Biostratigraphic correlation of Pleistocene marine deposits and sea levels, Atlantic Coastal Plain of the southeastern United States. *Quat. Res.* **13**, pp. 213–29.
——, Szabo, B. J., Ager, T. A., Hazel, J. E. and Owens, J. P. 1981: Quaternary climates and sea levels in the US Atlantic Coastal Plain. *Science* **211**, pp. 233–40.
Cruden, D. M. 1975: The influence of discontinuities on the stability of rock slopes. In *Mass Wasting*, ed. E. Yatsu, A. J. Ward, and F. Adams. 4th Guelph Symp. Geomorph., Geo. Abstr., Univ. East Anglia, Norwich, England, pp. 57–67.
Curray, J. R. 1956: The analysis of two-dimensional orientation data. *J. Geol.* **64**, pp. 117–31.
——. 1965: Late Quaternary history, continental shelves of the United States. In *The Quaternary of the United States*, ed. H. E. Wright and D. G. Frey (Princeton University Press, Princeton, NJ), pp. 723–35.
——, Shepard, F. P. and Veeh, H. H. 1970: Late Quaternary sea-level studies in Micronesia: Carmasel Expedition. *Geol. Soc. Am. Bull.* **81**, pp. 1865–80.
Curtis, C. D. 1976a: Stability of minerals in surface weathering reactions: a general thermochemical approach. *Earth Surface Processes* **1**, pp. 63–70.
——. 1976b: Chemistry of rock weathering, fundamental reactions and controls. In *Geomorphology and Climate*, ed. E. Derbyshire (Wiley-Interscience, London), pp. 25–57.
Cvijic, J. 1902: Lea crypto-depressions de l'Europe. *La Géogr.* **5**, pp. 247–54.
Dahl, E. 1946: On the origin of the strandflat. *Norsk Geol. Tidsskr.* **11**, pp. 159–72.
Dai Pra, G. 1982: The late Pleistocene marine deposits of Torre Castiglione (southern Italy). *Geogr. Fis. Dinam. Quat.* **5**, pp. 115–9.
Dalongeville, M. 1977: Formes littorales de corrosion dans les roches carbonatées au Liban. *Méditerranée* **30**, pp. 21–33.
—— and Guilcher, A. 1982: Les plates-formes à vasques en Méditerranée, notamment leur extension vers le nord. *Karsts Littoraux*, Comité National Français de Geographie (Actes du Colloqium of Perpignan) **1**, pp. 13–22.
Daly, R. A. 1920a: A general sinking of sea level in recent time. *Proc. National Acad. Sci.* **6**, pp. 246–50.
——. 1920b: A recent worldwide sinking of sea level. *Geol. Mag.* **57**, pp. 246–61.
——. 1925: Pleistocene changes of level. *Am. J. Sci.* **10**, pp. 281–313.

——. 1927: The geology of St Helena Island. *Proc. Am. Acad. Arts Sci.* **62**, pp. 31–92.

——. 1934: *The Changing World of the Ice Age* (Yale University Press, New Haven, Conn.).

Dana, J. D. 1849: *Geology* (Putnam, New York), Rept. US Exploration Exped. 1838–1842 **10**, pp. 109, 442.

——. 1880: *Manual of Geology* (Ivison, Blakeman, and Taylor, New York), 3rd ed.

Danel, P. 1952: On the limiting clapotis. *Gravity Waves, Nat. Bur. Std. Circ.* **521**, pp. 35–8.

Dansgaard, W. and Tauber, H. 1969: Glacier oxygen-18 content and Pleistocene ocean temperatures. *Science* **166**, pp. 499–502.

——, Johnsen, S. J., Clausen, H. B. and Langway, C. C. 1971: Climatic record revealed by the Cape Century ice core. In *The Late Cenozoic Glacial Ages*, ed. K. K. Turekian (Yale University Press, New Haven, Conn.), pp. 37–56.

——, ——, Møller, J. and Langway, C. C. 1969: One thousand centuries of climatic record from Cape Century on the Greenland ice sheet. *Science* **166**, pp. 377–81.

Dart, J. K. G. 1972: Echinoids, algal lawn, and coral recolonization. *Nature* **239**, pp. 50–1.

D'Auria, L. 1890: On the force of impact of waves and the stability of the superstructure of breakwaters. *J. Franklin Inst.* **130**, pp. 373–6.

Davidson-Arnott, R. G. D. and Askin, R. W. 1980: Factors controlling erosion of the nearshore profile in overconsolidated till, Grimsby, Lake Ontario. *Proc. Can. Coastal Conf. Burlington, Ont.* pp. 185–99.

Davies, G. L. H. and Stephens, N. 1978: *Ireland*: The Geomorphology of the British Isles (Methuen, London).

Davies, J. L. 1959: Sea level change and shoreline development in southeastern Tasmania. *Proc. Roy. Soc. Tasmania* **93**, pp. 89–95.

——. 1964: A morphogenic approach to world shorelines. *Z. Geomorph.* **8**, pp. 127–42.

——. 1972: *Geographical Variations in Coastal Development* (Oliver and Boyd, Edinburgh).

Davies, K. H. 1983: Amino acid analysis of Pleistocene marine molluscs from the Gower Peninsula. *Nature* **302**, pp. 137–9.

Davies, O. 1971: Sea level during the past 11,000 years (Africa). *Quaternaria* **14**, pp. 195–204.

——. 1981: A review of Wilson's theory that the last interglacial ended with an ice-surge, and the South African evidence therefor. *Ann. Natal. Mus.* **24**, pp. 701–20.

Davis, W. M. 1896: The outline of Cape Cod. *Proc. Am. Acad. Arts Sci.* **31**, pp. 303–32.

——. 1912: A geographical pilgrimage from Ireland to Italy. *Ann. Assoc. Am. Geogr.* **2**, pp. 73–100.

——. 1928: *The Coral Reef Problem* (Am. Geogr. Soc., New York).

——. 1933: Glacial epochs of the Santa Monica Mountains, California. *Geol. Soc. Am. Bull.* **44**, pp. 1041–133.

Dawson, A. G. 1980: Shore erosion by frost: an example from the Scottish late-glacial. In *Studies in the Late-Glacial of North-West Europe*, ed. J. J. Lowe, J. M. Gray and J. E. Robinson (Pergamon, Oxford), pp. 45–53.

Dean, W. R. 1945: On the reflexion of surface waves by a submerged plane barrier. *Proc. Cambridge Phil. Soc.* **41**, pp. 231–8.

Debrat, J. M. 1974: Étude d'un karst calcaire littoral Méditerranéen: exemple du littoral de Nice à Menton. *Méditerranée* **17**, pp. 63–85.

De Freitas, M. H. 1972: Some examples of cliff failure in south-west England. *Proc. Ussher Soc.* **2**, pp. 388–97.

—— and Watters, R. J. 1973: Some field examples of toppling failure. *Géotechnique* **23**, pp. 495–514.

Degens, E. T. 1965: *Geochemistry of Sediments* (Prentice-Hall, Englewood Cliffs, NJ).

De Jong, A. F. M. and Mook, W. G. 1981: Natural C-14 variations and consequences for sea-level fluctuations and frequency analysis of periods of peat growth. *Geol. Mijnbouw* **60**, pp. 331–6.

De la Beche, H. T. 1839: Report on the geology of Cornwall, Devon and west Somerset. HMSO London.

De Lamothe, L. 1911: Les anciennes lignes de rivage du Sahel d'Alger et d'une partie de la côte Algerienne. *Mém. Soc. Géol. France, Paris* **1** (series 4, memoir 6), pp. 1–288.

De Lapparent, A. 1898: *Leçons de Géographie Physique* (Masson, Paris), 2nd ed.

——. 1900: *Traité de gélogie: phénomènes actuels* (Masson, Paris), 4th ed.

Delibrias, G. and Guillier, M. T. 1971: The sea level on the Atlantic coast and the Channel for the last 10,000 years by the ^{14}C method. *Quaternaria* **14**, pp. 131–5.

—— and Laborel, J. 1971: Recent variations of the sea level along the Brazilian coast. *Quaternaria* **14**, pp. 45–9.

——, Giot, P. R., Gouletquer, P. L. and Morzadec-Kerfourn, M. T. 1971: Evolution de la ligne de rivage le long du littoral Armoricain depuis le Néolithique. *Quaternaria* **14**, pp. 175–9.

De Martonne, E. 1909: *Traité de géographie physique* (Armand Colin, Paris).

Denizot, G. 1934: Description des massifs de Marseilleveyre et de Puget. *Ann. Mus. Hist. Nat. Marseille* **26**.

Denness, B., Conway, B. W., McCann, D. M. and Grainger, P. 1975: Investigation of a coastal landslip at Charmouth, Dorset. *Quart. J. Eng. Geol.* **8**, pp. 119–40.

Denny, D. F. 1951: Further experiments on wave pressure. *J. Inst. Civil Eng.* **35**, pp. 330–45.

Denton, G. H., Armstrong, R. L. and Stuiver, M. 1971: The Late Cenozoic glacial history of Antarctica. In *Late Cenozoic Glacial Ages*, ed. K. K. Turekian (Yale University Press, New Haven, Conn.), pp. 267–306.

Depéret, C. 1906: Les anciennes lignes de rivage de la côte francaise de la Méditerranée. *Bull. Soc. Géol. France* **4**, pp. 207–30.

——. 1918: Essai de coordination chronologique générale des temps quaternaires. *Comptes Rendus Acad. Sci.* **167**, pp. 418–22.

Derbyshire, E., Cooper, R. G. and Page, L. W. F. 1979: Recent movements on the cliff at St Mary's Bay, Brixham, Devon. *Geogr. J.* **145**, pp. 86–96.

——, Page, L. W. F. and Burton, R. 1975: Integrated field mapping of a dynamic land surface: St Mary's Bay, Brixham. In *Environment, Man, and Economic*

Change: Essays Presented to S. H. Beaver, ed. A. D. M. Phillips and B. J. Turton (Longman, London), pp. 48–77.

De Swardt, A. M. J. and Casey, O. P. 1963: The coal resources of Nigeria. *Bull. Geol. Surv. Nigeria* **28**, pp. 1–28.

De Virville, A. D. 1934, 1935: Recherches écologiques sur la flore des flaques du littoral de l'Océan Atlantique et de la Manche. *Rev. Gén. Botanie* **46** (1934), pp. 705–21; and 47 (1935) pp. 26–43, 96–114, 160–77, 230–43, and 309–23.

Devoto, G. and Oli, G. C. 1971: *Disionario della lingua italiana* (Monnier, Florence).

Devoy, R. J. N. 1977: Flandrian sea-level changes in the Thames Estuary and the implications for land subsidence in England and Wales. *Nature* **270**, pp. 712–5.

——. 1979: Flandrian sea-level changes and vegetational history of the lower Thames estuary. *Phil. Trans. Roy. Soc. London* **B285**, pp. 355–410.

——. 1982: Analysis of the geological evidence for Holocene sea-level movements in southeast England. *Proc. Geol. Assoc.* **93**, pp. 65–90.

——. 1983: Late Quaternary shorelines in Ireland: an assessment of their implications for isostatic land movement and relative sea-level changes. In *Shorelines and Isostasy*, ed. D. E. Smith and A. G. Dawson (Academic Press, London), Inst. Brit. Geogr. Spec. Publ. **16**, pp. 227–3.

Dewey, H. 1908: In discussion of . . . Barrow's 1908 paper. *Quart. J. Geol. Soc. London* **64**, p. 397.

Dibb, T. E., Hughes, D. W. and Poole, A. B. 1983: The identification of critical factors affecting rock durability in marine environments. *Quart. J. Eng. Geol. London* **16**, pp. 149–61.

Dicken, S. N. 1961: Some recent physical changes on the Oregon coast. Rept. Proj. NR 388–062 (Off. Naval Res.).

Dietz, R. S. 1963: Wave-base marine profile of equilibrium, and wave-built terraces: a critical appraisal. *Geol. Soc. Am. Bull.* **74**, pp. 971–90.

—— and Menard, H. W. 1951: Origin of abrupt change in slope at continental shelf margins. *Am. Assoc. Petrol. Geol. Bull.* **35**, pp. 1994–2016.

Digerfeldt, G. 1975: A standard profile for Littorina transgressions in western Skane, south Sweden. *Boreas* **4**, pp. 125–42.

Dill, R. F. 1977: The blue holes: geologically significant submerged sink-holes and caves off British Honduras and Andros, Bahama Islands. *Proc. 3rd Int. Coral Reef Symp.* (Miami, Florida), **2**, pp. 237–42.

Dillon, W. P. and Oldale, R. N. 1978: Late Quaternary sea-level curve: reinterpretation based on glaciotectonic influence. *Geology* **6**, pp. 56–60.

Dionne, J.-C. 1964: Notes sur les marmites littorales. *Rev. Géogr. Montreal* **18**, pp. 249–77.

——. 1972: Les plates-formes rocheuses littorales de l'estuaire du Saint Laurent. Unpubl. mss. presented to the Assoc. Canadienne–Française pour l'avancement des sciences, 40th congress, Ottawa, pp. 1–9.

——. 1978: Le glaciel en Jamésie et en Hudsonie, Québec subartique. *Géogr. Phys. Quat.* **32**, pp. 3–70.

——. 1980: Les glaces comme agent littoral sur la côte orientale de la Baie de James, Québec. *Proc. Can. Coastal Conf. Burlington, Ont.*, pp. 80–92.

DiSalvo, L. H. 1969: Isolation of bacteria from the corralum of *Porites lobata* (Vaughn) and its possible significance. *Am. Zool.* **9**, pp. 735–40.

Dodge, R. E., Fairbanks, R. G., Benninger, L. K. and Maurrasse, F. 1983: Pleistocene sea levels from raised coral reefs of Haiti. *Science* **219**, pp. 1423–5.

Dominick, T. F., Wilkins, B. and Roberts, H. H. 1973: A one-dimensional mathematical model for beach groundwater fluctuations. *Louisiana State University Coastal Studies Inst. Tech. Rept.* **152**, pp. 1–78.

Domzig, H. 1955: Wellendruck und druckerzeugender seegang. *Mitt. Hannover Versuchsanstalt fur Grundbau und Wasserbau (Franzius-Institut der Technischen Hochschule Hannover)* **8**, pp. 1–79.

Donn, W. L., Farrand, W. R. and Ewing, M. 1962: Pleistocene ice volumes and sea-level lowering. *J. Geol.* **70**, pp. 206–14.

Donner, J. J. 1970: Land/sea-level changes in Scotland. In *Studies in the Vegetational History of the British Isles*, ed, D. Walker and R. G. West (Cambridge University Press, Cambridge), pp. 23–39.

Donner, J. J. 1980: The determination and dating of synchronous late Quaternary shorelines in Fennoscandia. In *Earth Rheology, Isostasy, and Eustasy*, ed. N.-A. Mörner (Wiley, Chichester), pp. 285–93.

—— and Jungner, H. 1975: Radiocarbon dating of shells from marine deposits in the Disko Bugt area, west Greenland. *Boreas* **4**, pp. 25–45.

Doty M. S. 1957: Rock intertidal surfaces. In *Treatise on Marine Ecology and Paleoecology*, ed. J. W. Hedgpeth. Mem. Geol. Soc. Am. **67**, pp. 535–85.

Douglas, G. R. 1980: Magnitude frequency study of a rockfall in Co. Antrim, N. Ireland. *Earth Surface Processes* **5**, pp. 123–9.

Dragovich, D. 1967: Flaking and weathering process operating on cavernous rock surfaces. *Geol. Soc. Am. Bull.* **78**, pp. 801–4.

——. 1969: The origin of cavernous surfaces (tafoni) in granitic rocks of southern South Australia. *Z. Geomorph.* **13**, pp. 163–81.

Dreimanis, A. and Goldthwait, R. P. 1973: Wisconsin glaciation in the Huron, Erie, and Ontario lobes. *Geol. Soc. Am. Mem.* **136**, pp. 71–106.

Drewry, D. J. 1980: Pleistocene bimodal response of Antarctic ice. *Nature* **287**, pp. 214–6.

Driscoll, E. M. 1958: The denudation chronology of the Vale of Glamorgan. *Trans. Inst. Brit. Geogr.* **25**, pp. 45–57.

Drouhin, G., Gautier, M. and Dervieux, F. 1948: Slide and subsidence of the hills of St Raphael–Telemly. *Proc. 2nd Int. Conf. Soil Mech. Foundation Eng. (Rotterdam)* **5**, pp. 104–6.

Duckmanton, N. M. 1974: The shore platforms of the Kaikoura Peninsula. Univ. Canterbury (NZ), MA thesis.

Duerden, J. E. 1902: Boring algae as agents in the destruction of corals. *Am. Mus. Nat. Hist. Bull.* **16**, pp. 323–32.

Dunn, E. J. 1915: Geological notes, Northern Territory, Australia. *Proc. Roy. Soc. Vict.* **28**, pp. 113–4.

Dunn, J. R. and Hudec, P. P. 1965: The influence of clays on water and ice in rock pores, part II. *Dept. Geol. Rensselaer Polytechnic Inst. Contrib.* **65–8**, pp. 1–149.

——. ——. 1966: Water, clay, and rock soundness. *Ohio J. Sci.* **66**, pp. 153–68.

——. ——. 1972: Frost and sorption effects in argillaceous rocks. *Highway Res. Record* **393**, pp. 65–78.

Duplessy, J. C., Labeyrie, J., Lalou, C. and Nguyen, H. V. 1971: La mésure des variations climatiques continentales application à la période comprise entre 130,000 et 90,000 ans BP *Quat. Res.* **1**, pp. 162–74.

Durgin, P. B. 1974: Discussion of Easton (1973). *Bull. Assoc. Eng. Geol.* **11**, pp. 161–6.

——. 1977: Landslides and the weathering of granite rocks. In *Reviews in Engineering Geology, Landslides*, ed. D R. Coates. Geol. Soc. Am., Boulder, Colorado, III, pp. 127–31.

Eardley, A. J. 1964: Polar rise and equatorial fall of sea level. *Am. Scientist* **52**, pp. 488–97.

Easton, W. H. 1973: Earthquakes, rain, and tides at Portuguese Bend landslide, California. *Bull. Assoc. Eng. Geol.* **10**, pp. 173–94.

—— and Ku, T.-L. 1980: Holocene sea-level changes in Palau, west Caroline Islands. *Quat. Res.* **14**, pp. 199–209.

—— and Olson, E. A. 1976: Radiocarbon profile of Hanauma Reef, Oahu, Hawaii. *Geol. Soc. Am. Bull.* **87**, pp. 711–9.

Eddy, J. A. 1977: The case of the missing sunspots. *Scientific Am.* **236**, pp. 80–92.

Edwards, A. B. 1941: Storm wave platforms. *J. Geomorph.* **4**, pp. 223–36.

——. 1942: The San Remo Peninsula. *Proc. Roy. Soc. Vict.* **54**, pp. 59–74.

——. 1945: The geology of Philipp Island. *Proc. Roy. Soc. Vict.* **57**, pp. 1–16.

——. 1951: Wave action in shore platform formation. *Geol. Mag.* **88**, pp. 41–9.

——. 1958: Wave cut platforms at Yampi Sound in the Buccaneer Archipelago. *J. Roy. Soc. West. Austr.* **41**, pp. 17–21.

Eggler, D. H., Larson, E. E. and Bradley, W. C. 1969: Granites, grusses and the Sherman erosion surface, southern Laramie Range, Colorado–Wyoming. *Am. J. Sci.* **267**, pp. 510–22.

Einsele, G., Herm, D. and Schwarz, H. U. 1974: Sea level fluctuation during the past 6,000 years at the coast of Mauritania. *Quat. Res.* **4**, pp. 282–9.

Embleton, C. and King, C. A. M. 1975: *Periglacial Geomorphology* (Edward Arnold, London).

Emery, K. O. 1941: Rate of surface retreat of sea cliffs based on dated inscriptions. *Science* **93**, pp. 617–8.

——. 1946: Marine solution basins. *J. Geol.* **54**, pp. 209–28.

——. 1954: The Painted Cave, Santa Cruz Island. *Sea and Pacific Motorboat* **46**, pp. 37–9, 91–2.

——. 1960: *The Sea of Southern California* (Wiley, New York).

——. 1962: Marine geology of Guam. *US Geol. Surv. Prof. Paper* **403–B**, pp. 1–76.

——. 1963: Organic transportation of marine sediments. In *The Sea*, ed. M. N. Hill (Wiley-Interscience, New York), pp. 776–93.

——. 1967: The activity of coastal landslides related to sea level. *Rev. Géogr. Phys. Géol. Dyn.* **9**, pp. 177–80.

—— and Foster, H. L. 1956: Shoreline nips in tuff at Matsushima, Japan. *Am. J. Sci.* **254**, pp. 380–5.

—— and Kuhn, G. G. 1980: Erosion of rock coasts at La Jolla, California. *Marine Geol.* **37**, pp. 197–208.

——. ——. 1982: Sea cliffs: their processes, profiles, and classification. *Geol. Soc. Am. Bull.* **93**, pp. 644–54.

Emiliani, C. 1955: Pleistocene temperatures. *J. Geol.* **63**, pp. 538–78.

——. 1966: Paleotemperature analysis of Caribbean cores P6304–8 and P6304–9 and a generalized temperature curve for the past 425,000 years. *J. Geol.* **74**, pp. 109–26.

——. 1971: The amplitude of Pleistocene climatic cycles at low latitudes and the isotopic composition of glacial ice. In *The Late Cenozoic Glacial Ages*, ed. K. K. Turekian (Yale University Press, New Haven, Conn.), pp. 183–97.

—— and Shackleton, N. J. 1974: The Brunhes Epoch: isotopic paleotemperatures and geochronology. *Science* **183**, pp. 511–14.

Endean, R. 1977: *Ancanthaster planci* infestation of reefs of the Great Barrier Reef. *Proc. 3rd Int. Coral Reef Symp. (Miami, Florida)* 1, pp. 185–91.

England, J. H. and Andrews, J. T. 1973: Broughton Island—a reference area for Wisconsin and Holocene chronology and sea level changes on eastern Baffin Island. *Boreas* **2**, pp. 17–32.

Epstein, H. and Tyrrell, F. C. 1949: Design of rubblemound breakwaters. *17th Int. Congr. Navig.* pp. 81–98.

Epstein, S., Sharp, R. P. and Gow, A. J. 1970: Antarctic ice sheet: stable isotope analysis of Byrd Station cores and interhemispheric climatic implications. *Science* **168**, pp. 1570–2.

Erez, J. 1979: Modification of the oxygen-isotope record in deep-sea cores by Pleistocene dissolution cycles. *Nature* **281**, pp. 535–8.

Ericson, D. B. and Wollin, G. 1968: Pleistocene climates and chronology in deep-sea sediments. *Science* **162**, pp. 1227–34.

——. ——. 1970: Pleistocene climates in the Atlantic and Pacific Oceans: a comparison based on deep sea sediments. *Science* **167**, pp. 1483–5.

——, Ewing, M. and Wollin, G. 1964: The Pleistocene epoch in deep-sea sediments. *Science* **146**, pp. 723–32.

Erinc, S. 1978: Changes in the physical environment in Turkey since the end of the last glacial. In *The Environmental History of the Near and Middle East since the Last Ice Age*, ed. W. C. Brice (Academic Press, London), pp. 87–110.

Eronen, M. 1983: Late Weichselian and Holocene shore displacement in Finland. In *Shorelines and Isostasy*, ed. D. E. Smith and A. G. Dawson (Academic Press, London), Inst. Brit. Geogr. Spec. Publ. **16**, pp. 183–207.

Evans, I. S. 1970: Salt crystallization and rock weathering: a review. *Rev. Géomorph. Dyn.* **19**, pp. 153–77.

Evans, J. W. 1968: The role of *Penitella penita* (Conrad 1837) (family Pholadidae) as eroders along the Pacific coast of North America. *Ecol.* **49**, pp. 156–9.

——. 1970: Palaeontological implications of a biological study of rock boring clams (family Pholadidae). In *Trace Fossils*, ed. T. P. Crimes and J. C. Harper (Seel House Press, Liverpool), pp. 127–40.

Evans, R. S. 1981: An analysis of secondary toppling rock failures—the stress redistribution method. *Quart. J. Eng. Geol. London* **14**, pp. 77–86.

Everard, C. E., Lawrence, R. H., Witherick, M. E. and Wright, L. W. 1964: Raised beaches and marine geomorphology. In *Present Views on Some Aspects of the*

Geology of Cornwall and Devon, ed. K. F. G. Hosking, and G. J. Shrimpton (Roy. Geol. Soc. Cornwall, Penzance), pp. 283–310.

Everett, D. H. 1961: The thermodynamics of frost damage to porous solids. *Trans. Faraday Soc.* **57**, pp. 1541–51.

Evteev, S. A. 1960: At what speed does wind 'erode' stones in Antarctica? *Soviet Antarctic Expedition* **2**, p. 211.

Fagerlund, G. 1975: The significance of critical degrees of saturation at freezing of porous and brittle materials. In *Durability of Concrete* (Am. Concrete Inst., Detroit), paper SP 47–2, pp. 13–65.

Fahey, B. D. 1983: Frost action and hydration as rock weathering mechanisms on schist: a laboratory study. *Earth Surface Processes and Landforms* **8**, pp. 535–45.

——. 1985: Salt weathering as a mechanism of rock breakup in cold climates: an experimental approach. *Z. Geomorph.* **29**, pp. 99–111.

Fairbairn, H. W. 1943: Packing in ionic minerals. *Geol. Soc. Am. Bull.* **54**, pp. 1305–74.

Fairbanks, R. G. and Matthews, R. K. 1978: The marine oxygen isotope record in Pleistocene coral, Barbados, West Indies. *Quat. Res.* **10**, pp. 181–96.

Fairbridge, R. W. 1948: Notes on the geomorphology of the Pelsart group of the Houtman's Abrolhos Islands. *J. Roy. Soc. West. Austr.* **33**, pp. 1–43.

——. 1950: The geology and geomorphology of Point Peron, Western Australia. *J. Roy. Soc. West. Austr.* **34**, pp. 35–72.

——. 1952: Marine erosion. *Proc. 7th Pacific Sci. Congr.* pp. 347–58.

——. 1958: Dating the latest movements of the Quaternary sea level. *Trans. NY Acad. Sci.* Series ii, **20**, pp. 471–82.

——. 1961a: Eustatic changes in sea level. In *Physics and Chemistry of the Earth*, ed. L. H. Ahrens *et al.* (Pergamon Press, London), **4**, pp. 99–185.

——. 1961b: Radiation solaire et variations cycliques du niveau marin. *Rev. Géogr. Phys. Géol. Dyn.* **4**, pp. 2–14.

——. 1966: Mean sea level changes, long-term—eustatic and other. In *The Encyclopedia of Oceanography*, ed. R. W. Fairbridge (Reinhold, New York), pp. 479–85.

——. 1968a: Exudation. In *The Encyclopedia of Geomorphology*, ed. R. W. Fairbridge (Dowden, Hutchinson, and Ross, Stroudsburg, Penn.), pp. 342–3.

——. 1968b: Limestone coastal weathering. In *The Encyclopedia of Geomorphology*, ed. R. W. Fairbridge (Dowden, Hutchinson, and Ross, Stroudsburg, Penn.), pp. 653–7.

——. 1968c: Atlantic and Pacific type coasts. In *The Encyclopedia of Geomorphology*, ed. R. W. Fairbridge (Dowden, Hutchinson, and Ross, Stroudsburg, Penn.), pp. 34–5.

——. 1971: Quaternary shoreline problems at Inqua 1969. *Quaternaria* **15**, pp. 1–18.

——. 1972: Quaternary sedimentation in the Mediterranean region controlled by tectonics, paleoclimates and sea level. In *The Mediterranean Sea*, ed. E. O. Stanley (Dowden, Hutchinson, and Ross, Stroudsburg, Penn.), pp. 99–113.

——. 1976a: Shellfish-eating pre-ceramic Indians in coastal Brazil. *Science* **191**, pp. 353–9.

——. 1976b: Effects of Holocene climatic change on some tropical geomorphic processes. *Quat. Res.* **6**, pp. 529–56.
——. 1977: Rates of sea-ice erosion of Quaternary littoral platforms. *Studia Geol. Polonica* **52**, pp. 135–41.
—— and Krebs, O. A. 1962: Sea Level and the Southern Oscillation. *Geophys. J. Roy. Astron. Soc.* **6**, pp. 532–45.
—— and Newman, W. S. 1968: Postglacial crustal subsidence of the New York area. *Z. Geomorph.* **12**, pp. 296–317.
—— and Richards, H. G. 1967: The Inqua Shorelines Commission. *Z. Geomorph.* **11**, pp. 205–15.
Faniran, A. and Jeje, L. K. 1983: *Humid Tropical Geomorphology* (Longman, London).
Farrell, W. E. and Clark, J. A. 1976: On postglacial sea level. *Geophy. J. Roy. Astron. Soc.* **46**, pp. 647–67.
Faure, H. 1980: Late Cenozoic vertical movements in Africa. In *Earth Rheology, Isostasy, and Eustasy*, ed. N.-A Mörner (Wiley, Chichester), pp. 465–9.
—— and Élouard, P. 1967; Schema des variations du niveau de l'océan Atlantique sur la côte de l'ouest de l'Afrique depuis 40,000 ans. *Comptes Rendus Acad. Sci.* **265**, pp. 784–7.
——, Fontes, J. C., Hebrard, L., Monteillet, J. and Pirazzoli, P. A. 1980: Geoidal change and shore-level tilt along Holocene estuaries: Senegal River area, West Africa. *Science* **210**, pp. 421–3.
Fenneman, N. M. 1902: Development of the profile of equilibrium of the subaqueous shore terrace. *J. Geol.* **10**, pp. 1–32.
Fenton, J. D. 1985: Wave forces on vertical walls. *Am. Soc. Civil. Eng. J. Waterways, Port, Coastal and Ocean Eng.* **111**, pp. 693–718.
Ferrar, H. T. 1925: The geology of the Whangarei Bay of Island subdivision, Kaipara Division. *NZ Geol. Surv. Bull.* **27**, pp. 25–6.
Fevret, M. and Sanlaville, P. 1966: L'utilisation des vermets dans la détermination des anciens niveaux marins. *Méditerranée* **4**, pp. 357–64.
Fewkes, J. W. 1890: On excavations made in rocks by sea urchins. *Am. Naturalist* **14**, pp. 1–21.
Fisher, O. 1866: On the disintegration of a chalk cliff. *Geol. Mag.* **3**, pp. 354–6.
Flack, H. L. 1957: Freezing-and-thawing resistance of concrete as affected by the method of test. *Am. Soc. Test. Mat. Proc.* **57**, pp. 1077–95.
Fleming, B. W. 1977: Langebaan Lagoon: a mixed carbonate–siliciclastic tidal environment in a semi-arid climate. *Sed. Geol.* **18**, pp. 61–95.
Fleming, C. A. 1965: Two-storied cliffs at the Auckland Islands. *Trans. Roy. Soc. N.Z. Geol.* **3**, pp. 171–4.
Flemming, N. C. 1965: Form and relation to present sea level of Pleistocene marine erosion features. *J. Geol.* **73**, pp. 799–811.
——. 1969: Archaeological evidence for eustatic change of sea level and Earth movements in the western Mediterranean during the last 2,000 yrs. *Geol. Soc. Am. Spec. Paper* **109**, pp. 1–125.
—— and Roberts, D. G. 1973: Tectono-eustatic changes in sea level and sea floor spreading. *Nature* **243**, pp. 19–22.

Flett, J. S. and Hill, J. B. 1912: Geology of the Lizard and Meneage. *Mem. Geol. Surv. GB* (2nd ed published 1946 as New Series Sheet Memoir 359).

Flint, R. F. 1957: *Glacial and Pleistocene Geology* (John Wiley, New York).

Flohn, H. 1981: Climatic change, ice sheets and sea level. In *Sea Level, Ice, and Climatic Change*, Proc. Canberra Symp. Dec. 1979. Int. Assoc. Hydrol. Sci. Publ. **131**, pp. 431–40.

Flores Silva, E. 1952: Observaciones de costas en La Antàrtida Chilena. *Inf. Géogr.* **3**, pp. 85–93.

Focke, J. W. 1977: The effect of a potentially reef-building vermetid/coralline algal community on an eroding limestone coast, Curaçao, Netherlands Antilles. *Proc. 3rd Int. Coral Reef Symp.* (Miami), **1**, pp. 239–45.

——. 1978: Limestone cliff morphology on Curaçao (Netherlands Antilles), with special attention to the origin of notches and vermetid/coralline algal surf benches (corniches, trottoirs). *Z. Geomorph.* **22**, pp. 329–49.

Fogg, G. E. 1973: Physiology and ecology of marine blue algae. In *The Biology of Blue-Green Algae*, ed. N. G. Carr and B. A. Whitton (University of California Press. Berkeley, Calif.) Botanical Monogr. **9**, pp. 368–78.

Fookes, P. G. and Poole, A. B. 1981: Some preliminary considerations on the selection and durability of rock and concrete materials for breakwaters and coastal protection works. *Quart. J. Eng. Geol. London* **14**, pp. 97–128.

Fraser, J. K. 1959: Freeze–thaw frequencies and mechanical weathering in Canada. *Arctic* **12**, pp. 40–53.

Freeze, R. A. and Cherry, J. A. 1979: *Groundwater* (Prentice-Hall, Englewood Cliffs, NJ).

Frémy, P. 1945: Contribution à la physiologie des Thallophytes marines perforant et cariant les roches calcaires et les coquilles. *Ann. Inst. Océanogr.* **22**, pp. 107–44.

Froget, G. 1963: La morphologie et les mécanismes d'erosion du littoral rocheux de la provence occidentale. *Rec. Travaux de la Station Marine d'Endoume*, **30**, pp. 165–243.

Frydl, P. and Stearn, C. W. 1978: Rate of bioerosion by parrotfish in Barbados reef environments. *J. Sed. Petrol*. **48**, pp. 1149–58.

Fuhrboter, A. 1970: Air entrainment and energy dissipation in breakers. *Proc. 12th Conf. Coastal Eng*., pp. 391–8.

Fujii, S., Lin, C. C. and Tjia, H. D. 1971: Sea level changes in Asia during the past 11,000 years. *Quaternaria* **14**, pp. 211–16.

Gabet, C. 1971: La phase terminale de la transgression flandrienne sur le littoral charentais. *Quaternaria* **14**, pp. 181–8.

Gaillard, D. D. 1904: Wave action in relation to engineering structures. *US Army Corps. Eng. Prof. Paper* 31.

Galvin, C. J. 1968: Breaker type classification of three laboratory beaches. *J. Geophys. Res*. **73**, pp. 3651–9.

——. 1972: Waves breaking in shallow water. In *Waves on Beaches and Resulting Sediment Transport*, ed. R. E. Meyer (Academic Press, New York), pp. 413–56.

Gardner, J. S. 1969: Snowpatches: their influence on mountain wall temperatures and the geomorphic implications. *Geogr. Annlr*. **51A**, pp. 114–20.

Garrels, R. M., Thompson, M. E. and Siever, R. 1961: Control of carbonate solubility by carbonate complexes. *Am. J. Sci.* **259**, pp. 24–45.

Gascoyne, M., Ford, D. C. and Schwarcz, H. P. 1981: Late Pleistocene chronology and paleoclimate of Vancouver Island determined from cave deposits. *Can. J. Earth Sci.* **18**, pp. 1643–52.

——, Schwarcz, H. P. and Ford, D. C. 1980: A palaeotemperature record for the mid-Wisconsin in Vancouver Island. *Nature* **285**, pp. 474–6.

Gaunt, G. D. and Tooley, M. J. 1974: Evidence for Flandrian sea-level changes in the Humber Estuary and adjacent areas. *Bull. Geol. Surv. GB* **48**, pp. 25–41.

Gautier, F. 1971: Les processus de l'attaque des falaises sur le littoral continental de la baie de Bourgneuf. *Norois* **70**, pp. 221–36.

Geikie, A. 1903: *Textbook of Geology* (Macmillan, London).

George, P. 1948: Quelques formes karstiques de la Croatie occidentale et de la Slovénie méridionale (Yougoslavie). *Ann. Géogr.* **57**, no. 308, pp. 298–307.

George, T. N. 1933: The coast of Gower. *Proc. Swansea Field Naturalist Soc.* **1**, pp. 192–206.

——. 1955: Drainage in the Southern Uplands: Clyde, Nith, Annan. *Trans. Geol. Soc. Glasgow* **22**, pp. 1–34.

——. 1960: The stratigraphical evolutions of the Midland Valley. *Trans. Geol. Soc. Glasgow* **24**, pp. 32–107.

Gessner, F. 1970: Lithothamnium terrassem im Karibischen Meer. *Int. Rev. Ges. Hydro-biol.* **55**, pp. 757–62.

Geyh, M. A., Kudrass, H.-R. and Streif, H. 1979: Sea level changes during the Late Pleistocene and Holocene in the Strait of Malacca. *Nature* **278**, pp. 441–3.

Gigout, M. 1959: Ages, par radiocarbone, de deux formations des environs de Rabat (Maroc). *Comptes Rendus Acad. Sci.* **249**, pp. 2802–3.

——. 1971: Les niveaux marins mondiaux depuis 11,000 ans. *Quaternaria* **14**, pp. 63–4.

Gill, A. F. and Thomas, J. F. J. 1939: Some observations regarding frost action. *Highway Res. Board Proc.* **19**, pp. 281–3.

Gill, E. D. 1961: Changes in the level of the sea relative to the land in Australia during the Quaternary era. *Z. Geomorph.* Suppl. Band **3**, pp. 73–9.

——. 1964: Quaternary shorelines in Australia. *Aust. J. Sci.* **26**, pp. 388–91.

——. 1965: Radiocarbon dating of past sea levels in SE Australia. *Inqua VII Congress (Boulder, Colorado) Abstr.*, p. 167.

——. 1967: The dynamics of the shore platform process and its relation to changes in sea-level. *Proc. Roy. Soc. Vict.* **80**, pp. 183–92.

——. 1968: Eustasy. In *The Encyclopedia of Geomorphology*, ed. R. W. Fairbridge (Dowden, Hutchinson, and Ross, Stroudsburg, Penn.), pp. 333–66.

——. 1971: The Paris symposium on world sea levels of the past 11,000 years. *Quatenaria* **14**, pp. 1–6.

——. 1972a: Ramparts on shore platforms. *Pacific Geol.* **4**, pp. 121–33.

——. 1972b: The relationship of present shore platforms to past sea levels. *Boreas* **1**, pp. 1–25.

——. 1972c: Sanders' wave tank experiments and shore platforms. *Papers Proc. Roy. Soc. Tasmania* **106**, pp. 17–20.

——. 1973a: Rate and mode of retrogradation on rocky coasts in Victoria, Australia, and their relationship to sea level changes. *Boreas* **2**, pp. 143–71.

——. 1973b: Application of recent hypotheses to changes of sea level in Bass Strait, Australia. *Proc. Roy. Soc. Vict.* **85**, pp. 117–24.

——. 1981: Rapid honeycomb weathering (tafoni formation) in greywacke, SE Australia. *Earth Surface Processes and Landforms* **6**, pp. 81–3.

—— and Clarke, R. 1979: Clarke's slip at Eastern View, Otway coast, SE Australia. *Vict. Naturalist* **96**, pp. 3–7.

—— and Lang, J. G. 1982: The peak of the Flandrian transgression in Victoria, SE Australia—faunas and sea level changes. *Proc. Roy. Soc. Vict* **94**, pp. 23–34.

——, ——. 1983: Microerosion meter measurements of rock wear on the Otway Coast of southeast Australia. *Marine Geol.* **52**, pp. 141–56.

——, Segnit, E. R. and McNeill, N. H. 1981: Rate of formation of honeycomb weathering features (small scale tafoni) on the Otway coast, SE Australia. *Proc. Roy. Soc. Vict.* **92**, pp. 149–54.

Ginsburg, R. N. 1953a: Beachrock in south Florida. *J. Sed. Petrol.* **23**, pp. 85–92.

——. 1953b: Intertidal erosion on the Florida Keys. *Bull. Marine Sci. Gulf Carib.* **3**, pp. 55–69.

——, Rezak, R. and Wray, J. 1972: Recent reef algae. Atlantic reefs. In *Geology of Calcareous Algae*, ed. R. N. Ginsburg, R. Rezak and J. Wray (Univ. Miami, Florida, Comparative Sed. Lab.), pp. 9.4–9.5.

Givago, A. V. 1950: On solution and abrasion forms in limestones of the west coast of Crimea. *Bull. Geogr. Soc. USSR*. **82**, pp. 615–8 (in Russian).

Gjorv, O. E. 1965: Investigation of Norwegian marine concrete structures. *Int. Symp. Behaviour of Concretes Exposed to Sea Water (Palermo, Italy)*, pp. 150–7.

Glenie, R. C., Schofield, T. C. and Ward, W. T. 1968: Tertiary sea levels in Australia and New Zealand. *Palaeogeogr., Palaeoclim., Palaeoecol.* **5**, pp. 141–63.

Glynn, P. W., Wellington, G. M. and Birkeland, C. 1979: Coral reef growth in the Galapagos: limitation by sea urchins. *Science* **203**, pp. 47–9.

Goda, Y. 1970: Pressures of standing waves on vertical wall. *Am, Soc. Civil Eng. Proc. Waterways, Harbors and Coastal Eng. Div*. **96**, pp. 155–9.

—— and Abe, Y. 1968: Apparent coefficient of partial reflection of finite amplitude waves. *Rept. Port Harbour Res. Inst.* **7**.

—— and Kakizaki, S. 1966: Study of finite amplitude standing waves and their pressure upon a vertical wall (in Japanese). *Rept. Port Harbour Res. Inst.* **5**, pp. 1–57.

Godard, A. and Houel-Gangloff, F. 1965: Essais de gélifraction artificielle pratiqués sur des calcaires et des grès Lorrains. *Revue Géogr. Alpine* **5**, pp. 125–39.

Godwin, H., Suggate, R. P. and Willis, E. H. 1958: Radiocarbon dating of the eustatic rise in ocean level. *Nature* **181**, pp. 1518–9.

Goede, A., Harmon, R. and Kiernan, K. 1979: Sea caves of King Island. *Helictite* **17**, pp. 51–64.

Goldich, S. S. 1938: A study in rock weathering. *J. Geol.* **46**, pp. 17–58.

Goldthwait, J. W. 1911: The twenty-foot terrace and sea-cliff of the lower Saint Lawrence. *Am. J. Sci.* **32**, pp. 291–317.

——. 1933: The sawing of cliffs and platforms by batture ice. Originally unpublished ms, reprinted in 1971 in *Pleistocene Geology of the Central St Lawrence*, *Geol. Surv. Canada. Mem.* **359**, pp. 144–5.

Golubic, S. 1969: Distribution, taxonomy, and boring patterns of marine endolithic algae. *Am. Zool.* **9**, pp. 747–51.

——, Perkins, R. D. and Lukas, K. J. 1975: Boring micro-organisms and microborings in carbonate substrates. In *The Study of Trace Fossils*, ed. R. W. Frey (Springer-Verlag, Berlin), pp. 229–59.

Goodell, H. G., Watkins, N. D., Mather, T. T. and Koster, S. 1968: The Antarctic glacial history recorded in sediments of the Southern Ocean. *Palaeogeogr., Palaeoclim., Palaeoecol.* **5**, pp. 41–62.

Goodman, R. E. and Bray, J. W. 1977: Toppling of rock slopes. In *Rock Engineering* (9th Conf. Rock Eng. for Foundations and Slopes), Am. Soc. Civil Eng., Boulder, Colorado, vol. ii, pp. 201–34.

Goreau, T. J. 1980: Frequency sensitivity of the deep-sea climatic record. *Nature* **287**, pp. 620–2.

Goskar, K. L. and Trueman, A. E. 1934: The coastal plateaux of south Wales. *Geol. Mag.* **71**, pp. 468–77.

Goudie, A. S. 1974: Further experimental investigation of rock weathering by salt and other mechanical processes. *Z. Geomorph.* suppl. band 21, pp. 1–12.

——. 1977: Sodium sulphate weathering and the disintegration of Mohenjo-Dara, Pakistan. *Earth Surface Processes* **2**, pp. 75–86.

——, Cooke, R. U. and Evans, I. S. 1970: Experimental investigation of rock weathering by salts. *Area* **4**, pp. 42–8.

Gourret, M. 1935: Sur un mouvement approche de clapotis: application au calcul des digues maritimes verticales. *Ann. Ponts Chauss.* **105**, pp. 337–451.

Grant, D. R. 1970: Recent coastal submergence of the Maritime Provinces, Canada. *Can. J. Earth Sci.* **7**, pp. 676–89.

——. 1980: Quaternary sea-level change in Atlantic Canada as an indication of crustal delevelling. In *Earth Rheology, Isostasy, and Eustasy*, ed. N. -A. Mörner (Wiley, Chichester), pp. 201–14.

Grasty, R. L. 1967: Orogeny, a cause of world-wide regression of the seas. *Nature* **216**, pp. 779–80.

Grawe, O. R. 1936: Ice as an agent of rock weathering: a discussion. *J. Geol.* **44**, pp. 173–82.

Gray, J. M. 1974: Lateglacial and postglacial shorelines in western Scotland. *Boreas* **3**, pp. 129–38.

——. 1978: Low-level shore platforms in the southwest Scottish highlands: altitude, age, and correlation. *Trans. Inst. Brit. Geogr.* **3**, pp. 151–64.

Green, A. H. 1887: *Geology, Part I: Physical Geology* (Rivingtons, London), 2nd ed.

Green, J. F. N. 1936: The terraces of southernmost England. *Quart. J. Geol. Soc. London* **92**, pp. 58–88.

Greenly, E. 1919: The geology of Anglesey. *Mem. Geol. Surv. (GB)*.

Greensmith, J. T. and Tucker, E. V. 1973: Holocene transgressions and regressions on the Essex coast, outer Thames Estuary. *Geol. Mijnbouw* **52**, pp. 193–202.

——.——. 1980: Evidence for differential subsidence on the Essex coast. *Proc. Geol Assoc.* **91**, pp. 169–75.

Greenwald, I. 1941: The dissociation of calcium and magnesium carbonates and bicarbonates. *J. Biol. Chem.* **141**, pp. 789–96.

Gregory, J. W. 1912: The structural and petrographical classification of coast types. *Scientia* **2**, pp. 36–63.

Grenier, P. 1968: Observations sur les taffonis du désert Cihien. *Bull. Assoc. Géogr. Francais* 364–5, pp. 193–211.

Greslou, L. and Mahe, Y. 1955: Étude du coefficient de réflexion d'une houle sur un obstacle constitué par un plan incliné. *Proc. 5th Conf. Coastal Eng.*, pp. 68–84.

Grisez, L. 1960: Alvéolisation littorale de schistes métamorphiques. *Rev. Géomorph. Dyn.* **11**, pp. 164–7.

Grønlie, O. T. 1924: Contributions to the Quaternary geology of Novaya Zemlya. In *Report of the Scientific Results of the Norwegian Expedition to Novaya Zemlya 1921*, ed. O. Holtedahl (Videnskapsselskapet, Kristiania), pp. 1–124.

——. 1951: On the rise of sea and land and the forming of strandflats on the west coast of Fennoscandia. *Norsk Geol. Tidsskr.* **29**, pp. 26–63.

Groot, R. A. de 1977: Boring sponges (Clionidae) and their trace fossils from the coast near Rovinj (Yugoslavia). *Geol Mijnbouw* **56**, pp. 168–30.

Gruner, J. W. 1950: An attempt to arrange silicates in the order of reaction energies at relatively low temperatures. *Am. Mineralogist* **35**, pp. 137–48.

Gubkin, N. M. 1969: The effect of tides on the dynamics of the banks of the Shantarskie Islands. *Vestnik Moskovskogo Universiteta Geografiya* **2**, pp. 111–3.

Guilcher, A. 1948: *Le Relief de la Bretagne Meridionale* (Henri Potier, La Roche-sur-Yon).

——. 1950: Cryoplanation et solifluction quaternaires dans les collines de Bretagne occidentale et nord du Devonshire. *Rev. Géomorph. Dyn.* **1**, pp. 53–76.

——. 1952a: Formes de décomposition chimique dans la zone des embruns et des marees sur les côtes britanniques et bretonnes. *Vol. Jubil. Cinquant. Anniv. Labor. Geogr. Univ. Rennes*, pp. 167–81.

——. 1952b: Formes et processes d'erosion sur les recifs coralliens du nord du banc Farsan (Mer Rouge). *Rev. Géomorph. Dyn.* **6**, pp. 261–74.

——. 1953: Essai sur la zonation et la distribution des formes littorales de dissolution du calcaire. *Ann. Géogr.* **62**, pp. 161–79.

——. 1954: Morphologie littorale du calcaire en Méditerranée occidentale (Catalogne et environs d'Alger). *Bull. Assoc. Géogr. Français* **241**, pp. 50–8.

——. 1955: Géomorphologie de l'extrémité septentrionale du banc Farsan (Mer Rouge). *Ann. Inst. Océanogr.* **30**, pp. 55–100.

——. 1957: Formes de corrosion littorale du calcaire sur les côtes du Portugal. *Tijd. Kon. Nederl. Aardr. Gen.* **74**, pp. 263–9.

——. 1958a: *Coastal and Submarine Morphology* (Methuen, London).

——. 1958b: Coastal corrosion forms in limestones around the Bay of Biscay. *Scot. Geogr. Mag.* **74**, pp. 137–49.

——. 1966: Les grandes falaise et megafalaise des côtes sud-ouest et ouest de l'Irlande. *Ann. Géogr.* **75**, pp. 26–38.

——. 1969: Pleistocene and Holocene sea level changes. *Earth Sci. Rev.* **5**, pp. 69–97.
——. 1974a: Les rasas: un problème de morphologie littorale generale. *Ann. Géogr.* **83**, pp. 1–33.
——. 1974b: Coral reefs of the New Hebrides, Melanesia. with special reference to open-sea, not fringing reefs. *Proc. 2nd Int. Coral Reef Symp.* (Brisbane), **2**, pp. 523–35.
——. 1980: Observations géomorphologiques sur des littoraux subarctiques de la pointe nord de l'île Sakhaline (extreme-orient Soviétique). *Rev. Géomorph. Dyn.* **29**, pp. 101–15.
——. 1981: Cryplanation littorale et cordons glaciels de basse mer dans la région de Rimouski, côte sud de l'estuaire du Saint-Laurent, Québec. *Géogr. Phys. Quat.* **35**, pp. 155–69.
—— and Bodere, J. C. 1975: Formes de corrosion littorale dans les roches volcaniques aux moyennes et hautes latitudes dans l'Atlantique. *Bull. Assoc. Géogr. Français* **52**, pp. 179–85.
—— and Joly, F. 1954: Recherches sur la morphologie de la côte Atlantique du Maroc. *Trav. Inst. Sci. Chérifien, Tanger* (série géol. et géogr. phys.) **2**, pp. 1–140.
——, Berthois, L. and Battistini, R. 1962: Fromes de corrosion littorale dans les roches volcaniques particulièrement à Madagascar et au Cap Vert (Sénégal). *Cah. Océanogr.* **14**, pp. 209–40.
Guillien, Y. 1949: Gel et dégel du sol: les mécanismes morphologiques. *Inf. Géogr.* **13**, pp. 104–16.
—— and Lautridou, J. P. 1970: Recherches de gelifraction experimentale du Centre de Geomophologie: 1–Calcaires des Charentes. *Centre de Géomorphologie, CRNS Bull.* **5**, pp. 1–45.
Guimont, P. and Laverdiere, C. 1980: The sud-est de la mer d'Hudson: un relief de cuesta. In *The Coastline of Canada* ed. S. B. McCann, Geol. Surv. Can. Paper 80–10, pp. 303–9.
Gullick, C. F. W. R. 1936: A physiographical survey of west Cornwall. *Trans. Roy. Soc. Cornwall* **16**, pp. 380–99.
Gulliver, F. 1899: Shoreline topography. *Proc. Am. Acad. Arts Sci.* **34**, pp. 151–258.
Gygi, R. A. 1969: An estimate of the erosional effect of *Sparisoma viride* (Bonnaterre), the Green Parrot-fish, on some Bermuda reefs. Rept. Res. 7th Graduate Res. Sem., ed. R. N. Ginsburg and P. Garrett (Bermuda Biol. Stat., Bermuda), pp. 137–44.
——. 1975: *Sparisoma viride* (Bonnaterre), the Stoplight Parrot-fish, a major sediment producer on coral reefs of Bermuda. *Eclogae Geol. Helvetiae* **68**, pp. 327–59.
Hafsten, U. 1983: Biostratigraphical evidence for late Weichselian and Holocene sea-level changes in southern Norway. In *Shorelines and Isostasy*, ed. D. E. Smith and A. G. Dawson (Academic Press, London), Inst. Brit. Geogr. Spec. Publ. **16**, pp. 161–81.
—— and Tallantire, P. A. 1978: Palaeoecology and post-Weichselian shore-level changes on the coast of Møre, western Norway. *Boreas* **7**, pp. 109–22.

Haigler, S. A. 1969: Boring mechanism of *Polydora websteri* inhabiting *Crassostrea virginica*. *Am. Zool.* **9**, pp. 821–8.
Hails, J. R. 1965: A critical review of sea-level changes in eastern Australia since the last glacial. *Austr. Geogr. Studies* **3**, pp. 63–78.
——. 1968: The Late Quaternary of part of the mid-north coast, New South Wales. *Trans. Inst. Brit. Geogr.* **44**, pp. 139–45.
—— and Hoyt, J. H. 1971: The question of late Quaternary changes of sea level in New South Wales, Australia. *Quaternaria* **14**, pp. 255–64.
Hall, K. J. 1980: Freeze–thaw activity at a nivation site in northern Norway. *Arctic Alpine Res.* **12**, pp. 183–94.
Hallam, A. 1963: Major epeirogenic and eustatic changes since the Cretaceous, and their possible relationship to crustal structure. *Am. J. Sci.* **261**, pp. 397–423.
——. 1981: Plate tectonics, biogeography, and evolution. *Nature* **293**, pp. 31–2.
Hammer, C. U., Clausen, H. B. and Dansgaard, W. 1980: Greenland ice sheet evidence of postglacial volcanism and its climatic impact. *Nature* **288**, pp. 230–5.
Hamner, W. M. and Jones, M. S. 1976: Distribution, burrowing, and growth rates of the clam *Tridacna crocea* on interior reef flats. *Oecologia* **24**, pp. 207–27.
Hanor, J. S. 1978: Precipitation of beachrock cements; mixing of marine and meteoric waters vs. CO_2 degassing. *J. Sed. Petrol.* **48**, pp. 489–501.
Hansen, 1940: Molen und wellenkrafte an molen im deutschen ostseegebiet. *Zentralblatt der Bauverwaltung* **60**, pp. 423–31.
Hansom, J. D. 1983: Shore-platform development in the South Shetland Islands, Antarctica. *Marine Geol.* **53**, pp. 211–29.
Hardy, H. R. and Jayaraman, N. I. 1970: An investigation of methods for the determination of the tensile strength of rock. *Proc. 2nd Congr. Int. Soc. Rock Mech.* (*Belgrade*) **3**, paper 5–12, pp. 85–92.
Harmon, R. S., Land, L. S., Mitterer, R. M., Garrett, P., Schwarcz, H. P. and Larson, G. J. 1981: Bermuda sea level during the last interglacial. *Nature* **289**, pp. 481–3.
——, Mitterer, R. M., Kriausakul, N., Land, L. S., Schwarcz, H. P., Garrett, P., Larson, G. J., Vacher, H. L. and Rowe, M. 1983: U-series and amino-acid racemization geochronology of Bermuda: implications for eustatic sea-level fluctuation over the past 250,000 years. *Palaeogeogr., Palaeoclim. Palaeoecol.* **44**, pp. 41–70.
——. Schwarcz, H. P. and Ford, D. C. 1978a: Late Pleistocene sea level history of Bermuda. *Quat. Res.* **9**, pp. 205–18.
——. Thompson, P., Schwarcz, H. P. and Ford, D. C. 1978b: Late Pleistocene paleoclimates of N. America as inferred from stable isotope studies of speleothems. *Quat. Res.* **9**, 54–70.
Hawkshaw, C. 1878: On the action of limpets in sucking pits in and abrading the surface of chalk at Dover. *J. Linn. Soc.* (*Zool.*) **14**, pp. 406–11.
Hawley, D. L. 1965: Rock platforms, their formation, modification, and destruction. Univ. Sydney, BA thesis.
Hayashi, T. and Hattori, M. 1958: Pressure of the breaker against a vertical wall. *Coastal Eng. Japan 1*, pp. 25–37.
Hays, J. D. and Pitman, W. C. 1973: Lithospheric plate motion, sea-level changes and climatic and ecological consequences. *Nature* **246**, pp. 18–22.

Healy, J. 1953: Wave damping effect of beaches. *Proc. Minn. Int. Hydr. Conv. (Minneapolis, Minnesota)*, pp. 213–20.
Healy, T. R. 1968a: Shore platform morphology on the Whangaparaoa Peninsula, Auckland. *Conf. Ser. NZ Geogr. Soc.* **5**, pp. 163–8.
——. 1968b: Bioerosion on shore platforms developed in the Waitemata Formation, Auckland. *Earth Sci. J.* **2**, pp. 26–37.
Hedley, C. 1924: Differential elevation near Sydney. *J. Proc. Roy. Soc. NSW* **58**, pp. 61–6.
Hein, J. and Risk, M. J. 1975: Bioerosion of coral heads: inner patch reefs, Florida reef tract. *Bull. Marine Sci.* **25**, pp. 133–8.
Helmuth, R. A. 1960: Discussion of frost action in concrete. In *Chemistry of Cement,* Proc. 4th Int. Symp. Nat.Bur.Std. Monogr. **43**, pp. 829–33.
Hendy, C. H., Rayner, E. M., Shaw, J. and Wilson, A. T. 1979: Late Pleistocene glacial chronology, Taylor Valley, Antarctica, and the global climate. *Quat. Res.* **11**, 172–84.
Henkrel, L. 1906: A study of tide-pools on the west coast of Vancouver Island. *Postelsia* (Yearbook of the Minnesota Seaside Station), pp. 277–304.
Hernández-Pacheco, F. 1950: Las rasas de la costa Cantabrica en su segmento Asturiano. *Comptes Rendus Congr. Géogr. Inst. Lisbonne* **2**, pp. 29–86.
Hetu, B. and Gray, J. T. 1980: Évolution postglaciaire des versants de la région de Mont-Louis, Gaspésie, Québec. *Géogr. Phys. Quat.* **34**, pp. 187–208.
Heusser, C. J., Heusser, L. E. and Streeter, S. S. 1980: Quaternary temperatures and precipitation for the north-west coast of North America. *Nature* **286**, pp. 702–4.
Hey, R. W. 1978: Horizontal Quaternary shorelines of the Mediterranean. *Quat. Res.* **10**, pp. 197–203.
Heyworth, A. and Kidson, C. 1982: Sea-level in south-west England and Wales. *Proc. Geol. Assoc.* **93**, pp. 91–111.
Hiatt, R. W. and Strasburg, D. W. 1960: Ecological relationships of the fish fauna on coral reefs of the Marshall Islands. *Ecol. Monogr.* **30**, pp. 65–127.
Higgens, C. G. 1956: Rock-boring isopod. *Geol. Soc. Am. Bull.* **67**, p. 1770.
Higgins, C. G. 1969: Isostatic effects of sea level changes. In *Quaternary Geology and Climate* (ed. H. E. Wright) Proc. VII Congr. Int. Assoc. Quat. Res. **16**, pp. 141–5.
Hillaire-Marcel, C. 1980: Multiple component postglacial emergence, eastern Hudson Bay, Canada. In *Earth Rheology, Isostasy and Eustasy* (ed. N. -A. Mörner (Wiley, Chichester), pp. 215–30.
—— and Fairbridge, R. W. 1978: Isostasy and eustasy of Hudson Bay. *Geology* **6**, pp. 117–22.
—— and Occhietti, S. 1977: Fréquence des datations au ^{14}C de faunes marines post-glaciaires de l'est du Canada et variations paléoclimatiques. *Palaeogeogr., Palaeoclim., Palaeoecol.* **21**, pp. 17–54.
Hills, E. S. 1949: Shore platforms. *Geol. Mag.* **86**, pp. 137–52.
——. 1971: A study of cliffy coastal profiles based on examples in Victoria, Australia. *Z. Geomorph.* **15**, pp. 137–80.
——. 1972: Shore platforms and wave ramps. *Geol. Mag.* **109**, pp. 81–8.

Hinds, N. E. A. 1930: The geology of Kauai and Niihau. *Bernice P. Bishop Mus. Bull.* **71**, pp. 1–103.
Hiroi, I. 1919: On a method of estimating the force of waves. *J. Coll. Eng.* (*Tokyo Imperial Univ.*) **10**, pp. 1–19.
——. 1920: The force and power of waves. *Engineer* **130**, pp. 184–5.
Hodgkin, E. P. 1964: Rate of erosion of intertidal limestone. *Z. Geomorph.* **8**, pp. 385–92.
——. 1970: Geomorphology and biological erosion of limestone coasts in Malaysia. *Geol. Soc. Malaysia Bull.* **3**, pp. 27–51.
Hoel, A. 1909: Geologiske iagttagelser pa Spitsbergen ekspeditionerne 1906 og 1907. *Norsk Geol. Tidsskr.* **1**, pp. 1–28.
Hoffman, E. J. and Duce, R. A. 1974: The organic carbon content of marine aerosols collected in Bermuda. *J. Geophys. Res.* **79**, pp. 4474–7.
Hoffmeister, J. E. and Wentworth, C. K. 1942: Data for the recognition of changes of sea level. *6th Pacific Sci. Congr.* pp. 839–48.
Hogbom, A. G. 1899: Ragundadelens geologie. *Sver. Geol. Unders.* Ser. C, no. 182.
Hollin, J. T. 1965: Wilson's theory of ice ages. *Nature* **208**, pp. 12–6.
——. 1969: Ice sheet surges and the geological record. *Can. J. Earth Sci.* **6**, pp. 903–10.
——. 1972: Interglacial climates and Antarctic ice surges. *Quat. Res.* **2**, pp. 401–8.
——. 1977: Thames interglacial sites, Ipswichian sea levels and Antarctic ice surges. *Boreas* **6**, pp. 33–52.
——. 1980: Climate and sea level in isotope stage 5: an east Antarctic ice surge at 95,000 BP? *Nature* **283**, pp. 629–33.
Hollingworth, S. E. 1938: The recognition and correlation of high-level erosion surfaces in Britain: a statistical study. *Quart. J. Geol. Soc. London* **94**, pp. 55–84.
Holmes, A. 1965: *Principles of Physical Geology* (Nelson, London).
Holtedahl, H. 1960: Mountain, fiord, strandflat; geomorphology and general geology of western Norway. In *Guide to Excursions, 21st International Geol. Congr.*, ed. J. A. Dons (Oslo, Norway), pp. 1–29.
——. 1962: The problem of coastal genesis with special reference to the strandflat, the banks or grounds, and deep channels of the Norwegian and Greenland coasts: a discussion. *J. Geol.* **70**, pp. 631–3.
——. 1984: High pre-late Weichselian sea-formed caves and other marine features on the Møre–Romsdal coast, west Norway. *Norsk Geol. Tidssk.* **64**, pp. 75–85.
Holtedahl, O. 1929: On the geology and physiography of some Antarctic and sub-Antarctic islands with notes on the character and origin of fjords and strandflats of some northern lands. In *Scientific Results of the Norwegian Antarctic Expedition, 1927–8*, ed. O. Holtedahl, Det Norske Vid.-Akad. i Oslo, no. 3.
Homma, M. and Horikawa, K. 1965a: Experimental study on total wave force against sea wall. *Coastal Eng. Japan* **8**, pp. 119–29.
——, ——. 1965b: Wave forces against sea wall. *Proc. 9th Conf. Coastal Eng.*, pp. 490–503.

Honeyborne, D. B. 1965: Weathering processes affecting inorganic building materials. UK Building Res. Station 141/65, pp. 1–16.

Hopkins, D. M. 1973: Sea level history in Beringia during the past 250,000 years. *Quat. Res.* **3**, pp. 520–40.

Hopley, D. 1963: The coastal geomorphology of Anglesey. Univ. Manchester, MA thesis.

——. 1971: World sea levels during the past 11,000 years. Evidence from Australia and New Zealand. *Quaternaria* **14**, pp. 265–6.

——. 1977: The age of the outer Ribbon Reef surface, Great Barrier Reef, Australia: implications for hydro-isostatic models. *Proc. 3rd Int. Coral Reef Symp. (Miami, Florida)* pp. 23–8.

——. 1978: Sea level changes on the Great Barrier Reef: an introduction. *Phil. Trans. Roy. Soc. London* **291A**, pp. 159–66.

——. 1980: Mid-Holocene high sea levels along the coastal plain of the Great Barrier Reef Province: a discussion. *Marine Geol.* **35**, pp. M1–M9.

Horikawa, K. 1978: *Coastal Engineering* (Halsted Press, Wiley, New York).

—— and Kuo, C. T. 1966: A study of wave transformation inside surf zone. *Proc. 10th Conf. Coastal Eng.*, pp. 217–33.

—— and Sunamura, T. 1966: A study on coastal cliff erosion. *Mem. 21st Conv. J. Soc. Civ. Engr.*, pp. 1–88.

——, ——. 1967: A study on erosion of coastal cliffs by using aerial photographs. *Coastal Eng. Japan* **10**, pp. 67–83.

——, ——. 1970: A study on erosion of coastal cliffs and of submarine bedrocks. *Coastal Eng. Japan* **13**, pp. 127–39.

Hotta, S., Muzuguchi, M. and Isobe, M. 1982: A field study of waves in the nearshore zone. *Proc. 18th Conf. Coastal Eng.*, pp. 38–57.

Howard, A. D. 1968: Terraces–marine. In *The Encyclopedia of Geomorphology*, ed. R. W. Fairbridge (Dowden, Hutchinson, and Ross, Stroudsburg), pp. 1140–2.

Howarth, P. J. and Bones, J. G. 1972: Relationships between process and geometrical form on high Arctic slopes, south-west Devon Island, Canada. In *Polar Geomorphology*, ed. R. J. Price and D. E. Sugden, Inst. Brit. Geogr. Spec. Publ. **4**, pp. 139–53.

Hoyt, J. H. 1967: Intercontinental correlation of late Pleistocene sea levels. *Nature* **215**, pp. 612–4.

—— and Hails, J. R. 1967: Pleistocene shoreline sediments in coastal Georgia: deposition and modification. *Science* **155**, pp. 1541–3.

——, Henry, V. J. and Weimar, R. J. 1968: Age of late-Pleistocene shoreline deposits, coastal Georgia. In *Means of Correlation of Quaternary Successions*, ed. R. B. Morrison and H. E. Wright (University of Utah Press, Salt Lake City pp. 381–93.

Huang, W. H. and Keller, W. C. 1973: Organic acids as agents of chemical weathering of silicate minerals. *Nat. Phys. Sci.* **239**, pp. 149–51.

—— and Kiang, W. C. 1972: Laboratory dissolution of plagioclase feldspars in water and organic acids at room temperature. *Am. Mineralogist* **57**, pp. 1849–59.

Hudec, P. P. 1973: Weathering of rocks in Arctic and Sub-arctic environment. In

Canadian Arctic Geology, ed. J. D. Aitken and D. J. Glass, Geol. Soc. Assoc. Can.–Can. Soc. Petrol. Geol. Symp. (Saskatoon, Saskatchewan), pp. 313–35.
——. 1977: Deterioration and dimensional changes of carbonate rocks as function of adsorbed water content. *Proc. 2nd Int. Symp. Water—Rock Interaction (Strasbourg, France)* **4**, pp. 38–45.
——. 1978a: Rock weathering on the molecular level. *Geol. Soc. Am. Eng. Case Hist.* **11**, pp. 47–51.
——. 1978b: Standard engineering tests for aggregate: what do they actually measure? *Geol. Soc. Am. Eng. Case Hist.* **11**, pp. 3–6.
——. 1980a: Durability of carbonate rocks as function of their thermal expansion, water sorption, and mineralogy. In *Durability of Building Materials and Components* (ed. P. J. Sereda and G. G. Litvan, Am. Soc. Test. Mat. Spec. Tech. Publ. **691**, pp. 497–508.
——. 1980b: Effect of de-icing salts on deterioration and dimensional changes of carbonate rocks. In *Durability of Building Materials and Components*, ed. P. J. Sereda and G. G. Litvan, Am. Soc. Test. Mat. Spec. Tech. Publ. **691**, pp. 629–40.
——. 1982: Overconsolidated clays: shales. *Transport Res. Record* **873**, pp. 28–35.
—— and Rigbey, S. G. 1976: The effect of sodium chloride on water sorption characteristics of rock aggregates. *Bull. Assoc. Eng. Geol.* **13**, pp. 199–211.
—— and Sitar, N. 1975: Effect of water sorption on carbonate rock expansivity. *Can. Geotech. J.* **12**, pp. 179–86.
Hudson, J. H. 1977: Long-term bioerosion rates on a Florida reef: a new method. *Proc. 3rd Int. Coral Reef Symp. (Miami, Florida)* **2**, pp. 491–7.
Hudson, R. Y. 1952: Wave forces on breakwaters. *Proc. Am. Soc. Civil Eng.* **113**, pp. 1–22.
Hume, J. D. 1965: Sea-level changes during the last 2,000 years at Point Barrow, Alaska. *Science* **150**, pp. 1165–6.
—— and Schalk, M. 1967: Shoreline processes near Barrow, Alaska. *Arctic* **25**, pp. 272–8.
Hunt, B. G. 1979: The effects of past variations of the Earth's rotation rate on climate. *Nature* **281**, pp. 188–91.
Hunt, M. 1969: A preliminary investigation of the habits and habitat of the rock-boring urchin *Echinometra lucunter* near Devonshire Bay, Bermuda. Repts. Res. 7th Graduate Res. Sem., ed. R. N. Ginsburg and P. Garrett (Bermuda Biol. Stat., Bermuda), pp. 35–40.
Hutchinson, J. N. 1968a: Mass movement. In *The Encyclopedia of Geomorphology*, ed. R. W. Fairbridge (Dowden, Hutchinson, and Ross, Stroudsburg, Penn.), pp. 688–95.
——. 1968b: Field meeting on the coastal landslides of Kent. *Proc. Geol. Assoc.* **79**, pp. 227–37.
——. 1969: A reconsideration of the coastal landslides at Folkestone Warren, Kent. *Géotechnique* **19**, pp. 6–38.
——. 1972: Field and laboratory studies of a fall in Upper Chalk at Joss Bay, Isle of Thanet. In *Stress Strain Behaviour of Soils*, ed. R. H. G. Parry, Proc. Roscoe Mem. Symp., Cambridge (G. T. Foulis, Henley, Oxfordshire), pp. 692–706.
——. 1980: Various forms of cliff instability arising from coast erosion in south-east

England. *Fjellsprengningsteknikk Bergmekanikk/Geoteknikk (Trondheim, Norway)*, pp. 19.1–19.32.

——, Bromhead, E. N. and Lupini, J. F. 1980: Additional observations on the Folkestone Warren landslides. *Quart. J. Eng. Geol.* **13**, pp. 1–31.

Idnurm, M. and Cook, P. J. 1980: Palaeomagnetism of beach ridges in South Australia and the Milankovitch theory of ice ages. *Nature* **286**, pp. 699–702.

Imbrie, J. and Imbrie, J. Z. 1980: Modelling the climatic response to orbital variations. *Science* **207**, pp. 943–53.

——, Van Donk, J. and Kipp, N. G. 1973: Paleoclimate investigation of a Late Pleistocene Caribbean deep-sea core: comparison of isotopic and faunal methods. *Quat. Res.* **3**, pp. 10–38.

Inman, D. L. and Nordstrom, C. E. 1971: On the tectonic and morphologic classification of coasts. *J. Geol.* **79**, pp. 1–21.

Iribarren, R. C. 1938: Una formula para el calcula de los digues de escollera. *Revista de Obras Publicas (Madrid)*. Translated copy Tech. Rept. HE–116–295, Dept. Eng. Univ. Calif. Berkeley, 1948.

—— and Nogales y Olano, C. C. 1949: Protection des ports. *17th Int. Congr. Navig*. pp. 31–80.

Isakov, I. S. 1953: *Morskoi Atlas* (Ministry of the Navy, USSR), **2**, pl. 13.

Iseki, H. 1978: Review of studies on sea level in Japan (in Japanese). *Geogr. Rev. Japan* **51**, pp. 188–96.

Iverson, H. W. 1951: Studies of wave transformation in shoaling water, including breaking. In *Gravity Waves*, National Bur. Stand., circ. no. 521, pp. 9–32.

Jackson, M. L., Tyler, S. A., Willis, A. L., Bourbeau, G. A. and Pennington, R. P. 1948: Weathering sequence of clay size minerals in soils and sediments. *J. Physics and Colloid Chem*. **52**, pp. 1237–60.

Jacoby, E. 1936: Die berechnung der standsicherheit von seehafendammen. *Werft Reed*. **17**, pp. 96–8, 119–22.

Jacoby, W. R. 1972: Plate theory, epeirogenesis and eustatic sea-level changes. *Tectonophys*. **15**, pp. 187–96.

Jahn, A. 1960a: Some remarks on evolution of slopes on Spitsbergen. *Z. Geomorph*. **1**, pp. 49–58.

——. 1960b: Quantitative analysis of some periglacial processes. *19th Int. Geogr. Congr. Abstr*. pp. 134–5.

——. 1961: Quantitative analysis of some periglacial processes in Spitsbergen. *Uniwersytet Wroclawski Im Boleslawa Bieruta, Zeszyty Nauk. Nauki Przyrodnicze* **11**, pp. 1–34.

James, N. P., Mountjoy, E. W. and Omura, A. 1971: An Early Wisconsin reef terrace at Barbados, West Indies, and its climatic implications. *Geol. Soc. Am. Bull*. **82**, pp. 2011–8.

Jardetzky, W. C. 1962: Aperiodic pole shift and deformation of the Earth's crust. *J. Geophys. Res*. **67**, pp. 4461–72.

Jardine, F. 1925: The development and significance of benches in the littoral of eastern Australia. *Trans. Roy. Geogr. Soc. Austr. Queensland* **1**, pp. 111–30.

Jardine, W. G. 1959: River development in Galloway. *Scot. Geogr. Mag*. **75**, pp. 65–74.

——. 1964: Postglacial sea levels in south-west Scotland. *Scot. Geogr. Mag.* **80**, pp. 5–11.
——. 1966: Landscape evolution in Galloway. *Trans. J. Proc. Dumfries Galloway Nat. Hist. Antiq. Soc.* **43**, pp. 1–13.
——. 1971: Form and age of Late Quaternary shore-lines and coastal deposits of south-west Scotland: critical data. *Quaternaria* **14**, pp. 103–14.
——. 1975: Chronology of Holocene marine transgression and regression in south-western Scotland. *Boreas* **4**, pp. 173–96.
——. 1978: Radiocarbon ages of raised-beach shells from Oronsay. Inner Hebrides, Scotland: a lesson in interpretation and deduction. *Boreas* **7**, pp. 183–96.
——. 1981: Holocene shorelines in Britain: recent studies. *Geol. Mijnbouw* **60**, pp. 297–315.
——. 1982: Sea-level changes in Scotland during the last 18,000 years. *Proc. Geol. Assoc.* **93**, pp. 25–41.
Jehu, T. J. 1918: Rockboring organisms as agents in coast erosion. *Scot. Geogr. Mag.* **34**, pp. 1–11.
Jeletzky, J. A. 1978: Causes of Cretaceous oscillations of sea level in western and Arctic Canada and some general geotectonic implications. *Can. Geol. Surv. Paper* 77–18, pp. 1–44.
Jelgersma, S. 1966: Sea level changes during the last 10,000 years. In *Proc. Int. Symp. World Climate, 8000–0 BC.*, Roy. Meteorol. Soc. London. Reprinted in 1971 in *Introduction to Coastline Development*, ed. J. A. Steers (Macmillan, London), pp. 25–48.
——. 1980: Late Cenozoic sea level changes in the Netherlands and the adjacent North Sea basin. In *Earth Rheology, Isostasy, and Eustasy*, ed. N.-A. Mörner (Wiley, Chichester), pp. 435–47.
Jennings, J. N. 1961: Sea level changes in King Island, Bass Strait. *Z. Geomorph.* Suppl. Band 3, pp. 80–4.
——. 1968: Tafoni. In *The Encyclopedia of Geomorphology*, ed. R. W. Fairbridge (Dowden, Hutchinson, and Ross, Stroudsburg, Penn.), pp. 1103–4.
——. 1971: *Karst* (MIT Press, Cambridge, Mass.).
Johannessen, C. L., Feiereisen, J. J. and Wells, A. N. 1982: Weathering of ocean cliffs by salt expansion in a mid-latitude coastal environment. *Shore and Beach* **50**, pp. 26–34.
John, B. S. and Sugden, D. E. 1971: Raised marine features and phases of glaciation in the South Shetland Islands. *Brit. Antarctic Surv. Bull.* **24**, pp. 45–111.
——.——. 1975: Coastal geomorphology of high latitudes. *Prob. Phys. Geogr.* **7**, pp. 53–132.
Johnsen, S. J., Dansgaard, W., Clausen, H. B. and Langway, C. C. 1972: Oxygen isotope profiles through the Antarctic and Greenland ice sheets. *Nature* **235**, pp. 429–34.
Johnson, A. R. M. 1974: Cavernous weathering at Berowra, NSW. *Austr. Geogr.* **12**, pp. 531–5.
Johnson, D. W. 1919: *Shore Processes and Shoreline Development* (Wiley, New York).

——. 1925: *The New England—Acadian Shoreline* (Wiley, New York).
——. 1931: Supposed two-metre eustatic bench of the Pacific shores. *Congr. Int. Geogr. Comptes Rendus.* **2**, pp. 158–63.
——. 1933: Role of analysis in scientific investigation. *Geol. Soc. Am. Bull.* **44**, pp. 461–94.
——. 1938: Shore platforms. *J. Geomorph.* **1**, pp. 266–8.
Johnson, J. H. 1961: *Limestone Building Algae and Algal Limestones* (Johnson Publ. Co., Colorado School of Mines, Boulder, Colorado).
Johnson, T. W. and Sparrow, F. K. 1961: *Fungi in Oceans and Estuaries* (Hafner, New York).
Jones, M. L. 1969: Boring of shell by *Caobangia* in freshwater snails of southeast Asia. *Am. Zool.* **9**, pp. 829–35.
Jones, O. T. 1911: *The Physical Features and Geology of Central Wales*. National Union of Teachers Handbook (Aberystwyth Conference).
Jongsma, D. 1970: Eustatic sea level changes in the Arafura Sea. *Nature* **228**, pp. 150–1.
Joyce, E. B. and Evans, R. S. 1976: Some areas of landslide activity in Victoria, Australia. *Proc. Roy. Soc. Vict.* **88**, pp. 95–108.
Joyce, J. R. F. 1950: Notes on ice foot development. Neny Fjord, Graham Land, Antarctica. *J. Geol.* **58**, pp. 646–9.
Jukes-Browne, A. J. 1893: *Geology, an Elementary Textbook* (Bell and Sons, London).
Junge, C. E. 1963: *Air Chemistry and Radioactivity* (Academic Press, New York).
Jutson, J. T. 1918: The influence of salts in rock weathering in sub-arid Western Australia. *Proc. Roy. Soc. Vict.* **30**, pp. 165–72.
——. 1927: Notes on the coastal physiography of Port Campbell, Victoria. *Proc. Roy. Soc. Victoria* **40**, pp. 45–56.
——. 1939: Shore platforms near Sydney, New South Wales. *J. Geomorph.* **2**, pp. 237–50.
——. 1940: The shore platforms of Mt. Martha, Port Phillip Bay, Victoria, Australia. *Proc. Roy. Soc. Vict.* **52**, pp. 164–74.
——. 1949a: The shore platforms of Lorne, Victoria. *Proc. Roy. Soc. Vict.* **61**, pp. 43–59.
——. 1949b: The shore platforms of Point Lonsdale, Victoria. *Proc. Roy. Soc. Vict.* **61**, pp. 105–11.
——. 1950: The shore platforms of Flinders, Victoria, Australia. *Proc. Roy. Soc. Vict.* **60**, pp. 57–72.
——. 1954: The shore platforms of Lorne, Victoria, and the processes operating thereon. *Proc. Roy. Soc. Vict.* **65**, pp. 125–34.
Kaitera, P. 1966: Sea pressure as a cause of crustal movements. *Ann. Acad. Sci. Fennicae* AIII, 90, pp. 191–200.
Kamel, A. M. 1970: Shock pressures on coastal structures. *Am. Soc. Civil Eng. Proc. Waterways, Harbors and Coastal Eng. Div.* **96**, pp. 689–99.
Karlstrom, T. N. V. 1968: The Quaternary time scale–a current problem of correlation and radiometric dating. In *Means of Correlation of Quaternary Successions*, ed. R. B. Morrison and H. E. Wright (University of Utah Press, Salt Lake City), pp. 121–50.

Karrow, P. F. and Bada, J. L. 1980: Amino acid racemization dating of Quaternary raised marine terraces in San Diego County, California. *Geology* **8**, pp. 200–4.

Kaufman, A., Broecker, W. S., Ku, T. L. and Thurber, D. L. 1971: The status of U-series methods of mollusk dating. *Geochim. Cosmochim. Acta*. **35**, pp. 1155–83.

Kawasaki, I. 1954: Geomorphological study of the Byobugaura seacliff in the vicinity of Iioka-Machi, Chiba Prefecture. *Geogr. Rev. Japan* **27**, pp. 213–7 (in Japanese).

Kaye, C. A. 1957: The effect of solvent motion on limestone solution . *J. Geol.* **65**, pp. 35–46.

——. 1959: Shoreline features and Quaternary shoreline changes Puerto Rico. *U.S. Geol. Survey Prof. Paper* 317–B, pp. 49–140.

——. and Barghoorn, E. S. 1964: Late Quaternary sea-level change and crustal rise at Boston, Massachusetts, with notes on the autocompaction of peat. *Geol. Soc. Am. Bull.* **75**, pp. 63–80.

Keatch, D. R. A. 1965: Geomorphology of the coast Cardiff to Porthcawl with special reference to the transportation of beach material. Univ. Wales, Ph.D. thesis.

Keeble, A. B. 1971: Freeze–thaw cycles and rock weathering in Alberta. *Alberta Geogr.* **7**, pp. 34–42.

Keen, D. H., Harmon, R. S. and Andrews, J. T. 1981: U series and amino acid dates from Jersey. *Nature* **289**, pp. 162–4.

Keen, M. 1961: A proposed reclassification of the gastropod family Vermetidae. *Bull. Brit. Mus. (Nat. Hist.) Zool.* **7**, pp. 181–213.

Keller, W. D. 1957: *The Principles of Chemical Weathering* (Lucas Bros., Columbia, Missouri).

Kelletat, D. 1980: Studies on the age of honeycombs and tafoni features. *Catena* **7**, pp. 317–25.

——. 1982: Hohlkehlen sowie rezente organische gesteinsbildungen an den kösten und ihre beziehungen zum meeresniveau. *Essener Geogr. Arb.* **1**, pp. 1–27.

Kelly, W. C. and Zumberge, J. H. 1961: Weathering of a quartz diorite at Marble Point, McMurdo Sound, Antarctica. *J. Geol.* **69**, pp. 433–46.

Kempf, M. and Laborel, J. 1968: Formations de vermets et d'algues calcaires sur les côtes de Brésil. *Rec. Travaux de Station Marine d'Endoume* **43**, pp. 9–23.

Kennedy, T. B. and Mather, K. 1953: Correlation between laboratory freezing and thawing and weathering at Treat Island, Maine. *Proc. Am. Concrete Inst.* **50**, pp. 141–72.

Kennett, J. P. and Thunell, R. C. 1975: Global increase in Quaternary explosive volcanism. *Science* **187**, pp. 497–503.

Kenney, T. C. 1975: Weathering and changes in strength as related to landslides. In *Mass Wasting*, ed. E. Yatsu, A. J. Ward and F. Adams, 4th Guelph Symp. Geomorph., Geo Abstr. (University of East Anglia, Norwich), pp. 69–78.

Kern, J. P. 1977: Origin and history of upper Pleistocene marine terraces, San Diego, California. *Geol. Soc. Am. Bull.* **88**, pp. 1553–66.

Kerr, R. A. 1981a: Milankovitch climatic cycles: old and unsteady. *Science* **213**, pp. 1095–6.

——. 1981b: Staggered Antarctic ice formation supported. *Science* **213**, pp. 427–8.

Keulegan, G. and Patterson, G. 1940: Mathematical theory of irrotational translation waves. *J. Res. Nat. Bur. Std.* **24**, pp. 47–101.

Kidson, C. 1962: Coastal cliffs: report of a symposium. *Geogr. J.* **128**, pp. 303–20.

—— and Heyworth, A. 1976: The Quaternary deposits of the Somerset Levels. *Quart. J. Eng. Geol.* **9**, pp. 217–35.

——. ——. 1978: Holocene eustatic sea level change. *Nature* **273**, pp. 748–50.

Kiernan, K. 1979: Sea caves and morphological karst on the Tasmanian coastline. *Southern Caver* **11**, pp. 2–23.

Kieslinger, A. 1959: Rahmenverwitterung. *Geologie und Bauwesen* **24**, pp. 171–86.

Kinahan, G. H. 1866: Ancient sea margins in the counties Clare and Galway. *Geol. Mag.* **3**, pp. 337–43.

King, C. A. M. 1959: *Beaches and Coasts* (Edward Arnold, London).

——. 1963: Some problems concerning marine planation and the formation of erosion surfaces. *Trans. Inst. Brit. Geogr.* **33**, pp. 29–43.

——. 1969: Some Arctic coastal features around Foxe Basin and in eastern Baffin Island NWT, Canada. *Geogr. Annlr.* **51A**, pp. 207–18.

——. 1972: *Beaches and Coasts* (Edward Arnold, London), 2nd edn.

Kino, Y. 1958: The geological map of Hyuga-Aoshima and its explanatory text. *Geol. Surv. Japan*, pp. 1–5.

Kirk, R. M. 1977: Rates and forms of erosion on intertidal platforms at Kaikoura Peninsula, South Island, New Zealand. *NZ J. Geol. Geophys.* **20**, pp. 571–613.

Kirkby, M. J. 1983: Modelling cliff development in south Wales: Savigear reviewed. School of Geogr. Univ. Leeds, Working Paper **351**, pp. 1–17.

——. 1984: Modelling cliff development in south Wales: Savigear re-viewed. *Z. Geomorph.* **28**, pp. 405–26.

Kirkgoz, M. S. 1981: A theoretical study of plunging breakers and their run-up. *Coastal Eng.* **5**, pp. 353–70.

——. 1982: Shock pressure of breaking waves on vertical walls. *Am. Soc. Civil Eng. J. Waterways Port Coastal and Ocean Div.* **108**, pp. 81–95.

——. 1983: Secondary pressures of waves breaking on seawall. *Am. Soc. Civil Eng. J. Waterways, Port, Coastal, and Ocean Div.* **109**, pp. 487–90.

Kitano, Y., Kanamori, N. and Tokuyama, A. 1969: Effects of organic matter on solubilities and crystal form of carbonates. *Am. Zool.* **9**, pp. 681–8.

Klaer, W. 1956: Verwitterungsformen in granit auf Korsika. *Pet. Geogr. Mitt. Ergh.* **261**, pp. 1–146.

Kleemann, K. 1973: Der gesteinsabbau durch atzmuscheln an kalkkusten. *Oecologia* **13**, pp. 377–95.

Klemsdal, T. 1982: Coastal classification and the coast of Norway. *Norsk Geogr. Tidsskr.* **36**, pp. 129–52.

Koba, M., Takashi, N. and Takahashi, T. 1982: Late Holocene eustatic sea-level changes deduced from geomorphological features and their ^{14}C dates in the Ryukyu Islands, Japan. *Palaeogeogr., Palaeoclim., Palaeoecol.* **39**, pp. 231–60.

Kobluk, D. R. and Risk, M. J. 1977: Rate and nature of infestation of a carbonate substratum by a boring algae. *J. Exper. Marine Biol. Ecol.* **27**, pp. 107–15.

Koch, B. and Disteche, A. 1984: Pressure effect on magnesian calcite coating and synthetic magnesian calcite seeds added to sea water in a closed system. *Geochim. Cosmochim. Acta* **48**, pp. 583–9.

Kohlmeyer, J. 1969: The role of marine fungi in the penetration of calcareous substances. *Am. Zool.* **9**, pp. 741–6.

Komar, P. D. 1976: *Beach Processes and Sedimentation* (Prentice-Hall, Englewood Cliffs, NJ).

Konishi, K., Omura, A. and Nakamichi, O. 1974: Radiometric coral ages and sea level records from the Late Quaternary reef complexes of the Ryukyu Islands. *Proc. 2nd Int. Coral Reef Symp. (Queensland, Austr.)* **2**, pp. 595–613.

——, Schlanger, S. O. and Omura, A. 1970: Neotectonic rates in the central Ryukyu Islands derived from Th^{230} ages. *Marine Geol.* **9**, pp. 225–40.

Koster, R. 1971: Postglacial sea-level changes on the German North Sea and Baltic shorelines. *Quaternaria* **14**, pp. 97–100.

Kowsmann, R. O. and Costa, M. P. A. 1974: Sea levels on the northern and southern parts of the Brazilian shelf. *Revista Brasileira Geociências* **4**, pp. 215–22.

Kranck, K. 1972: Geomorphological development and post-Pleistocene sea level changes, Northumberland Strait, Maritime Provinces. *Can. J. Earth Sci.* **9**, pp. 835–44.

Krumbein, W. E. and Pers, J. N. C. van der 1974: Diving investigations on biodeterioration by sea urchins in the sublittoral of Helgoland. *Helgolander Wiss. Meeresunters* **26**, pp. 1–17.

Ku, T.-L. and Kern, J. P. 1974: Uranium-series age of the upper Pleistocene Nestor terrace, San Diego, California. *Geol. Soc. Am. Bull.* **85**, pp. 1713–6.

Kuenen, P. H. 1933: Geology of coral reefs. *The Snellius Expedition* **5**, part 2, pp. 1–125.

——. 1950: *Marine Geology* (Wiley, New York).

Kukla, G. J., Berger, A., Lotti, R. and Brown, J. 1981: Orbital signature of interglacials. *Nature* **290**, pp. 295–300.

Kuznetsov, A. 1970: Wave loads against the vertical wall at a random angle of approaching waves. *Abstr. 12th Conf. Coastal Eng.*, pp. 227–32.

Kwaad, F. J. P. M. 1970: Experiments on the granular disintegration of granite by salt action. In *Univ. Amsterdam Fysisch Geografisch en Bodemkundig Lab. Publ.* **16**, pp. 67–80.

Labeyrie, J., Lalou, C. and Delibrias, G. 1969: Étude des transgressions marines sur l'atoll de Mururoa par la datation des différents niveaux de corail. *Cah. Pacifique* **13**, pp. 59–68.

Laborel, J. 1966: Contribution à l' étude des Madréporaires des Bermudes (systématique et répartition). *Bull. Mus. Nat. Hist. National*, 2nd series, 38.

—— and Delibrias, G. 1976: Niveaux marins recents à vermetidae du littoral ouest Africain. *Assoc. Sénégal Quat. Afr. Bull. Liaison* **47**, pp. 97–110.

Lalou, C. and Duplessey, J. C. 1977: Sea level variations: interest for neotectonic studies. In *International Symposium on Geodynamics in the South-West Pacific (Noumea, New Caledonia)*, (Editions Technip, Paris), pp. 405–12.

——, ——, and Nguyen, H. V. 1971: Données géochronologiques actuelles sur les niveaus de l'interglaciaire Riss/Wurm. *Rev. Géogr. Phys. Dyn.* **13**, pp. 447–61.

——, Nguyen, H. Van, Faure, H. and Moreira, L. 1970: Datation par la méthode uranium–thorium des hauts niveaux de coraux de la dépression de l'Afar (Ethiopie). *Rev. Géogr. Phys. Géol. Dyn.* **12**, pp. 3–8.

Land, L. S., MacKenzie, T. F. and Gould, S. J. 1967: Pleistocene history of Bermuda. *Geol. Soc. Am. Bull.* **78**, pp. 993–1006.

Langford-Smith, T. and Hails, J. R. 1966: New evidence of Quaternary sea levels from the north coast of New South Wales. *Austr. J. Sci.* **28**, pp. 352–3.

Lappo, D. D. and Zagryadskaya, N. N. 1977; Studies of pressure and energy of standing waves. *Am. Soc. Civil Eng. J., Waterways, Port, Coastal and Ocean Div.* **103**, pp. 335–47.

Larras, J. 1937a: Solution rigoureuse du problème du clapotis. *Ann. Ponts Chauss.* **107**, pp. 392–9.

——. 1937b: La résistance des jetées verticales aux houles obliques. *Ann. Ponts Chauss.* **107**, pp. 167–81.

——. 1937c: Le déferlement des lames sur les jetées verticales. *Ann. Ponts Chauss.* **107**, pp. 643–80.

Larsen, T. C. and Cady, P. D. 1967: Identification of aggregates exhibiting frost susceptibility. NCHRP Final Rept. Project 4–3, Dept. Civil Eng., Penn. State University, Pennsylvannia.

——, ——. 1969: Identification of frost-susceptible particles in concrete aggregates. Nat. Co-operative Highway Res. Progr. Rept. 66, pp. 1–62.

Launay, J. and Recy, J. 1972: Variations relatives de la mer et néo-tectonique en Nouvelle-Calédonie au Pléistocene supérieur et a l'Holocène. *Rev. Géogr. Phys. Géol. Dyn.* **14**, pp. 47–65.

Laurent, J. and Devimeux, W. 1951: Étude expérimentale de la réflexion de la houle sur des obstacles accorés. *Rev. Gén Hydraul.* **17**, pp. 235–46.

Lautridou, J. P. 1971: Conclusions générales des recherches de gélifraction expérimentale du Centre de Géomorphologie. *Centre de Géomorphologie, CRNS Bull.* **10**, pp. 63–84.

——. 1978: Principaux resultats des experiences de gelifraction expérimentale effectuées au Centre de Géomorphologie. *Inter-Nord* **15**, pp. 5–13.

—— and Coutard, J.-P. 1971: Recherches de gélifraction expérimentale du Centre de Géomorphologie. *Centre de Géomorphologie, CRNS Bull.* **10**, pp. 1–84.

—— and Ozouf, J. C. 1982: Experimental frost shattering. *Progr. Phys. Geogr.* **6**, pp. 215–32.

Lawson, A. C. 1893: The post-Pliocene diastrophism of the coast of southern California. *Univ. Calif. Dept. Geol. Sci. Bull.* **1**, pp. 115–60.

——. 1940: Isostatic control of fluctuations of sea level. *Science* **92**, pp. 162–4.

Leach, A. L. 1933: The geology and scenery of Tenby and the south Pembrokeshire coast. *Proc. Geol. Assoc.* **44**, pp. 187–216.

Le Campion-Alsumard, T. 1975: Étude expérimentale de la colonisation d'éclats de calcite par les Cyanophycées endolithes marine. *Cah. Biol. Marine* **16**, pp. 177–85.

——. 1979: Les cyanophycées endolithes marines. Systématique, ultrastructure, écologie et biodestruction. *Oceanologica Acta* **2**, pp. 143–56.

Lee, L., Pinckney, C. J. and Bemis, C. 1976: Sea cliff base erosion, San Diego,

California. *Am. Soc. Civ. Eng. National Water Resource Ocean Eng. Conf. April 5–8, 1976* preprint 2708, pp. 1–13.

Lelong, F. 1966: Régime des nappes phréatiques contenues dans les formations d'altération tropicale. Conséquences pour la pédogenèse. *Sci. Terre* **11**, pp. 203–44.

—— and Millet, G. 1966: Sur l'origine des minéraux micaces des altérations latéritiques. Diagenèse régressive minéraux en transit. *Bull. Ser. Carte. Géol. Alsace Lorraine* **19**, pp. 271–87.

Lewis, D. W., Dolch, W. L. and Woods, K. P. 1953: Porosity determinations and the significance of pore characteristics of aggregates. *Am. Soc. Test. Mat. Proc.* **53**, pp. 949–58.

Lewis, J. R. 1964: *The Ecology of Rocky Shores* (English Univ. Press, London).

Ley, R. G. 1977: The influence of lithology on marine karren. *Abh. Karst-und Hohlenkunde, A-Spelaologie* **15**, pp. 81–100.

——. 1979: The development of marine karren along the Bristol Channel coastline. *Z. Geomorph*. Suppl. Band 32, pp. 75–89.

Lietz, J. and Schmincke, H.-U. 1975: Miocene–Pliocene sea-level changes and volcanic phases on Gran Canaria (Canary Islands) in the light of new K–Ar ages. *Palaeogeogr., Palaeoclim., Palaeoecol*. **18**, pp. 213–39.

Lind, A. O. 1969: *Coastal Landforms of Cat Island, Bahamas* (University of Chicago, Chicago) Dept. Geog. Res. Paper 122, pp. 1–156.

Linton, D. L. 1955: The problem of tors. *Geogr. J*. **121**, pp. 470–87.

——. 1964: Tertiary landscape evolution. In *The British Isles*, ed. J. B. Sissons (Nelson, London), pp. 110–30.

Lira, J. 1926: Breakwaters or jetties in tideless seas. *Int. Congr. Navig. Rept. (Cairo)* **29**.

——. 1927: Le calcul des brise-lames à parement vertical. *Genie Civil* **90**, pp. 140–5.

Litvan, G. G. 1972a: Phase transitions of adsorbates: III. Heat effects and dimensional changes in nonequilibrium temperature cycles. *J. Colloid and Interface Sci*. **38**, pp. 75–83.

——. 1972b: Phase transitions of adsorbates: IV. Mechanism of frost action in hardened cement paste. *J. Am. Ceramic Soc*. **55**, pp. 38–42.

——. 1973a: Frost action in cement paste. *Matériaux et Constructions* **6**, pp. 293–8.

——. 1973b: Phase transitions of adsorbates: V. Aqueous sodium chloride solutions adsorbed of porous silica glass. *J. Colloid and Interface Sci*. **45**, pp. 154–69.

——. 1975: Phase transitions of adsorbates: VI. Effect of de-icing agents on the freezing of cement paste. *J. Am. Ceramic Soc*. **58**, pp. 26–30.

——. 1976: Frost action in cement in the presence of de-icers. *Cement and Concrete Res*. **6**, pp. 351–6.

Lockwood, J. G. 1980: Milankovitch theory and ice ages. *Progr. Phys. Geogr*. **4**, pp. 79–87.

Loney, R. A. 1964: Stratigraphy and petrography of the Pybus–Gambier area, Admiralty I., Alaska. *US Geol. Surv. Bull.* **1178**, pp. 1–103.

Longuet-Higgins, M. S. 1980: The unsolved problem of breaking waves. *Proc. 17th Conf. Coastal Eng*., pp. 1–28.

Longwell, C. R. and Flint, R. F. 1955: *Physical Geology* (Wiley, New York).
Lorius, C., Merlivat, L., Jouzel, J. and Pourchet, M. 1979: A 30,000-yr isotope climatic record from Antarctic ice. *Nature* **280**, pp. 644–8.
Loughnan, F. 1969: *Chemical Weathering of Silicate Minerals* (Elsevier, New York).
Luckman, B. H. 1976: Rockfalls and rockfall inventory data: some observations from Surprise Valley, Jasper National Park, Canada. *Earth Surface Processes* **1**, pp. 287–98.
Luiggi, L. 1922: Discussion on paper no. 4312. The improvement of the port of Valparaiso. *Min. Proc. Inst. Civil Eng.* **214**, pp. 40–5.
Lukas, K. J. 1969: An investigation of the filamentous endolithic algae in shallow-water corals from Bermuda. Repts. Res. 7th Graduate Res. Sem., ed. R. N. Ginsburg and P. Garrett (Bermuda Biol. Stat., Bermuda), pp. 145–52.
Lundgren, H. 1969: Wave shock forces: an analysis of deformations and forces in the wave and in the foundation. *Symp. Res. Wave Action (Delft Hydraulic Lab., Netherlands)* **2**, pp. 1–20.
Lyell, C. 1873: *Principles of Geology* (D. Appleton, New York), vol. 1, 11th ed.
Mabbutt, J. A. 1977: *Desert Landforms* (MIT Press, Cambridge, Mass.).
McCann, S. B. and Owens, E. H. 1969: The size and shape of sediments in three Arctic beaches, southwest Devon Island, NWT, Canada. *Arctic and Alpine Res.* **1**, pp. 267–78.
——, ——. 1970: Plan and profile characteristics of beaches in the Canadian Arctic archipelago. *Shore and Beach* **38**, pp. 26–50.
—— and Richards, A. 1969: The coastal features of the island of Rhum in the Inner Hebrides. *Scot. J. Geol.* **5**, pp. 15–25.
McCowan, J. 1891: On the solitary wave. *Phil. Mag.* **32**, pp. 45–58.
Macfadyen, W. A. 1930: The undercutting of coral reef limestone on the coasts of some islands of the Red Sea. *Geogr. J.* **75**, pp. 27–37.
McFarlan, E. 1961: Radiocarbon dating of late Quaternary deposits, south Louisiana. *Geol. Soc. Am. Bull.* **72**, pp. 129–58.
MacGeachy, J. K. 1977: Factors controlling sponge boring in Barbados reef corals. *Proc. 3rd Int. Coral Reef Symp. (Miami, Florida)* **2**, pp. 477–83.
McGreevy, J. P. 1981: Some perspectives on frost shattering. *Progr. Phys. Geogr.* **5**, pp. 56–75.
——. 1982: 'Frost and salt' weathering: further experimental results. *Earth Surface Processes and Landforms* **7**, pp. 475–88.
——. 1985a: A preliminary scanning electron microscope study of honeycomb weathering of sandstone in a coastal environment. *Earth Surface Processes and Landforms* **10**, pp. 509–18.
——. 1985b: Thermal properties as controls on rock surface temperature maxima, and possible implications for rock weathering. *Earth Surface Processes and Landforms* **10**, pp. 125–36.
—— and Whalley, W. B. 1982: The geomorphic significance of rock temperature variations in cold environments: a discussion. *Arctic Alpine Res.* **14**, pp. 157–62.
——, ——. 1984: Weathering. *Progr. Phys. Geogr.* **8**, pp. 543–69.
Machado, F. 1967: Geological evidence for a pulsating gravitation. *Nature* **214**, pp. 1317–18.

MacInnes, C. and Whiting, J. D. 1979: The frost resistance of concrete subjected to a de-icing agent. *Cement and Concrete Res.* **9**, pp. 325–36.

McIntire, W. G. 1961: Mauritius: river mouth terraces and present eustatic sea stand. *Z. Geomorph.* Suppl. Band 3, pp. 39–47.

—— and Walker, H. J. 1964: Tropical cyclones and coastal morphology in Mauritius. *Ann. Assoc. Am. Geogr.* **54**, pp. 582–96.

MacIntyre, I. G., Pilkey, O. H. and Stuckenrath, R. 1978: Relict oysters on the United States Atlantic continental shelf: a reconsideration of their usefulness in understanding late Quaternary sea-level history. *Geol. Soc. Am. Bull.* **89**, pp. 277–82.

Mackay, J. R. 1963: The Mackenzie Delta area, N. W. T. Canada. *Mines and Tech. Surv. Geogr. Branch Mem.* **8**, pp. 1–202.

Maclaren, C. 1842: The glacial theory of Prof. Agassiz. *Am. J. Sci.* **42**, pp. 346–65.

McLean, J. F. 1967: Objective description of shore platforms on the northeast coast of the South Island. Univ. Canterbury (NZ), MA thesis.

McLean, R. F. 1964: Mechanical and biological erosion of beach rock in Barbados, West Indies. McGill Univ. Montreal, Ph.D. thesis.

——. 1967a: Measurements of beachrock erosion by some tropical marine gastropods. *Bull. Marine Sci.* **17**, pp. 551–61.

——. 1967b: Erosion of burrows in beachrock by the tropical sea urchin *Echinometra lucunter. Can. J. Zool.* **45**, pp. 586–8.

——. 1974: Geological significance of bioerosion of beachrock. *Proc. 2nd Int. Coral Reef Symp. (Queensland, Australia)* **2**, pp. 401–8.

—— and Davidson, C. F. 1968: The role of mass-movement in shore platform development along the Gisborne coastline, New Zealand. *Earth Sci. J.* **2**, pp. 15–25.

——, Stoddart, D. R., Hopley, D. and Polach, H. 1978: Sea level change in the Holocene on the northern Great Barrier Reef. *Phil. Trans. Roy. Soc. London* 291A, pp. 167–86.

Malin, M. 1974: Salt-weathering on Mars. *J. Geophys. Res.* **79**, pp. 3888–94.

Malmgren, B. A. and Kennett, J. P. 1978: Late Quaternary paleoclimatic applications of mean size variations in *Globigerina bulloides* d'Orbigny in the southern Indian Ocean. *J. Palaeont.* **52**, pp. 1195–207.

Mariette, H. 1971: L'archéologie des dépôts flandriens du Boulonnais. *Quaternaria* **14**, pp. 137–50.

Markov, K. K. 1934: Criteria of transgression and regression. *Proc. 1st All-Union Geogr. Congr.* **3**.

Marshall, J. F. and Launay, J. 1978: Uplift rates of the Loyalty Islands as determined by $^{230}Th/^{234}U$ dating of raised coral terraces. *Quat. Res.* **9**, pp. 186–92.

—— and Thom, B. G. 1976: The sea level in the last interglacial. *Nature* **263**, pp. 120–1.

Martens, C. S., Wesolowski, J. J., Harris, R. C. and Kaifer, R. 1973: Chlorine loss from Puerto Rican and San Francisco Bay area marine aerosols. *J. Geophys. Res.* **78**, pp. 8778–92.

Martin, L., Suguio, K., Flexor, J.-M., Bittencourt, A. and Vilas-Boas, G. 1980: Le

Quaternaire marin brêsilien (littoral pauliste sud fluminense et bahianais) *Cah. ORSTOM* (sér. géol.) **11**, pp. 95–124.

Martini, A. 1967: Preliminary experimental studies on frost weathering of certain rock types from the west Sudetes. *Biul. Peryglac.* **16**, pp. 147–94.

——. 1973: Experimental investigations of frost weathering on granites. *Studia Geographica Brno* **33**, pp. 61–6.

——. 1975: The weathering of beach pebbles in Hornsund. *Acta Universitatis Wratislaviensis* (Studia Gogaficzne) 251, pp. 187–93.

Martini, I. P. 1978: Tafoni weathering, with examples from Tuscany, Italy. *Z. Geomorph.* **22**, pp. 44–67.

——. 1981: Ice effect on erosion and sedimentation on the Ontario shores of James Bay, Canada. *Z. Geomorph.* **25**, pp. 1–16.

Mason, B. J. 1976: Towards the understanding and prediction of climatic variations. *Quart. J. Roy. Meteor. Soc.* **102**, pp. 473–98.

Massel, S. R., Oleskiewicz, M. and Trapp, W. 1978: Impact wave forces on vertical and horizontal plate. *Proc. 16th Conf. Coastal Eng.*, pp. 2340–59.

Masseport, J. 1959: Premiers résultats d'expériences de laboratorie sur les roches. *Revue Géogr. Alpine* **48**, pp. 531–7.

Mather, B. 1969: Behavior of concrete exposed to the sea. In *Civil Engineering in the Oceans,* Proc. Am. Soc. Civil Eng. Conf. (Miami Beach), pp. 987–98.

Matthews, J. A., Dawson, A. G. and Shakesby, R. A. 1986: Lake shoreline development, frost weathering and rock platform erosion in an alpine environment, Jotunheimen, southern Norway. *Boreas* **15**, pp. 33–50.

Matthews, R. K. 1973: Relative elevation of Late Pleistocene high sea level stands: Barbados uplift rates and their implications. *Quat. Res.* **3**, pp. 147–53.

—— and Poore, R. Z. 1980: Tertiary δ^{18}O record and glacio-eustatic sea-level fluctuations. *Geology* **8**, pp. 501–4.

May, V. J. 1971: The retreat of chalk cliffs. *Geogr. J.* **137**, pp. 203–6.

Mellor, M. 1970: Phase composition of pore water in cold rocks. US Army Corps Eng. *CRREL Res. Rept.* **292**, pp. 1–61.

——. 1971: Strength and deformability of rocks at low temperatures. US Army Corps. Eng. *CRREL Res. Rept.* **294**, pp. 1–73.

——. 1973: Mechanical properties of rocks at low temperatures. In *Permafrost* (North American Contribution to the 2nd Int. Conf., Nat. Acad. Sci. Washington, DC), pp. 334–44.

Menard, H. W. 1969: Elevation and subsidence of oceanic crust. *Earth Planet. Sci. Lett.* **6**, pp. 275–84.

Mensching, H. 1965: Beobachtungen zum formenschatz des küstenkarstes an der kantabrischen kuste bei Santander und Llanes (Nordspanien). *Erdkunde* **19**, pp. 24–31.

Mercer, J. H. 1968a: The discontinuous glacio-eustatic fall in Tertiary sea level. *Palaeogeogr., Palaeoclim., Palaeoecol.* **5**, pp. 77–85.

——. 1968b: Antarctic ice and Sangamon sea level. *Int. Assoc. Hydrol. Sci. Publ.* **79**, pp. 217–25.

——. 1981: West Antarctic ice volume: the interplay of sea level and temperature, and a strandline test for absence of the ice sheet during the last interglacial. In *Sea Level, Ice, and Climatic Change*, Proc. Canberra Symp. Dec. 1979. Int. Assoc. Hydrol. Sci. Publ. **131**, pp. 323–30.

Merriam, R. 1960: Portuguese Bend landslide, Palos Verdes Hills, California. *J. Geol.* **68**, pp. 140–53.

Mesolella, K. J., Matthews, R. K., Broecker, W. S. and Thurber, D. L. 1969: The astronomical theory of climatic change: Barbados data. *J. Geol.* **77**, pp. 250–74.

——, Sealy, H. A. and Matthews, R. K. 1970: Facies geometries within Pleistocene reefs of Barbados, West Indies. *Am. Assoc. Petrol. Geol. Bull.* **54**, pp. 1899–917.

Meszaros, A. and Vissy, K. 1974: Concentration, size distribution, and chemical nature of atmospheric aerosol particles in remote oceanic areas. *Aerosol Sci.* **5**, pp. 101–9.

Miche, R. 1944: Mouvements endulatoires de la mer en profondeur constante ou décroissante. *Ann. Ponts Chauss.* **114**, pp. 25–78, 131–64, 270–92, 369–406.

——. 1951: Le pouvoir réfléchissant des ouvrages maritimes exposés à l'action de la houle. *Ann. Ponts Chauss.* **121**, pp. 285–319.

Middlemiss, F. A. 1983: Instability of Chalk cliffs between the South Foreland and Kingsdown, Kent, in relation to geological structure. *Proc. Geol. Assoc.* **94**, pp. 115–22.

Mii, H. 1962: Coastal geology of Tanabe Bay. *Sci. Rept. Tohoku Univ. (series 2)* **34**, pp. 1–93.

——. 1963: Relation of shore erosion to sea level. *J. Marine Geol.* **2**, pp. 8–17.

Milankovitch, M. 1938: Astronomische mittel zur erforschung der erdgeschichtlichen klimate. *Handb. d. Geophys.* **9**, pp. 593–698.

Miller, A. A. 1935: The entrenched meanders of the Herefordshire Wye. *Geogr. J.* **85**, pp. 160–78.

——. 1937: The 600′ plateau in Pembrokeshire and Carmarthenshire. *Geogr. J.* **90**, pp. 148–59.

——. 1939a: River development in southern Ireland. *Proc. Roy. Ir. Acad.* **45B**, pp. 32–54.

——. 1939b: Attainable standards of accuracy in the determination of preglacial sea levels by physiographic methods. *J. Geomorph.* **2**, pp. 95–115.

——. 1955: The origin of the south Ireland peneplane. *Ir. Geogr.* **3**, pp. 79–86.

Miller, D. J. 1960: Giant waves in Lituya Bay, Alaska. *US Geol. Surv. Prof. Paper* **354-C**, pp. 51–86.

Miller, G. H., Locke, W. W. and Locke, C. W. 1980: Physical characteristics of the southeastern Baffin Island coastal zone. In *The Coastline of Canada*, ed. S. B. McCann, Geol. Surv. Can. Paper **80–10**, pp. 251–65.

Miller, R. L. 1972: Study of air entrainment in breaking waves. *Eos* **53**, p. 426.

——, Leverette, S., O'Sullivan, J., Tochko, J. and Theriault, K. 1974: Field measurements of impact pressures in surf. *Proc. 14th Conf. Coastal Eng.*, pp. 1761–77.

Miller, W. J. 1931: The landslide at Point Firmin, California. *Scientific Monthly* **32**, pp. 464–9.

Milliman, J. D. and Emery, K. O. 1968: Sea levels during the past 35,000 years. *Science* **162**, pp. 1121–3.

Milojevic, B. Z. 1952: Les formes karstiques de la côte Dinarique. *Vol. Jubil. Cinquant. Anniv. Labor. Geogr. Univ. Rennes*, pp. 198–207.

Minikin, R. R. 1950: *Winds, Waves, and Maritime Structures* (Charles Griffin, London).

Minty, E. J. 1965: Preliminary report on an investigation into the influence of several factors on the sodium sulphate soundness test for aggregate. *Austr. Road Res.* **2**, pp. 49–52.

Mitchell, G. F. 1977: Raised beaches and sealevels. In *British Quaternary Studies, Recent Advances*, ed. F. W. Shotton (Clarendon Press, Oxford), pp. 169–86.

Mitsuyasu, H. 1963: Wave pressure and wave run-off (in Japanese). In *Handbook of Hydraulics*, ed. S. Yokota, *(Japan. Soc. Civil Eng. Tokyo)*, pp. 505–28.

——. 1966: Shock pressure of breaking wave. *Proc. 10th Conf. Coastal Eng.*, pp. 268–83

Mogi, A., Tsuchide, M. and Fukushima, M. 1980: Coastal erosion of the new volcanic island Nishinoshima. *Geogr. Rev. Japan* **53**, pp. 449–62.

Mogridge, G. R. and Jamieson, W. W. 1980: Wave impact pressures on composite breakwaters. *Proc. 17th Conf. Coastal Eng.*, pp. 1829–48.

Moign, A. 1974a: Strandflats immergés et émergés du Spitsberg central et nord-occidental (Thesis Brest, Lille III).

——. 1974b: Géomorphologie du strandflat au Svalbard; problèmes (âge, origine, processes) méthodes de travail. *Inter-Nord* **13–14**, pp. 57–72.

——. 1976: L'action des glaces flottantes sur le littoral et les fonds marins du Spitsberg central et nord-occidental. *Rev. Géogr. Montreal* **30**, pp. 51–64.

Moign, Y. and Moign, A. 1970: Les îles Heimaey and Surtsey de l'archipel volcanique Vestmannaejar (Islande). *Norois* **67**, pp. 305–34.

Molinier, R. 1954: Première contribution à l'étude des peuplements marins superficiels des Iles Pithyuses (Baléares). *Vie et Milieu* **5**, pp. 226–42.

——. 1955a: Les plate-formes et corniches récifales de vermets (*Vermetus cristatus* Biondi) en Méditerranée occidentale. *Comptes Rendus Acad. Sci.* **240**, pp. 361–3.

——. 1955b: Deux nouvelles formations organogènes construites en Méditerranée occidentale. *Comptes Rendus Acad. Sci.* **240**, pp. 2166–8.

——. 1960: Étude des biocénoses marines du Cap Corse. *Végétatio* **9**, pp. 121–312.

——. and Picard, J. 1953: Notes biologiques à propos d'un voyage d'études sur les côtes de Sicile. *Ann. Inst.Ocèanogr.* **28**, pp. 164–87.

——. 1954: Eléments de bionomie marine sur les côtes de Tunisie. *Ann. Station Océanogr. Salammbô* **48**.

——, ——. 1957: Un nouveau type de plateforme organogène dans l'étage médiolittoral sur les côtes de L'île de Majorque (Baléares). *Comptes Rendus Acad. Sci.* **244**, pp. 674–5.

Molitor, D. 1934: Wave pressures on sea-walls and breakwaters. *Am. Soc. Civil Eng. Papers* **60**, pp. 653–71.

Montaggioni, L. 1974; Coral reefs and Quaternary shore-lines in the Mascarene Archipelago (Indian Ocean). *Proc. 2nd Coral Reef Symp. (Queensland, Austr.)* **2**, pp. 579–93.

Moore, C. H. and Shedd, W. W. 1977: Effective rates of sponge bioerosion as a function of carbonate production. *Proc. 3rd Int. Coral Reef Symp. (Miami, Florida)* **2**, pp. 499–505.

Moore, D. G. 1954: Origin and development of sea caves. *Bull. Am. Caver* **16**, pp. 71–6.

Moore, W. S. and Somayajulu, B. L. K. 1974: Age determinations of fossil corals using ^{230}Th/^{234}Th and ^{230}Th/^{227}Th. *J. Geophys. Res.* **79**, pp. 5065–8.

Moraes, C. de C. 1970: Experiments on wave reflexion on impermeable slopes. *Proc. 12th Conf. Coastal Eng.*, pp. 509–21.

Morison, J. R. 1948: Wave pressures on a vertical wall. *Dept. Eng. Univ. Calif. Berkeley Tech. Rept.* **HE-116-298**, pp. 1–11.

——, Johnson, J. W. and O'Brien, M. P. 1954: Experimental studies of wave forces on piles. *Proc. 4th Conf. Coastal Eng.*, pp. 340–70.

Mörner, N.-A. 1969: The late Quaternary history of the Kattegatt Sea and the Swedish west coast. *Sverig. Geol. Unders*. Ser. C, no. 640, pp. 1–487.

——. 1971a: The position of the ocean level during the interstadial about 30,000 BP—a discussion from a climatic–glaciologic point of view. *Can. J. Earth Sci*. **8**, pp. 132–43.

——. 1971b: Eustatic and climatic oscillations. *Arctic Alpine Res.* **3**, pp. 167–71.

——. 1971c: The Holocene eustatic sea level problem. *Geol. Mijnbouw* **50**, pp. 699–702.

——. 1971d: Late Quaternary isostatic, eustatic, and climatic changes. *Quaternaria* **14**, pp. 65–83.

——. 1971e: Eustatic changes during the last 20,000 years and a method of separating the isostatic and eustatic factors in an uplifted area. *Palaeogeogr., Palaeoclim., Palaeoecol.* **9**, pp. 153–81.

——. 1971f: Relations between ocean, glacial, and crustal changes. *Geol. Soc. Am. Bull.* **82**, pp. 787–8.

——. 1972: Isostasy, eustasy, and crustal sensitivity. *Tellus* **24**, pp. 586–92.

——. 1973: Eustatic changes during the last 300 years. *Palaeogeogr., Palaeoclim., Palaeoecol.* **13**, pp. 1–14.

——. 1976: Eustasy and geoid changes. *J. Geol*. **84**, pp. 123–51.

——. 1980a: Relative sea level, tectono-eustasy, geoidal-eustasy, and geodynamics during the Cretaceous. *Cretaceous Res*. **1**, pp. 329–40.

——. 1980b: Eustasy and geoid changes as a function of core/mantle changes. In *Earth Rheology, Isostasy, and Eustasy*, ed. N.-A. Mörner (Wiley, Chichester), pp. 535–53.

——. 1981a: Eustasy, paleoglaciation, and paleoclimatology. *Geol. Rundschau* **70**, pp. 691–702.

——. 1981b: Eustasy, palaeogeodesy, and glacial volume changes. In *Sea Level, Ice and Climatic Change*, Proc. Canberra Symp. Dec. 1979, *Int. Assoc. Hydrol. Sci. Publ.* **131**, pp. 277–80.

——. 1983: Sea levels. In *Mega-Geomorphology*, ed. R. Garner and H. Scoging (Clarendon Press, Oxford), pp. 73–91.

Morse, J. W., Mucci, A. and Millero, F. J. 1980: The solubility of calcite and aragonite in seawater of 35‰ salinity at 25°C and atmospheric pressure. *Geochim. Cosmochim. Acta* **44**, pp. 85–94.

——, ——, Walter, L. M. and Kaminsky, M. S. 1979: Magnesium interaction with the surface of calcite in seawater. *Science* **205**, pp. 904–5.

Mortensen, H. 1933: Die salzsprengung und ihre bedeutung fur die regional klimatische gliederung der wusten. *Pet. Geogr. Mitt.* **79**, pp. 130–5.

Mortimer, C. and Saric, N. 1972: Landform evolution in the coastal region of Tarapaca province, Chile. *Rev. Géomorph. Dyn*. **21**, pp. 162–70.

——. ——. 1975: Cenozoic studies in northernmost Chile. *Geol. Rundschau* **64**, pp. 395–420.

Mottershead, D. N. 1981: The persistence of oil pollution on a rocky shore. *Applied Geogr.* **1**, pp. 297–304.

——. 1982a: Coastal spray weathering of bedrock in the supratidal zone of east Prawle, south Devon. *Field Studies* **5**, pp. 663–84.

——. 1982b: Rapid weathering of greenschist by coastal salt spray, east Prawle, south Devon: a preliminary report. *Proc. Ussher Soc.* **5**, pp. 347–53.

Mucci, A. and Morse, J. W. 1984: The solubility of calcite in seawater solutions of various magnesium concentration, $I_t = 0.697$ m at 25°C and one atmosphere total pressure. *Geochim. Cosmochim. Acta* **48**, pp. 815–22.

Muhs, D. R. 1979: Geomorphic implications of late Quaternary paleosols on the Palos Verdes Peninsula, California. *Geol. Soc. Am. Abstracts with Programs* **11**, p. 297.

——. 1983: Quaternary sea-level events on northern San Clemente Island, California. *Quat. Res.* **20**, pp. 322–41.

—— and Szabo, B. J. 1982: Uranium-series age of the Eel Point terrace, San Clemente Island, California. *Geology* **10**, pp. 23–6.

Muir-Wood, A. M. 1955: Folkestone Warren landslips: investigations 1948–1950. *Proc. Inst. Civil Eng.* **4**, pp. 410–28.

——. 1971: Engineering aspects of coastal landslides. *Proc. Inst. Civil Eng.* **50**, pp. 257–76.

Mullin, J. W. 1961: *Crystallization* (Butterworths, London).

Munk, W. H. 1949: The solitary wave theory and its application to surf problems. *Ann. N.Y. Acad. Sci.* **51**, pp. 376–424.

—— and Macdonald, G. F. 1960: *The Rotation of the Earth* (Cambridge University Press, London).

—— and Revelle, R. 1952: Sea level and the rotation of the Earth. *Am. J. Sci.* **250**, pp. 829–33.

Mustoe, G. E. 1982: The origin of honeycomb weathering. *Geol. Soc. Am. Bull.* **93**, pp. 108–15.

——. 1983: Cavernous weathering in the Capitol Reef Desert, Utah. *Earth Surface Processes and Landforms* **8**, pp. 517–26.

Nadson, G. 1927: Les algues perforantes, leur distribution, et leur rôle dans la nature. *Comptes Rendus Acad. Sci.* **184**, pp. 896–8, 1015–7.

Nagai, S. 1961: Shock pressures exerted by breaking waves on breakwaters. *Trans. Am. Soc. Civil Eng.* **126**, pp. 772–809.

——. 1969: Pressures of standing waves on vertical wall. *Am. Soc. Civil Eng. Proc. Waterways, Harbors Div.* **95**, pp. 53–76.

Nakamura, M., Shiraishi, H. and Sasaki, Y. 1966: Wave decay due to breaking. *Proc. 10th Conf. Coastal Eng.*, pp. 234–53.

Nansen, F. 1904: *The Norwegian North Polar Expedition 1893–1896* (Longman, Green and Co., London). Reprinted in 1969 by Greenwood Press, New York, see pp. 102–30.

——. 1922: *The Strandflat and Isostasy*. Videnskapssel skapets Skrifter 1, Mat.-Naturv. Klasse 1921, no. 11 (Kristiana).

Nardin, T. R., Osborne, R. N. and Bottjer, D. J. 1981: Holocene sea-level curves for Santa Monica Shelf, California continental borderland. *Science* **213**, pp. 331–3.

Neef, G. and Veeh, H. H. 1977: Uranium series ages and late Quaternary uplift in the New Hebrides. *Nature* **269**, pp. 682–3.

Nerenst, P. 1960: Frost action in concrete. In *Chemistry of Cement, Proc. 4th Int. Symp. Nat. Bur. Std. Monogr.* **43**, ii, pp. 807–28.

Nesteroff, V. 1955: Les récifs coralliens du banc Farsan Nord (Mer Rouge). In *Résultats Scientifiques des Campagne de la 'Calypse'*: I. *Campagne en Mer Rouge (1951–1952)* (Masson, Paris).

Neudecker, S. 1977: Transplant experiments to test the effect of fish grazing on coral distribution. *Proc. 3rd Int. Coral Reef Symp. (Miami, Florida)* **1**, pp. 317–23.

Neumann, A. C. 1966: Observations on coastal erosion in Bermuda and measurement of the boring rate of the sponge *Cliona lampa. Limnol. Oceanogr.* **11**, pp. 92–108.

——. 1968: Biological erosion of limestone coasts. In *Encyclopedia of Geomorphology*, ed. R. W. Fairbridge, (Dowden, Hutchinson, and Ross, Stroudsburg, Penn.), pp. 75–81.

Neumann, C. A. and Moore, W. S. 1975: Sea level events and Pleistocene coral ages in the northern Bahamas. *Quat. Res.* **5**, pp. 215–24.

Newell, N. D. 1956: Geological reconnaissance of Raroia (Kon Tiki) Atoll, Tuamotu Archipelago. *Am. Mus. Nat. Hist. Bull.* **109**, pp. 315–72.

——. 1961: Recent terraces of tropical limestone shores. *Z. Geomorph.* Suppl. Band 3, pp. 87–106.

—— and Bloom, A. L. 1970: The reef flat and 'two-metre eustatic terrace' of some Pacific atolls. *Geol. Soc. Am. Bull.* **81**, pp. 1881–94.

—— and Imbrie, J. 1955: Biogeological reconnaissance in the Bimini area, Great Bahama Bank. *Trans. NY Acad. Sci.* **18**, pp. 3–14.

Newman, W. S. 1968: Coastal stability. In *The Encyclopedia of Geomorphology*, ed. R. W. Fairbridge (Dowden, Hutchinson, and Ross, Stroudsburg, Penn.), pp. 150–6.

——, Cinquemani, L. J., Pardi, R. R. and Marcus, L. F. 1980a: Holocene delevelling of the United States' east coast. In *Earth Rheology, Isostasy and Eustasy*, ed. N.-A. Mörner (Wiley, Chichester), pp. 449–63.

——, Marcus, L. F. and Pardi, R. R. 1981: Palaeogeodesy: late Quaternary geoidal configurations as determined by ancient sea levels. In *Sea Level, Ice and Climatic Change*, Proc. Canberra Symp. Dec. 1979. Int. Assoc. Hydrol. Sci. Publ. **131**, pp. 263–75.

——, ——, ——, Paccione, J. A. and Tomecek, S. M. 1980b: Eustasy and deformation of the geoid: 1,000–6,000 radiocarbon years BP. In *Earth Rheology, Isostasy, and Eustasy*, ed. N.-A. Mörner (Wiley, Chichester), pp. 555–67.

Nichols, R. L. 1960: Geomorphology of Marguerite Bay area, Palmer Peninsula, Antarctica. *Geol. Soc. Am. Bull.* **71**, pp. 1421–50.

Nicod, J. 1951: Les problème de la classification des 'calanques' parmi les formes de côtes de submersion. *Rev. Géomorph. Dyn.* **2**, pp. 120–7.

——. 1963: Problèmes de morphologie karstique en Grèce. *Méditerranée* **4**, pp. 15–25.

——. 1967: *Recherches morphologiques en Basse-Provence calcaire*. (Ophrys Editions, Gap-05).

——. 1972: *Pays et paysages du calcaire* (Presses Universitaires de France, Paris).

Nielsen, N. 1979: Ice-foot processes. Observations of erosion on a rocky coast, Disko, west Greenland. *Z. Geomorph*. **23**, pp. 321–31.

Nilsson, T. 1983: *The Pleistocene* (D. Reidel, Arlov, Sweden).

Nordenskjold, O. 1928: The Geography of the Polar Regions. *Am. Geogr. Soc. Spec. Publ*. **8** (ed. W. L. G. Joerg), pp. 58–60.

Norrman, J. O. 1980: Coast erosion and slope development in Surtsey Island, Iceland. *Z. Geomorph*. Suppl. Band 34, pp. 20–38.

North, F. J. 1929: *The Evolution of the Bristol Channel* (National Museum of Wales, Cardiff).

North, G. R. and Coakley, J. A. 1979: Differences between seasonal and mean energy balance model calculations of climate and climatic sensitivity. *J. Atmosph. Sci*. **36**, pp. 1189–204.

North, W. J. 1954: Size distribution, erosive activities, and gross metabolic efficiency of the marine intertidal snails, *Littorina planaxis* and *L. scutulata. Biol. Bull*. **106**, pp. 185–97.

Nossin, J. J. 1965: Analysis of younger beach ridge deposits in eastern Malaya. *Z. Geomorph*. **9**, pp. 186–208.

Nunn, P. D. 1984: Review of evidence for late Tertiary shorelines occurring on South Atlantic coasts. *Earth Sci. Rev*. **20**, pp. 185–210.

Nye, P. H. 1955: Some soil forming processes in the humid tropics: pt.ii. The development of the upper slope member of the catena. *J. Soil Sci*. **6**, pp. 51–62.

Oaks, R. Q., Coch, N. K., Saders, J. E. and Flint, R. F. 1974: Post-Miocene shorelines and sea levels, southeastern Virginia. In *Post-Miocene Stratigraphy, Central and Southern Atlantic Coast Plain*, ed. R. Q. Oaks and J. R. Dunbar (Utah State University Press, Logan, Utah), pp. 53–87.

Oerlemans, J. 1981: Effect of irregular fluctuations in Antarctic precipitaion on global sea level. *Nature* **290**, pp. 770–2.

Olausson, E. 1965: Evidence of climatic changes in deep-sea cores, with remarks on isotope palaeotemperature analysis. In *Progress in Oceanography*, ed. M. Shears (Pergamon Press, London), vol. 3, pp. 221–52.

Oldale, R. N. and O'Hara, C. J. 1980: New radiocarbon dates from the inner continental shelf off southeastern Massachusetts and a local sea level rise curve for the past 12,000 yr. *Geology* **8**, pp. 102–6.

Ollier, C. D. 1969: *Weathering* (Oliver and Boyd, Edinburgh).

Olmsted, F. H. 1958: Geological reconnaissance of San Clemente Island, California. *US Geol. Surv. Bull*. **1071-B**, pp. 55–68.

Omura, A., Emerson, W. K. and Ku, T. L. 1979: Uranium series ages of echinoids and corals from the upper Pleistocene Magdalena Terrace, Baja California Sur, Mexico. *The Nautilus* **94**, pp. 184–9.

Ongley, M. 1940: Note on coastal benches formed by spray weathering. *NZ J. Sci. Tech*. **B22**, pp. 34–5.

Orme, A. R. 1962: Abandoned and composite sea cliffs in Britain and Ireland. *Irish Geogr*. **4**, pp. 279–91.

——. 1964: Planation surfaces in the Drum Hills, County Waterford, and their wider implications. *Ir. Geogr.* **5**, pp. 48–72.

——. 1972: Quaternary deformation of western Baja California, Mexico, as indicated by marine terraces and associated deposits. In *Tectonics* (Harpell's Press Co-operative, Gardenvale, Québec), 24th Int. Geol. Congr. Sect. **3**, pp. 627–34.

——. 1974: Quaternary deformation of marine terraces between Ensenada and El Rosario. In *Geology of Peninsula California*, Pacific sections of the Am. Assoc. Petrol. Geol., SEPM, and SEG, pp. 67–79.

Osman, C. W. 1917: The landslips at Folkestone Warren and the thickness of the Lower Chalk and Gault near Dover. *Proc. Geol. Assoc.* **28**, pp. 59–84.

Osmond, J. K., Carpenter, J. R. and Windom, H. 1965: Th^{230}/U^{234} age of the Pleistocene corals and oolites of Florida. *J. Geophys. Res.* **70**, pp. 1843–7.

——, May, J. P. and Tanner, W. F. 1970: Age of the Cape Kennedy barrier-and-lagoon complex. *J. Geophys. Res.* **75**, pp. 469–79.

Otter, G. W. 1932: Rock-burrowing echinoids. *Biol. Rev.* **7**, pp. 89–107.

——. 1937: Rock-destroying organisms in relation to coral reefs. *Rept. Great Barrier Reef Exped.* **1**, pp. 324–52.

Owen, H. G. 1976: Continental displacement and expansion of the Earth during the Mesozoic and Cenozoic. *Phil. Trans. Roy. Soc. London* **281A**, pp. 223–91.

Paepe, R. 1971: Quaternary marine formations of Belgium. *Quaternaria* **15**, pp. 99–104.

Palmer, H. S. and Powers, H. A. 1935: Pits in coastal pahoehoe lavas controlled by gas bubbles. *J. Geol.* **43**, pp. 639–43.

Panzer, W. 1933: Junge kustenhebung im Bismarck-Archipel und auf Neu-Guinea. *Z. Ges. Erdk.* (Berlin), pp. 175–90.

——. 1949: Brandungshohlen und brandungskehlen. *Erdkunde* **3**, pp. 29–41.

Paskoff, R. P. 1978a: Aspects géomorphologiques de L'île de Paques. *Bull. Assoc. Géogr. Franç.* **452**, pp. 147–57.

——. 1978b: Sur l'évolution géomorphologique du grand escarpement côtier du désert Chilien. *Géogr. Phys. Quat.* **32**, pp. 351–60.

——. 1980: Late Cenozoic crustal movements and sea level variations in the coastal area of northern Chile. In *Earth Rheology, Isostasy, and Eustasy*, ed. N.-A. Mörner (Wiley, Chichester), pp. 487–95.

—— and Sanlaville, P. 1978: Observations géomorphologiques sur les côtes de l'archipel Maltais. *Z. Geomorph.* **22**, pp. 310–28.

Peach, A. M. E. 1916: The preglacial platform and raised beaches of Prince Charles foreland. *Trans. Edin. Geol. Soc.* **10**, pp. 289–91.

Pedro, G. 1957: Nouvelles recherches sur l'influence des sels dans la désagrégation des roches. *Comptes Rendus Acad. Sci.* **244**, pp. 2822–4.

Peltier, L. C. 1950: The geographic cycle in periglacial regions. *Ann. Assoc. Am. Geogr.* **40**, pp. 214–36.

Peltier, W. R. 1980: Ice sheets, oceans, and the Earth's shape. In *Earth Rheology, Isostasy, and Eustasy*, ed. N.-A. Mörner (Wiley, Chichester), pp. 45–63.

—— and Andrews, J. T. 1976: Glacial-isostatic adjustment: I. The forward problem. *Geophys. J. Roy. Astron. Soc.* **46**, pp. 605–46.

——, ——. 1983: Glacial geology and glacial isostasy of the Hudson Bay region. In *Shorelines and Isostasy*, ed. D. E. Smith and A. G. Dawson (Academic Press, London), Inst. Brit. Geogr. Spec. Publ. **16**, pp. 285–319.

——. Farrell, W. E. and Clark, J. A. 1978: Glacial isostasy and relative sea level: a global finite element model. *Tectonophys*. **50**, pp. 81–110.

Peregrine, D. H. and Svendsen, I. A. 1978: Spilling breakers, bores and hydraulic jumps. *Proc. 16th Conf. Coastal Eng*., pp. 540–51.

Pérès, J. M. 1968: Trottoir. In *Encyclopedia of Geomorphology*, ed. R. W. Fairbridge (Dowden, Hutchinson, and Ross, Stroudsburg, Penn), pp. 1173–4.

—— and Picard, J. 1952: Les corniches d'origine biologique en Méditerranée occidentale. *Rec. Travaux Station Marine D'Endoume* **4**, pp. 1–34.

——, ——. 1964: Nouveau manuel de bionomie benthique de la mer Méditerranée. *Rec. Travaux Station Marine d'Endoume* **31**, pp. 1–137.

Pettersson, O. 1914: Climatic variations in historic and prehistoric time. *Svenska Hydrogr. Biol. Komm*. **5**.

Pheasant, D. R. and Andrews, J. T. 1973: Wisconsin glacial chronology and relative sea-level movements, Narpaing Fiord Broughton Island Area, eastern Baffin Island, NWT. *Can. J. Earth Sci.* **10**, pp. 1621–41.

Phillips, B. A. M. 1970a: The significance of inheritance in the interpretation of marine and lacustrine coastal histories. *Lakehead Univ. Rev*. **3**, pp. 36–45.

——. 1970b: Effective levels of marine planation on raised and present rock platforms. *Rev. Geogr. Montreal* **24**, pp. 227–40.

——. 1977: Shoreline inheritance in coastal histories. *Science 195*, pp. 11–16.

——. 1978: Coastal erosion: north and east shores of Lake Superior. *Proc. 2nd Workshop of Great Lakes Erosion and Sedimentation*, pp. 37–40.

——. 1980: The morphology and processes of the Lake Superior north shore. In *The Coastline of Canada*, (ed. S. B. McCann), *Geol. Surv. Can. Paper* **80–10**, pp. 407–15.

Pillans, B. 1983: Upper Quaternary marine terrace chronology and deformation, south Taranaki, New Zealand. *Geology* **11**, pp. 292–7.

Pinckney, C. J. and Lee, L. 1973: Sea cliff recession study, southern one-half of San Diego County. In *Field Trip Guidebook Engineering Problem Areas in the Southwest*, ed. W. J. Elliott (San Diego County, California), pp. 14–18.

Pirazzoli, P. A. 1976: Sea level variations in the northwest Mediterranean during Roman times. *Science* **194**, pp. 519–21.

——. 1977: Sea level relative variations in the world during the last 2,000 years. *Z. Geomorph*. **21**, pp. 284–96.

——. 1978: High stands of Holocene sea levels in the northwest Pacific. *Quat. Res*. **10**, pp. 1–29.

Pissart, A. 1970: Les phénomènes physiques essentielles liés au gel, les structures périglaciares qui en résultent et leur signification climatique. *Ann. Soc. Géol. Belg*. **93**, pp. 7–49.

Pitman, W. C. 1979: The effect of eustatic sea level changes on stratigraphic sequences at Atlantic margins. In *Geological and Geophysical Investigations of Continental Margins*, ed. J. S. Watkins, L. Montadert and P. W. Dickerson, Am. Assoc. Petrol. Geol. Mem. **29**, pp. 453–60.

Pitts, J. 1979: Morphological mapping in the Axmouth–Lyme Regis undercliffs, Devon. *Quart. J. Eng. Geol.* **12**, pp. 205–17.

——. 1983: The temporal and spatial development of landslides in the Axmouth–Lyme Regis Undercliffs national nature reserve. *Earth Surface Processes and Landforms* **8**, pp. 589–603.

Pitty, A. F. 1971: *Introduction to Geomorphology* (Methuen, London).

Plummer, L. N. and Mackenzie, F. T. 1974: Predicting mineral solubility from rate data: application to the dissolution of magnesian calcites. *Am. J. Sci.* **274**, pp. 61–83.

Pollack, J. B., Toon, O. B., Sagan, C., Summers, A., Baldwin, B. and Van Camp, W. 1976: Volcanic explosions and climatic change: a theoretical assessment. *J. Geophys. Res.* **81**, pp. 1071–83.

Polunin, N. V. C. 1974: Devastation of a fringing coral reef by *Acanthaster*. *Nature* **249**, pp. 589–90.

Polynov, B. B. 1937: *The Cycle of Weathering* (Murty, London).

Pomponi, S. A. 1977: Etching cells of boring sponges: an ultrastructural analysis. *Proc. 3rd Int. Coral Reef Symp. (Miami, Florida)* **2**, pp. 485–90.

Popov, B. A. 1957: Opyt analiticheskogo issedovaniya protsessa formirovaniya morskikh terrass. *Trudy Okeanogr. Kommiss. Akad. Nauk. SSSR.* **2**, pp. 111–15.

Porter, S. C. 1981: Recent glacier variations and volcanic eruptions. *Nature* **291**, pp. 139–42.

Potts, A. S. 1970: Frost action in rocks: some experimental data. *Trans. Inst. Brit. Geogr.* **49**, pp. 109–24.

Powers, H. A. 1961: The emerged shoreline at 2–3 metres in the Aleutian Islands. *Z. Geomorph.* Suppl. Band, **3**, pp. 36–8.

Powers, T. C. 1945: A working hypothesis for further studies of frost resistance of concrete. *J. Am. Concrete Inst.* **16**, pp. 245–72.

——. 1949: The air requirement of frost-resistant concrete. *Proc. Highway Res. Board* **29**, pp. 184–202.

——. 1955: Basic considerations pertaining to freezing and thawing tests. *Am. Soc. Test. Mat. Proc.* **55**, pp. 1132–55.

——. 1965: The mechanisms of frost action in concrete. *Stanton Walker Lecture Ser. Mat. Sci.* **3** (Nat. Sand and Gravel Assoc., University of Maryland), pp. 1–35.

——. 1975: Freezing effects in concrete. In *Durability of Concrete* (Am. Concrete Inst., Detroit), paper SP 47–1, pp. 1–11.

—— and Brownyard, T. L. 1947: Studies of the physical properties of hardened Portland cement paste. *J. Am. Concrete Inst.* **18**, pp. 101–32, 249–336, 469–504, 549–95, 669–712, 845–64, 865–80, 933–69, 971–92.

—— and Helmuth, R. H. 1953: Theory of volume change in hardened Portland cement paste during freezing. *Proc. Highway Res. Board* **32**, pp. 285–97.

Prat, H. 1935: Les formes d'érosion littorale dans l'archipel des Bermudes, et l'évolution des atolls et récifs coralliens. *Rev. Géogr. Phys. Géol. Dyn.* **8**, pp. 257–83.

Pratt, R. M. and Dill, R. F. 1974: Deep eustatic terrace levels: further speculations. *Geology* **2**, pp. 155–9.

Prebble, M. M. 1967: Cavernous weathering in the Taylor dry valley, Victoria Land, Antarctica. *Nature* **216**, pp. 1194–5.
Prêcheur, C. 1960: Le littoral de la Manche, de Saint-Adresse à Ault, étude morphologique. *Norois* 138p.
Prescott, G. W. 1968: *The Algae* (Houghton-Mifflin, Boston).
Proix-Noé, M. 1946: Étude d'un glissement de terrain dû à la présence de glauconie. *Comptes Rendus Acad. Sci.* **222**, pp. 403–5.
Proudfoot, V. B. 1954: Erosion surfaces in the Mourne Mountains. *Ir. Geogr.* **3**, pp. 26–35.
Purdy, E. G. and Kornicker, L. S. 1958: Algal disintegration of Bahamian limestone coasts. *J. Geol.* **66**, pp. 96–9.
Putnam, W. C. 1937: The marine cycle of erosion for a steeply sloping shoreline of emergence. *J. Geol.* **45**, pp. 844–50.
Pytkowicz, R. M. 1965: Rates of inorganic calcium carbonate precipitation. *J. Geol.* **73**, pp. 196–9.
——. 1969: Chemical solution of calcium carbonate in sea water. *Am. Zool.* **9**, pp. 673–9.
——. 1973: Calcium carbonate retention in super-saturated seawater. *Am. J. Sci.* **273**, pp. 515–22.
——. 1975: Some trends in marine chemistry and geochemistry. *Earth Sci. Rev.*, **11**, pp. 1–46.
—— and Hawley, J. E. 1974: Bicarbonate and carbonate ion-pairs and a model of seawater at 25°C. *Limnol. Oceanogr.* **19**, pp. 223–34.
Quatrefages, A. de 1854: *Souvenirs d'un naturaliste* (Charpentier, Paris) 1.
Quinif, Y. and Dupuis, C. 1985: Un karst en zone intertropicale: le Gunung Sewu à Java—aspects morphologiques et concepts evolutifs. *Rev. Géomorph. Dyn.* **34**, pp. 1–16.
Radtke, U., Hennig, G. J., Linke, W. and Mungersdorf, J. 1981: $^{230}Th/^{234}U$- and ESR-dating problems of fossil shells in Pleistocene marine terraces (northern Latium, central Italy). *Quaternaria* **23**, pp. 37–50.
Ramkema, C. 1978: A model law for wave impacts on coastal structures. *Proc. 16th Conf. Coastal Eng.*, pp. 2308–27.
Ramsey, A. C. 1846: On the denudation of south Wales and the adjacent counties of England. *Mem. Geol. Surv. GB.*, pp. 297–335.
Randall, J. E. 1974: The effect of fishes on coral reefs. *Proc. 2nd Int. Coral Reef Symp. (Queensland, Australia)* **1**, pp. 159–66.
Ranson, M. G. 1955a: Observations sur les principaux agents de la dissolution du calcaire sous-marin dans la zone côtière des îles coralliennes de l'Archipel des Tuamotu. *Comptes Rendus Acad. Sci.* **240**, pp. 806–8.
——. 1955b: Observations sur l'agent essentiel de la dissolution du calcaire dans les régions exondées des îles coralliennes de l'Archipel des Tuamotu: conclusions sur le processus de la dissolution du calcaire. *Comptes Rendus Acad. Sci.* **240**, pp. 1007–9.
——. 1959: Érosion biologique des calcaires côtiers et autres calcaires d'origine animale. *Comptes Rendus Acad. Sci.* **249**, pp. 438–40.
Rapp, A. 1960a: Recent development of mountain slopes in Karkwagge and surroundings in northern Scandanavia. *Geogr. Annlr.* **42A**, pp. 65–200.

——. 1960b: Talus slopes and mountain walls at Tempelfjorden, Spitsbergen. *Norsk Polarinstitutt Skrifter* **119**, pp. 1–96.

—— and Rudberg, S. 1964: Studies on periglacial phenomena in Scandanavia 1960–63. *Biul. Peryglac.* **14**, pp. 75–89.

Rasool, S. I. 1973: *Chemistry of the Lower Atmosphere* (Plenum Press, London).

Reese, E. S. 1977: Coevolution of corals and coral feeding fishes of the family Chaetodontidae. *Proc. 3rd Int. Coral Reef Symp. (Miami, Florida)* **1**, pp. 267–74.

Reffell, G. 1978: Descriptive analysis of the subaqueous extensions of subaerial rock platforms. Univ. Sydney, BA thesis.

Reiche, P. 1950: A survey of weathering processes and products. *University of New Mexico Publ. Geol.* **3**.

Reid, C. 1890: The Pliocene deposits of Britain. *Mem. Geol. Surv. GB.*

——. 1907. Memorandum on the geological conditions affecting the coast of Holderness, Lincolnshire, and Norfolk, the south coast generally, and the coast of Cornwall and N. Devon. In *The Royal Commission on Coast Erosion (UK)* Rept. 1, (vol. 1, part 2), London, pp. 165–72.

——, Barrow, G. and Dewey, H. 1910: The geology of the country around Padstow and Camelford. *Mem. Geol. Survey GB, New Series Sheet Memoirs* **335** and **336**.

Rekstad, J. 1915: Om strandlinjer og strandlinjedannelse. *Norsk Geol. Tidsskr.* **3**, pp. 3–18.

Reusch, H. 1894: The Norwegian coastal plain. *J. Geol.* **2**, pp. 347–9.

Revelle, R. 1955: On the history of the oceans. *J. Marine Res.* **14**, pp. 446–61.

—— and Emery, K. O. 1957: Chemical erosion of beach rock and exposed reef rock. *US Geol. Surv. Prof. Paper* 260–T, pp. 699–709.

—— and Fairbridge, R. 1957: Carbonates and carbon dioxide. In *Treatise on Marine Ecology and Paleoecology*, ed. J. W. Hedgpeth, Geol. Soc. Am. Mem. **67**, vol. 1, pp. 239–96.

Rezanov, I. A. 1979: Major sinkings of the ocean floor and the constancy of sea level. *Int. Geol. Rev.* **21**, pp. 509–16.

Rice, M. E. 1969: Possible boring structures of sipunculids. *Am. Zool.* **9**, pp. 803–12.

Richards, A. 1969: Some aspects of the evolution of the coastline of north east Skye. *Scot. Geogr. Mag.* **85**, pp. 122–31.

Richards, A. F. 1960: Rates of marine erosion of tephra and lava at Isla San Benedicto, Mexico. *Int. Geol. Congr. (Norden, Submarine Geol.)* **21**, pp. 59–63.

Richards, H. G. 1970: Annotated bibliography of Quaternary shorelines, supplement 1965–1969. *Nat. Acad. Sci. (US) Spec. Publ.* **10**.

——. 1974: Annotated bibliography of Quaternary shorelines, second supplement 1970–1973. *Nat. Acad. Sci. (US) Spec. Publ.* **11**.

——. and Broecker, W. 1963: Emerged Holocene South American shorelines. *Science* **141**, pp. 1044–5.

—— and Fairbridge, R. W. 1965: Annotated bibliography of Quaternary shorelines (1945–1964). *Nat. Acad. Sci. (US) Spec. Publ.* **6**.

Richert, G. 1968: Experimental investigation of shock pressures against breakwaters. *Proc. 11th Conf. Coastal Eng*, pp. 954–73.

——. 1974: Shock pressures of breaking waves. *Bull. Roy. Inst. Tech. Hydraulics Lab. (Stockholm)* **84**, pp. 1–93.

Richter, A. 1930: Neue methoden der standsicherheitsberechnung von aussenhafenwerken, einschliesslich der wirkung der meereswellen. *Berichte des II, Hydrol. Kongr. USSR.* **3**, pp. 508–13.

Richter, E. 1896: Die Norwegische strandebene und ihre entstehung. *Globus* **69**, pp. 313–19.

Ridlon, J. B. 1972: Pleistocene–Holocene deformation of the San Clemente Island crustal block, California. *Geol. Soc. Am. Bull.* **83**, pp. 1831–44.

Robinson, L. A. 1977a: Marine erosive processes at the cliff foot. *Marine Geol.* **23**, pp. 257–71.

——. 1977b: Erosive processes on the shore platform of northeast Yorkshire, England. *Marine Geol.* **23**, pp. 339–61.

——. 1977c: The morphology and development of the northeast Yorkshire shore platform. *Marine Geol.* **23**, pp. 237–55.

Rode, K. 1930: Geomorphogenie des Ben Lomond (Kalifornien) eine studie uber terrassenbildung durch marine abrasion. *Z. Geomorph.* **5**, pp. 16–78.

Rognon, P., Cussenot-Curien, M., Seyer, C. and Weisrock, A. 1967: Remarques sur le comportement des grès et granites vosgiens sous climat froid. *Revue Géogr. Est* **7**, pp. 403–18.

Rona, P. A. 1973: Relations between rates of sediment accumulation on continental shelves, sea floor spreading, and eustasy inferred from the central North Atlantic. *Geol. Soc. Am. Bull.* **84**, pp. 2851–72.

Rondeau, M. A. 1961: Les taffonis. In *Recherches geomorphologiques en Corse* (Armand Colin, Paris), pp. 159–82.

——. 1965: Formes d'érosion superficielles dans les grès Fontainebleau. *Assoc. Géogr. Français Bull.* 334–5, pp. 58–66.

Rosen, P. S. 1980: Coastal environments of the Makkovik region, Labrador. In *The Coastline of Canada*, ed. S. B. McCann, Geol. Surv. Can. Paper 80–10, pp. 267–80.

Ross, C. 1954: Shock pressure of breaking waves. *Proc. 4th Conf. Coastal Eng.*, pp. 323–32.

——. 1955: Laboratory study of shock pressures of breaking waves. *Beach Erosion Board Tech. Memo.* **59**, pp. 1–22.

Rossiter, J. R. 1962: Long-term variations in sea-level. In *The Sea*, ed. M. N. Hill (Wiley, New York), **1**, pp. 590–610.

Rougerie, G. 1960: Le Façonnement actuel des modèles en Côte D'Ivoire forestière. *Mem. Inst. Française Afrique Noire* **58**.

Round, F. E. 1973: *The Biology of the Algae* (Arnold, London).

Rouville, A. de, Besson, P. and Petry, P. 1938: État actuel des études internationales sur les efforts dûs aux lames. *Ann. Ponts Chauss.* **108**, pp. 5–113.

Rozier, I. T. and Reeves, M. J. 1979: Ground movement at Runswick Bay, north Yorkshire. *Earth Surface Processes* **4**, pp. 275–80.

Rudberg, S. 1967: The cliff coast of Gotland and the rate of cliff retreat. *Geogr. Annlr.* **49A**, pp. 283–99.

Rudwick, M. J. S. 1965: Ecology and paleoecology: Part H. Brachiopoda. In *Treatise on Invertebrate Paleontology*, ed. R. C. Moore (University of Kansas Press, Lawrence), pp. H199–214.

Ruggieri, G. and Spovieri, R. 1977: A revision of Italian Pleistocene stratigraphy. *Geologica Romana* **16**, pp. 131–9.

Ruitenberg, A. A., McCutcheon, S. R. and Venugopal, D. V. 1976: Recent gravity sliding and coastal erosion, Devil's Half Acre, Fundy Park, NB *Geoscience Can.* **3**, pp. 237–9.

Ruiz, C. L. 1962: Osmotic interpretation of the swelling of expansive soils. *Highway Res. Board Bull.* **313**, pp. 47–77.

Rundgren, L. 1958: Water wave forces. *Trans. Roy. Inst. Tech.* **122**, pp. 1–121 (Bull. 54, Inst. Hydraulics, Stockholm, Sweden).

Russell, F. S. and Yonge, C. M. 1949: *The Seas* (Warne, London).

Russell, K. L. 1968: Oceanic ridges and eustatic changes in sea level. *Nature* **218**, pp. 861–2.

Russell, R. D. 1939: Effects on transportation on sedimentary particles. *Recent Marine Sediments*, Am. Assoc. Petrol. Geol., pp. 32–7.

Russell, R. J. 1943: Freeze-and-thaw frequencies in the United States. *Trans. Am. Geophys. Union* **1**, pp. 125–33.

——. 1963: Recent recession of tropical cliffy coasts. *Science* **139**, pp. 9–15.

——. 1971: Water-table effects on sea coasts. *Geol. Soc. Am. Bull.* **82**, pp. 2343–8.

Rutford, R. H., Craddock, C. and Bastien, T. W. 1968: Late Tertiary glaciation and sea level changes in Antarctica. *Palaeogeogr., Palaeoclim., Palaeoecol.* **5**, pp. 15–39.

Rutland, R. W. R. 1971: Andean orogeny and ocean floor spreading. *Nature* **233**, pp. 252–5.

Rutzler, K. 1974: The burrowing sponges of Bermuda. *Smithson. Contrib. Zool.* **165**, pp. 1–32.

——. 1975: The role of burrowing sponges in bioerosion. *Oecologia* **19**, pp. 203–16.

Ruxton, B. P. and Berry L. 1957: The weathering of granite and associated erosional features in Hong Kong. *Geol. Soc. Am. Bull.* **68**, 1263–92.

Rzhevsky, V. and Novik, G. 1971: *The Physics of Rocks* (Mir Publishers, Moscow).

Safriel, U. N. 1966: Recent vermetid formation on the Mediterranean shore of Israel. *Proc. Malac. Soc. London* **37**, pp. 27–34.

——. 1974: Vermetid gastropods and intertidal reefs in Israel and Bermuda. *Science* **186**, pp. 1113–16.

Sainflou, M. 1928: Essai sur des digues maritimes verticales. *Ann. Ponts Chauss.* **98**, pp. 5–48.

——. 1935: Note sur le calcul des digues maritimes verticales. *Ann. Ponts Chauss.* **105**, pp. 735–53.

Saint-Venant, B. De and Flamant, A. 1888: De la houle et du clapotis. *Ann Ponts Chauss.* **15**, pp. 705–809.

Samuelsson, L. and Werner, M. 1978: Weathering pits in the Lake Mjorn area northeast of Goteborg. *Geogr. Annlr.* **60A**, pp. 9–21.

Sanders, N. K. 1968a: The development of Tasmanian shore platforms. Univ. Tasmania, Ph.D. thesis.

——. 1968b: Wave tank experiments on the erosion of rocky coasts. *Papers Proc. Roy. Soc. Tasmania* **102**, pp. 11–16.

——. 1970: The production of horizontal high-tidal shore platforms. *Austr. Nat. Hist.* **16**, pp. 315–19.

Sanlaville, P. 1972: Vermetus dating of changes in sea level. In *Underwater Archeology: A Nascent Discipline*, Unesco, Paris, pp. 185–91.
Sarjeant, W. A. S. 1975: Plant trace fossils. In *The Study of Trace Fossils*, ed. R. W. Frey (Springer-Verlag, Berlin), pp. 163–79.
Saumell, E. S., Belles, J. F. M. and Cueva, A. P. 1982: Carst marino: estado de la cuestion. *Estudios Geograficos* **43**, pp. 411–38.
Savage, C. N. 1968: Mass wasting. In *The Encyclopedia of Geomorphology*, ed. R. W. Fairbridge (Dowden, Hutchinson, and Ross, Stroudsburg, Penn.), pp. 696–700.
Savigear, R. A. G. 1952: Some observations on slope development in south Wales. *Trans. Inst. Brit. Geogr.* **18**, pp. 31–52.
——. 1956: Technique and terminology in the investigation of slope forms. *Slopes Comm. Rept.* **1**, pp. 66–75.
——. 1962: Some observations on slope development in north Devon and north Cornwall. *Trans. Inst. Brit. Geogr.* **31**, pp. 23–42.
Sawaragi, T. and Iwata, K. 1974: Turbulence effect on wave deformation. *Coastal Eng. Japan* **17**, pp. 39–49.
Schaffer, R. J. 1955: The weathering, preservation, and restoration of stone buildings. *J. Roy. Soc. Arts* **103**, pp. 843–67.
Scheidegger, A. E. 1957: *Physics of Flow through Porous Media* University of Toronto Press, Toronto).
——. 1970: *Theoretical Geomorphology* (Springer-Verlag, New York).
Scherrer, P. H. 1979: Solar variability and terrestrial weather. *Rev. Geophys. Space Phys.* **17**, pp. 724–31.
Schmalz, R. F. and Chave, K. E. 1963: Calcium carbonate: factors affecting saturation in ocean waters off Bermuda. *Science* **139**, pp. 1206–7.
—— and Swanson, F. J. 1969: Diurnal variations in the carbonate saturation of seawater. *J. Sed. Petrol.* **39**, pp. 255–67.
Schnable, J. E. and Goodell, H. G. 1968: Pleistocene–Recent stratigraphy evolution, and development of the Apalachicola coast, Florida. *Geol. Soc. Am. Spec. Paper* **112**, pp. 1–72.
Schneider, J. 1976: Biological and inorganic factors in the destruction of limestone coasts. *Contrib. Sedimentology* **6**, pp. 1–112.
—— and Torunski, H. 1983: Biokarst on limestone coasts, morphogenesis and sediment production. *Marine Ecol.* **4**, pp. 45–63.
Schneider, S. H. and Thompson, S. L. 1979: Ice ages and orbital variations: some simple theory and modelling. *Quat. Res.* **12**, pp. 188–203.
Schoemaker, H. J. and Thijsse, T. Th. 1949: Investigations of the reflection of waves. *Int. Assoc. Hydraul. Structures Res.* 3rd meeting (Grenoble), paper I–2.
Schofield, J. C. 1960: Sea level fluctuations during the last 4,000 years as recorded by a chenier plain, Firth of Thames, New Zealand. *NZ J. Geol. Geophys.* **3**, pp. 467–85.
——. 1970: Correlation between sea level and volcanic periodicities of the last millennium. *NZ J. Geol. Geophys.* **13**, pp. 737–41.
——. 1977: Late Holocene sea level, Gilbert and Ellice Islands, west central Pacific Ocean. *NZ J. Geol. Geophys.* **20**, pp. 503–29.
——. 1980: Postglacial transgressive maxima and second-order transgressions of

the southwest Pacific Ocean. In *Earth Rheology, Isostasy, and Eustasy*, ed. N.-A. Mörner (Wiley, Chichester), pp. 517–21.

—— and Suggate, R. P. 1971: Late Quaternary sea level records of the south-west Pacific. *Quaternaria* **14**, pp. 243–53.

Scholl, D. W., Craighead, F. C. and Stuiver, M. 1969: Florida submergence curve revised: its relation to coastal sedimentation rates. *Science* **163**, pp. 562–4.

Schubert, C. and Szabo, B. J. 1978: Uranium-series ages of Pleistocene marine deposits on the islands of Curaçao and La Blanquilla, Caribbean Sea. *Geol. Mijnbouw* **57**, pp. 325–32.

Schulke, H. 1968: Quelques types de dépressions fermeés littorales et supralittorales lieés a l'action destructive de la mer (Bretange, Corse, Asturies). *Norois* **57**, pp. 23–42.

Schwartz, M. L. 1972: Seamounts as sea-level indicators. *Geol. Soc. Am. Bull.* **83**, pp. 2975–80.

Scott, D. B. and Medioli, F. S. 1982: Micropaleontological documentation for early Holocene fall of relative sea level on the Atlantic coast of Nova Scotia. *Geology* **10**, pp. 278–81.

Scott, R. F. 1969: The freezing process and mechanics of frozen ground. US Army Corps Eng. *CRREL Monogr*. II-DI, pp. 1–65.

Sekyra, J. 1972: Forms of mechanical weathering and their significance in the stratigraphy of the Quaternary in Antarctica. In *Antarctica Geology and Geophysics*, ed. R. J. Adie (Universitetsforlaget, Oslo), pp. 669–76.

Sergin, V. Y. 1980: Origin and mechanisms of large-scale climatic oscillations. *Science* **209**, pp. 1477–82.

Shackleton, N. J. 1967: Oxygen isotope analysis and Pleistocene temperatures re-assessed. *Nature* **215**, pp. 15–17.

——. 1977a: The oxygen isotope stratigraphic record of the Late Pleistocene. *Phil. Trans. Roy. Soc. London* **280B**, pp. 169–82.

——. 1977b: Oxygen isotope stratigraphy of the Middle Pleistocene. In *British Quaternary Studies, Recent Advances*, ed. F. W. Shotton (Clarendon Press, Oxford), pp. 1–16.

—— and Matthews, R. K. 1977: Oxygen isotope stratigraphy of Late Pleistocene coral terraces in Barbados. *Nature* **268**, pp. 618–20.

—— and Opdyke, N. D. 1973: Oxygen isotope and palaeomagnetic stratigraphy of equatorial Pacific core V28-238: oxygen isotope temperatures and ice volumes on a 10^5-year and 10^6-year scale. *Quat. Res*. **3**, pp. 39–55.

——, ——. 1976: Oxygen-isotope and paleomagnetic stratigraphy of Pacific core V28-239: late Pliocene to latest Pleistocene. In *Investigations of Late Quaternary Paleoceanography and Paleoclimatology*, ed. R. M. Cline and J. D. Hays, Geol. Soc. Am. Mem. **145**, pp. 449–64.

——, ——. 1977: Oxygen isotope and palaeomagnetic evidence for early Northern Hemisphere glaciation. *Nature* **270**, pp. 216–19.

Sharaf, S. G. and Budnikova, N. A. 1969: Secular perturbations in the elements of the Earth's orbit and the astronomical theory of climate variations. *Trans. Inst. Teor. Astron*. **14**, pp. 48–85.

Sharpe, C. F. S. 1938: *Landslides and Related Phenomena* (Columbia Univ. Press, New York).

Shennan, I. 1982: Interpretation of Flandrian sea-level data from the Fenland, England. *Proc. Geol. Assoc.* **93**, pp. 53–63.

——. 1983: Flandrian and late Devensian sea-level changes and crustal movements in England and Wales. In *Shorelines and Isostasy*, ed. D. E. Smith and A. G. Dawson (Academic Press, London), Inst. Brit. Geogr. Spec. Publ. **16**, pp. 255–83.

Shepard, F. P. 1963: Thirty-five thousand years of sea level. In *Essays in Marine Geology in Honor of K. O. Emery*, ed. T. Clements (University of Southern California Press, Los Angeles), pp. 1–10.

—— and Curray, J. R. 1967: Carbon-14 determination of sea level changes in stable areas. In *Progress in Oceanography*, ed. M. Sears (Pergamon Press, London) vol. 4, pp. 283–91.

—— and Grant, U. S. 1947: Wave erosion along the southern California coast. *Geol. Soc. Am. Bull.* **58**, pp. 919–26.

—— and Kuhn, G. G. 1983: History of sea arches and remnant stacks of La Jolla, California, and their bearing on similar features elsewhere. *Marine Geol.* **51**, pp. 139–61.

—— and Wanless, H. R. 1971: *Our Changing Coastlines* (McGraw-Hill, New York).

——, Curray, J. R., Newman, W. A., Bloom, A. L., Newell, N. D., Tracey, J. I. and Veeh, H. H. 1967: Holocene changes in sea level in Micronesia. *Science* **157**, pp. 542–4.

Shtencel, V. K. 1972. Pressure upon vertical wall from standing waves. *Proc. 13th Conf. Coastal Eng.*, pp. 1649–59.

Siddiquie, H. N. 1980: The ages of the storm beaches of the Lakshadweep (Laccadives). *Marine Geol.* **38**, pp. M11–20.

Siffre, M. 1961: Niveau de base et formes karstiques submergeés. *Ann. Spéléo.* **16**, pp. 87–92.

Silen, L. 1946: On two new groups of Bryozoa living in shells of molluscs. *Ark. Zool.* 38B, pp. 1–7.

Silvester, R. 1974: *Coastal Engineering* (Elsevier, Amsterdam), **2** vols.

Simonett, D. S. 1968: Landslides. In *The Encyclopedia of Geomorphology*, ed. R. W. Fairbridge (Dowden, Hutchinson, and Ross, Stroudsburg, Penn.), pp. 639–41.

Sissons, J. B. 1960: Erosion surfaces, cyclic slopes and drainage systems in south Scotland and northern England. *Trans. Inst. Brit. Geogr.* **28**, pp. 23–38.

——. 1967: *The Evolution of Scotland's Scenery* (Oliver and Boyd, Edinburgh).

——. 1974: Late-glacial marine erosion in Scotland. *Boreas* **3**, pp. 41–8.

——. 1976: *The Geomorphology of the British Isles: Scotland* (Methuen, London).

——. 1981: British shore platforms and ice sheets. *Nature* **291**, pp. 473–5.

——. 1983: Shorelines and isostasy in Scotland. In *Shorelines and Isostasy*, ed. D. E. Smith and A. G. Dawson (Academic Press, London), Inst. Brit. Geogr. Spec. Publ. **16**, pp. 209–25.

—— and Brooks, C. L. 1971: Dating of early postglacial land and sea level changes in the western Forth Valley. *Nature (Phys. Sci.)* **234**, pp. 124–7.

Sjoberg, R. 1981: Tunnel caves in Swedish Archean rocks. *Trans. Brit. Cave Res. Assoc.* **8**, pp. 159–67.

Slosson, J. E. and Cilweck, B. A. 1966: Parson's Landing landslide—a case history

including the effects of eustatic sea level changes on stability. *Eng. Geol.* **3**, pp. 1–9.
Smale, D., Van der Lingen, G. J. and Bell, D. H. 1982: A bedding-plane landslide near Mt. Vulcan, north Canterbury. *NZ J. Geol. Geophys.* **25**, pp. 397–404.
Small, R. J. 1970: *The Study of Landforms* (Cambridge University Press, Cambridge).
Smart, J. 1977: Late Quaternary sea-level changes, Gulf of Carpentaria, Australia. *Geology* **5**, pp. 755–9.
Smith, B. J. 1978: The origin and geomorphic implications of cliff foot recesses and tafoni on limestone hamadas in the northwest Sahara. *Z. Geomorph.* **22**, pp. 21–43.
Smith, S. V., Dygas, J. A. and Chave, K. E. 1968: Distribution of calcium carbonate in pelagic sediments. *Marine Geol.* **6**, pp. 391–400.
Snead, R. E. 1969: Physical geography reconnaissance—west Pakistan coastal zone. *Univ. New Mexico Publ. Geogr. (Albuquerque)* **1**.
So, C. L. 1965: Coastal platforms of the Isle of Thanet, Kent. *Trans. Inst. Brit. Geogr.* **37**, pp. 147–56.
Soderman, G. 1980: Slope processes in cold environments of northern Finland. *Fennia* **158**, pp. 83–152.
Soliman, G. N. 1969: Ecological aspects of some coral-boring gastropods and bivalves on the northwestern Red Sea. *Am. Zool.* **9**, pp. 887–94.
Sollid, J. L., Andersen, S., Harme, N., Kjeldsen, O., Salvigsen, O., Sturd, S., Tveita, T. and Wilhelmsen, A. 1973: Deglaciation of Finmark, north Norway. *Norsk Geogr. Tidsskr.* **27**, pp. 233–325.
Sorensen, R. N. 1968: Recession of marine terraces—with special reference to the coastal area north of Santa Cruz, California. *Proc. 11th Conf. Coastal Eng.* **1**, pp. 653–69.
Soule, J. D. and Soule, D. F. 1969: Systematics and biogeography of burrowing bryozoans. *Am. Zool.* **9**, pp. 791–802.
Southward, A. J. 1964: Limpet grazing and control of vegetation on rocky shores. *Proc. 4th Symp. Brit. Ecol. Soc.* pp. 265–73.
Sparks, B. W. 1971: *Rocks and Relief* (Longman, London).
——. 1972: *Geomorphology* (Longman, London), 2nd edn.
—— and West, R. G. 1972: *The Ice Age in Britain* (Methuen, London).
Stearn, C. W. and Scoffin, T. P. S. 1977: Carbonate budget of a fringing reef, Barbados. *Proc. 3rd Int. Symp. Coral Reefs (Miami, Florida)* **2**, pp. 471–6.
Stearns, C. E. 1976: Estimates of the position of sea level between 140,000 and 75,000 years ago. *Quat. Res.* **6**, pp. 445–9.
—— and Thurber, D. L. 1967: Th^{230}/U^{234} dates of late Pleistocene marine fossils from the Mediterranean and Moroccan littorals. In *Progress in Oceanography*, ed. M. Sears (Pergamon Press, London), Vol. **4**, pp. 293–305.
Stearns, H. T. 1935: Shore benches on the island of Oahu, Hawaii. *Geol. Soc. Am. Bull.* **46**, pp. 1467–82.
——. 1945: Eustatic shore lines of the Pacific. *Geol. Soc. Am. Bull.* **56**, pp. 1071–8.
——. 1961: Eustatic shorelines on Pacific Islands. *Z. Geomorph.* Suppl. Band. **3**, pp. 1–16.

——. 1971: Geologic setting of an Eocene fossil deposit on Eua Island, Tonga. *Geol. Soc. Am. Bull.* **82**, pp. 2541–52.
——. 1978: Quaternary shorelines in the Hawaiian Islands. *B.P. Bishop Museum Bull. (Honolulu, Hawaii)* **237**, pp. 1–57.
Steers, J. A. 1929: The Queensland coast and the Great Barrier Reef. *Geogr. J.* **74**, pp. 341–70.
——. 1937: The coral islands and associated features of the Great Barrier Reefs. *Geogr. J.* **89**, pp. 1–28.
——. 1962a: *The Sea Coast* (Collins, London).
——. 1962b: Coastal cliffs: report of a symposium. *Geogr. J.* **128**, pp. 303–20.
——. 1964: *The Coastline of England and Wales* (Cambridge University Press, Cambridge).
——. 1973: *The Coastline of Scotland* (Cambridge University Press, Cambridge).
——. 1981: *Coastal Features of England and Wales* (Oleander Press, Cambridge).
Steinen, R. P., Harrison, R. S. and Matthews, R. K. 1973: Eustatic low stand of sea level between 125,000 and 105,000 BP: evidence from the subsurface of Barbados, West Indies. *Geol. Soc. Am. Bull.* **84**, pp. 63–70.
Stephens, N. 1957: Some observations on the 'interglacial' platform and early post-glacial raised beach on the east coast of Ireland. *Proc. Roy. Ir. Acad.* **58B**, pp. 129–49.
Stephens, N. and Synge, F. M. 1966: Pleistocene shorelines. In *Essays in Geomorphology*, ed. G. H. Dury (Heinemann, London), pp. 1–51.
Stephenson, T. A. and Stephenson, A. 1949: The universal features of zonation between tide marks on rocky coasts. *J. Ecol.* **37**, pp. 289–305.
Stephenson, W. 1961: Experimental studies on the ecology of intertidal environments at Heron Island: II. The effect of substratum. *Austr. J. Marine Freshwater Res.* **12**, pp. 164–76.
Stevenson, T. 1874: *The Design and Construction of Harbours* (A. & C. Black, Edinburgh), 2nd edn.
Stewart, A. D. 1981: The expanding Earth. *Nature* **290**, p. 627.
Stewart, N. B. 1970: Detailed petrography, an aid in the evaluation of gravel aggregates for freeze–thaw resistance. *Penn. Geol. Surv. Mineral Resources Rept. 4th Ser. Bull.* **M64**, pp. 55–89.
Stive, M. J. F. 1980: Velocity and pressure field of spilling breakers. *Proc. 17th Conf. Coastal Eng.*, pp. 547–66.
Strakhov, N. M. 1967: *Principles of Lithogenesis* (Oliver and Boyd, Edinburgh).
Strøm, K. M. 1948: The geomorphology of Norway. *Geogr. J.* **62**, pp. 19–27.
Stuiver, M. 1980: Solar variability and climatic change during the current millennium. *Nature* **286**, pp. 868–71.
—— and Daddario, J. J. 1963: Submergence of the New Jersey coast. *Science* **142**, p. 951.
—— and Quay, P. D. 1980: Changes in atmospheric carbon-14 attributed to a variable sun. *Science* **207**, pp. 11–19.
——, Heusser, C. J. and Yang, I. C. 1978: North American glacial history extended to 75,000 years ago. *Science* **200**, pp. 16–21.
Stumm, W. and Brauner, P. A. 1975: Chemical speciation. In *Chemical Oceanography*, ed. J. P. Riley and G. Skirrow (Academic Press, London), Vol. 1, pp. 173–239.

Stump, E., Sheridan, M. F. and Scott, G. B. 1980: Early Miocene subglacial basalts, the east Antarctic ice sheet, and uplift of the Transantarctic Mountains. *Science* **207**, pp. 757–9.
Suarez, M. J. and Held, I. M. 1979: The sensitivity of an energy-balance climate model to variations in orbital parameters. *J. Geophys. Res.* **84**, pp. 4825–36.
Suess, E. 1888: *The Face of the Earth II*, English translation by H. B. C. Sollas, 1906 (Oxford University Press, Oxford).
Suess, E. 1970: Interaction of organic compounds with calcium carbonate: I. Association phenomena and geochemical implications. *Geochim. Cosmochim. Acta* **34**, pp. 157–68.
—— and Futterer, D. 1972: Aragonitic ooids: experimental precipitation from seawater in the presence of humic acid. *Sedimentology* **19**, pp. 129–39.
Suggate, R. P. 1968: Postglacial sea-level rise in the Christchurch metropolitan area, New Zealand. *Geol. Mijnbouw* **47**, pp. 291–7.
Suguio, K., Martin, L. and Flexor. J.-M. 1980: Sea level fluctuations during the past 6,000 years along the coast of the State of Sao Paulo, Brazil. In *Earth Rheology, Isostasy and Eustasy*, ed. N.-A. Mörner (Wiley, Chichester), pp. 471–86.
Sunamura, T. 1973: Coastal cliff erosion due to waves–field investigations and laboratory experiments. *J. Faculty Eng. Univ. Tokyo* **32**, pp. 1–86.
——. 1975: A laboratory study of wave-cut platform formation. *J. Geol.* **83**, pp. 389–97.
——. 1976: Feedback relationship in wave erosion of laboratory rocky coast. *J. Geol.* **84**, pp. 427–37.
——. 1977: A relationship between wave-induced cliff erosion and erosive force of wave. *J. Geol.* **85**, pp. 613–18.
——. 1978a: Mechanisms of shore platform formation on the southeastern coast of the Izu Peninsula, Japan. *J. Geol.* **86**, pp. 211–22.
——. 1978b: A mathematical model of submarine platform development. *Math. Geol.* **10**, pp. 53–8.
——. 1978c: A model of the development of continental shelves having erosional origin. *Geol. Soc. Am. Bull.* **89**, pp. 504–10.
——. 1982a: A predictive model for wave-induced cliff erosion, with application to Pacific coasts of Japan. *J. Geol.* **90**, pp. 167–78.
——. 1982b: A wave tank experiment on the erosional mechanism at a cliff base. *Earth Surface Processes and Landforms* **7**, pp. 333–43.
—— and Tsujimoto, H. 1982: Long-term recession rate of Taitomisaki coastal cliff. *Ann. Rept. Inst. Geosci. Univ. Tsukuba* **8**, pp. 55–6.
Suzuki, T., Takahashi, K., Sunamura, Y. and Terada, M. 1970: Rock mechanisms on the formation of washboard-like relief on wave-cut benches at Arasaki, Miura Peninsula, Japan. *Geogr. Rev. Japan* **43**, pp. 211–22.
Svendsen, I. A., Madsen, P. A. and Hansen, J. B. 1978: Wave characteristics in the surf zone. *Proc. 16th Conf. Coastal Eng.*, pp. 520–39.
Swan, S. B. St. C. 1965: Coast erosion principles and a classification of south west Ceylon's beaches on the basis of their erosional stability. *Ceylon Geogr.* **19**, pp. 1–18.
——. 1968: Coastal classification with reference to the east coast of Malaya. *Z. Geomorph.* Suppl. Band 7, pp. 114–32.

——. 1971: Coastal geomorphology in a humid tropical low energy environment: the islands of Singapore. *J. Tropical Geogr.* **33**, pp. 43–61.

Swantesson, J. 1985: Experimental weathering studies of Swedish rocks. *Geogr. Annlr.* **67A**, pp. 115–8.

Sweet, H. S. 1948: Research on concrete durability as affected by coarse aggregate. *Am. Soc. Test. Mat. Proc.* **48**, pp. 988–1016.

Sweeting, M. M. 1955: The landforms of north-west county Clare, Ireland. *Trans. Inst. Brit. Geogr.* **21**, pp. 33–49.

——. 1973: *Karst Landforms* (Columbia University Press, New York).

Swinnerton, A. C. 1927: Observations on some details of wave erosion: wave furrows and shore potholes. *J. Geol.* **35**, pp. 171–9.

Synge, F. M. 1964: Some problems concerned with the glacial succession in south-east Ireland. *Ir. Geogr.* **5**, pp. 73–82.

Szabo, B. J. and Rosholt, J. N. 1969: Uranium series dating of Pleistocene molluscan shells from southern California—an open system model. *J. Geophys. Res.* **74**, pp. 3253–60.

—— and Vedder, J. G. 1971: Uranium-series dating of some Pleistocene marine deposits in southern California. *Earth Planetary Sci. Lett.* **11**, pp. 283–90.

——, Ward, W. C., Weidie, A. E. and Brady, M. J. 1978: Age and magnitude of the late Pleistocene sea level rise on the eastern Yucatan Peninsula. *Geology* **6**, pp. 713–15.

Taber, S. 1930: The mechanics of frost heaving. *J. Geol.* **38**, pp. 303–17.

——. 1950: Intensive frost action along lake shores. *Am. J. Sci.* **248**, pp. 784–93.

Taillefer, F. 1957: Le rivages des Bermudes et les formes littorales de dissolution de calcaire. *Cah. Géogr. Québec* **2**, pp. 115–37.

Taira, K. 1980: Radiocarbon dating of shell middens and Holocene sea-level fluctuations in Japan. *Palaeogeogr., Palaeoclim., Palaeoecol.* **32**, pp. 79–87.

Takahashi, T. 1977: *Shore Platforms in Southwestern Japan-Geomorphological Study*. Coastal Landform Study Soc. Southwestern Japan, Osaka.

Takenaga, K. 1968: The classification of notch profiles and the origin of notches. *Chigaku Zasshi (J. Geogr. Tokyo)* **77**, pp. 329–41.

Tanner, W. F. 1968: Multiple influences on sea-level changes in the Tertiary. *Palaeogeogr., Palaeoclim., Palaeoecol.* **5**, pp. 165–71. (See also his Introduction pp. 7–14.)

Tasch, P. and Gafford, E. L. 1969: Weathering of an Antarctic argillite: field, geochemical and mineralogical observations. *J. Sed. Petrol.* **39**, pp. 369–73.

Taylor, J. C. and Illing, L. V. 1969: Holocene intertidal calcium carbonate concentration, Qatar, Persian Gulf. *Sedimentology* **12**, pp. 69–107.

Taylor, J. D. and Way, K. 1976: Erosive activities of chitons at Aldabra Atoll. *J. Sed. Petrol.* **46**, pp. 974–7.

Taylor, R. B. 1980: Coastal environments along the northern shore of Somerset Island, District of Franklin In *The Coastline of Canada*, ed. S. B. McCann, *Geol. Surv. Can. Paper* **80–10**, pp. 239–50.

Teichert, C. 1947: Contributions to the geology of Houtman's Abrolhos, Western Australia. *Proc. Linn. Soc. NSW* **71**, pp. 145–96.

——. 1950: Late Quaternary sea-level changes at Rottnest Island, Western Australia. *Proc. Roy. Soc. Vict.* **59**, pp. 63–79.

Te Punga, M. T. 1957: Periglaciation in southern England. *Tidsskr. Kon. Ned. Aardrijksk Genootsch.* **73**, pp. 400–12.

Ters, M. 1975: Variations in sea level on the French Atlantic coast over the last 10,000 years. In *Quaternary Studies*, ed. R. P. Suggate and M. M. Cresswell, 9th Inqua Congr., Christchurch, Roy. Soc. NZ Bull. **13**, pp. 287–8.

—— and Peulvast, J. P. 1972: Un exemple d'érosion différentielle en milieu littoral: l'estran de Brétignolles, Vendée. *Bull. Assoc. Géogr. Franc.* **402–3**, pp. 319–25.

Terzaghi, K. 1950: Mechanism of landslides. *Bull. Geol. Soc. Am. Berkey Vol.*, pp. 83–122.

——. 1962: Stability of steep slopes on hard unweathered rock. *Géotechnique* **12**, pp. 251–70.

Thom, B. G. 1973: The dilemma of high interstadial sea levels during the last glaciation. *Progr. Geogr.* **5**, pp. 167–246.

—— and Chappell, J. 1975: Holocene sea levels relative to Australia. *Search* **6**, pp. 90–3.

——, ——. 1978: Holocene sea level change: an interpretation. *Phil. Trans. Roy. Soc. London* **291A**, pp. 187–94.

——, Hails, J. R. and Martin, J. R. H. 1969: Radiocarbon evidence against higher postglacial sea levels in eastern Australia. *Marine Geol.* **7**, pp. 161–8.

——, ——, ——, and Phipps, C. V. G. 1972: Postglacial sea levels in eastern Australia—a reply. *Marine Geol.* **12**, pp. 233–42.

Thomas, M. F. 1974: *Tropical Geomorphology* (Macmillan, London).

Thomas, R. H., Sanderson, T. J. O. and Rose, K. E. 1979: Effect of climatic warming on the west Antarctic ice sheet. *Nature* **277**, pp. 355–8.

Thomas, W. N. 1938: Experiments on the freezing of certain building materials. Sci. Ind. Res. Dept. (UK), *Building Res. Tech. Paper* **17**, pp. 1–146.

Thomson, K. W. 1979: Coastal erosion in part of the Bay of Fundy, Nova Scotia. Univ. Alberta, MSc thesis.

Thorarinsson, S., Einarsson, T. and Kjartansson, G. 1959: On the geology and geomorphology of Iceland. *Geogr. Annlr.* **41**, pp. 135–69.

Thorn, C. E. 1979: Bedrock freeze–thaw in an alpine environment, Colorado Front Range. *Earth Surface Processes* **4**, pp. 211–28.

——. 1982: Bedrock microclimatology and the freeze–thaw cycle: a brief illustration. *Ann. Assoc. Am. Geogr.* **72**, pp. 131–7.

Thornbury, W. D. 1954: *Principles of Geomorphology* (Wiley, New York).

Thornton, E. B. and Calhoun, R. J. 1972: Spectral resolution of breakwater reflected waves. *Am. Soc. Civil Eng. Proc. Waterways, Harbors and Coastal Eng. Div.* **98**, pp. 443–60.

Thorstenson, D. C. and Plummer, L. N. 1977: Equilibrium criteria for two component solids reacting with fixed composition in an aqueous phase—example: the magnesian calcites. *Am. J. Sci.* **277**, pp. 1203–23.

Thurber, D. L., Broecker, W. S., Blanchard, R. L. and Potranz, H. A. 1965: Uranium series ages of Pacific atoll corals. *Science* **149**, pp. 55–8.

Tietze, W. 1962: A contribution to the geomorphological problem of strandflats. *Pet. Geogr. Mitt.* **106**, pp. 1–20.

Tjia, H. D. 1985: Notching by abrasion on a limestone coast. *Z. Geomorph.* **29**, pp. 367–72.

Tomlinson, J. T. 1953: A burrowing barnacle of the genus *Trypetesa* (Order Acrothoracica). *J. Wash. Acad. Sci.* **43**, pp. 373–81.

——. 1969: Shell-burrowing barnacles. *Am. Zool.* **9**, pp. 837–40.

Toms, A. H. 1946: Folkestone Warren landslips: research carried out in 1939 by the Southern Railway. *Proc. Inst. Civil Eng.* (*Railway Paper*) **19**, pp. 3–25.

——. 1953: Recent research into the coastal landslips at Folkestone Warren, Kent, England. *Proc. 3rd Int. Conf. Soil Mech. Foundation Eng. (Zurich)* **2**, pp. 288–93.

Tooley, M. J. 1974: Sea level changes during the last 9,000 years in north-west England. *Geogr. J.* **140**, pp. 18–42.

——. 1979: *Proc. 1978 Int. Symp. Coastal Evolution in the Quaternary* (Sao Paulo, Brazil), pp. 502–33.

——. 1982: Sea-level changes in northern England. *Proc. Geol. Assoc.* **93**, pp. 43–51.

Topley, W. 1893: The landslip at Sandgate. *Proc. Geol. Assoc.* **13**, pp. 40–7.

Torunski, H. 1979: Biological erosion and its significance for the morphogenesis of limestone coasts and for nearshore sedimentation (northern Adriatic). *Senckenbergiana Marit.* **11**, pp. 193–265.

Tracey, J. L. and Ladd, H. S. 1974: Quaternary history of Eniwetok and Bikini Atolls, Marshall Islands. *Proc. 2nd Int. Coral Reef Symp. (Queensland, Austr.)* **2**, pp. 537–50.

——, Schlanger, S. O., Stark, J. T., Doan, D. B. and May, H. G. 1964: General geology of Guam. *US Geol. Surv. Prof. Paper* **403A**, pp. 1–104.

Tratman, E. K. 1971: The formation of the Gibraltar caves. *Trans. Cave Res. Group Great Britain* **13**, pp. 135–43.

Trenhaile, A. S. 1969: A geomorphological investigation of shore platforms and high-water rock ledges in the Vale of Glamorgan. Univ. Wales Ph.D. thesis.

——. 1971: Lithological control of high-water rock ledges in the Vale of Glamorgan, Wales. *Geogr. Annlr.* **53A**, pp. 59–69.

——. 1972: The shore platforms of the Vale of Glamorgan, Wales. *Trans. Inst. Brit. Geogr.* **56**, pp. 127–44.

——. 1974a: The geometry of shore platforms in England and Wales. *Trans. Inst. Brit. Geogr.* **62**, pp. 129–42.

——. 1974b: The morphology and classification of shore platforms in England and Wales. *Geogr. Annlr.* **56A**, pp. 103–10.

——. 1978: The shore platforms of Gaspé, Québec. *Annals Assoc. Am. Geogr.* **68**, pp. 95–114.

——. 1980: Shore platforms: a neglected coastal feature. *Progr. Phys. Geogr.* **4**, pp. 1–23.

——. 1983a: The development of shore platforms in high latitudes. In *Shorelines and Isostasy*, ed. D. E. Smith and A. G. Dawson (Academic Press, London), Inst. Brit. Geogr. Special Publ. **16**, pp. 77–93.

——. 1983b: The width of the shore platforms; a theoretical approach. *Geogr. Annlr.* **65A**, pp. 147–58.

——. (in press): Sea level oscillations and the development of rock coasts. In *Coastal Modelling: Techniques and Applications*, ed. V. C. Lakhan and A. S. Trenhaile (Elsevier, Amsterdam).

—— and Byrne, M.-L. 1986: A theoretical investigation of the Holocene development of rock coasts, with particular reference to shore platforms. *Geogr. Annlr.* **68A**, pp. 1–14.

—— and Layzell, M. G. J. 1980: Shore platform morphology and tidal-duration distributions in storm wave environments. In *The Coastline of Canada*, ed. S. B. McCann, Geol. Surv. Can. Paper **80–10**, pp. 207–14.

——, ——. 1981: Shore platform morphology and the tidal duration factor. *Trans. Inst. Brit. Geogr.* NS **6**, pp. 82–102.

—— and Mercan, D. W. 1984: Frost weathering and the saturation of coastal rocks. *Earth Surface Processes and Landforms* **9**, pp. 321–31.

—— and Rudakas, P. A. 1981: Freeze–thaw and shore platform development in Gaspé, Québec. *Géogr. Phys. Quat.* **35**, pp. 171–81.

Treves, S. B. 1967: The geology of Cape Evans and Cape Royds, Ross Island, Antarctica. In *Antarctic Research*, ed. H. Wexler, Am. Geophys. Union, Geophys. Monogr. **7**, pp. 40–6.

Tricart, J. 1956: Étude expérimentale du problème de la gélivation. *Biul. Peryglac.* **4**, pp. 285–318.

——. 1957: Aspects et problèmes géomorphologiques du littoral occidental de la Côte de Ivoire. *Bull. Inst. Franc. Afr. Noire* **19A**, pp. 1–20.

——. 1959: Problèmes géomorphologiques du littoral oriental du Brésil. *Cah. Océanogr. COEC* **11**, pp. 276–308.

——. 1960: Expèriences de désagrégation de roches granitiques par la cristallisation du sel marin. *Z. Geomorph.* Suppl. Band **1**, pp. 239–40.

——. 1962: Observations de géomorphologie littorale à Mamba Point (Monrovia, Libéria). *Erdkunde* **16**, pp. 49–57.

——. 1969a: *Le Modéle des régions sèches*. (Soc. d'edition d'enseignement supérieur, Paris).

——. 1969b: *Geomorphology of Cold Environments* (Macmillan, London).

——. 1970: Convergence de phénomènes entre l'action du gel et celle du sel. *Acta Geographica Lodziensia* **24**, pp. 425–36.

——. 1972: *Landforms of the Humid Tropics: Forests and Savannas* (Longman, London).

Trudgill, S. T. 1976a: Rock weathering and climate: quantitative and experimental aspects. In *Geomorphology and Climate*, ed. E. Derbyshire (Macmillan, London), pp. 59–99.

——. 1976b: The marine erosion of limestone on Aldabra Atoll, Indian Ocean. *Z. Geomorph*, Suppl. Band 26, pp. 164–200.

——. 1983: Preliminary estimates of intertidal limestone erosion, One Tree Island, southern Great Barrier Reef, Australia. *Earth Surface Processes and Landforms* **8**, pp. 189–93.

Trueman, A. E. 1922: The Liassic rocks of Glamorgan. *Proc. Geol. Assoc.* **33**, pp. 245–84.

——. 1949: *Geology and Scenery in England and Wales* (Penguin, Harmondsworth, Middx.)

Tsuchiya, Y. and Yamaguchi, M. 1970: Limiting conditions for standing wave theories by perturbation method. *Proc. 12th Conf. Coastal Eng.*, pp. 523–42.

Twenhofel, W. H. 1945: The rounding of sand grains. *J. Sed. Petrol.* **15**, pp. 59–71.

Twidale, C. R. 1982: *Granite Landforms* (Elsevier, Amsterdam).
——, Bourne, J. A. and Twidale, N. 1977: Shore platforms and sea level changes in the Gulfs Region of South Australia. *Trans. Roy. Soc. South Austr.* **101**, pp. 63–74.
Tylor, A. 1869: On the formation of deltas, and on the evidence and cause of great changes in the sea-level during the glacial period. *Quart. J. Geol. Soc. London* **25**, pp. 7–11.
Umgrove, J. H. F. 1931: Note on 'negroheads' (coral boulders) in the East Indian Archipelago. *Proc. Kon. Akad. Wetensch.* **34**, pp. 485–7.
Upson, J. E. 1949: Late Pleistocene and recent changes of sea level along the coast of Santa Barbara County, California. *Am. J. Sci.* **247**, 94–115.
——. 1951: Former marine shore lines of the Gaviota Quadrangle, Santa Barbara County, California. *J. Geol.* **59**, pp. 415–46.
Urien, C. M. and Ottmann, F. 1971: Histoire de Rio de la Plata au Quaternaire. *Quaternaria* **14**, pp. 51–9.
Ursell, F. 1947: The effect of a fixed vertical barrier on surface waves in deep water. *Proc. Cambridge Phil. Soc.* **43**, pp. 374–82.
Uusinoka, R. and Matti, E. 1979: On weathering depressions and their occurrence in Finland. *Terra* **91**, pp. 81–6.
Vail, P. R. and Hardenbol, J. 1979: Sea-level changes during the Tertiary. *Oceanus* **22**, pp. 71–9.
——, Mitchum, R. M. and Thompson, S. 1977: Seismic stratigraphy and global changes of sea level: Part 4: global cycles of relative changes of sea level. In *Seismic Stratigraphy—Applications to Hydrocarbon Exploration*, ed. C. E. Payton, *Am. Assoc. Petrol. Geol. Mem.* **26**, pp. 83–97.
Valentine, J. W. and Moores, E. M. 1970: Plate-tectonics regulation of faunal diversity and sea level: a model. *Nature* **228**, pp. 657–9.
—— and Veeh, H. H. 1969: Radiometric ages of Pleistocene terraces from San Nicolas Island, California. *Geol. Soc. Am. Bull.* **80**, pp. 1415–18.
Van Andel, T. H. and Laborel, J. 1964: Recent high relative sea level stand near Recife, Brazil. *Science* **145**, pp. 580–1.
Van Autenboer, T. 1964: The geomorphology and glacial geology of the Sør-Rondane, Dronning Maud Land. In *Antarctic Geology*, ed. R. J. Adie (North Holland, Amsterdam), pp. 81–103.
Van de Plassche, O. 1981: Sea level, groundwater, and basal peat growth—a reassessment of data from the Netherlands. *Geol. Mijnbouw* **60**, pp. 401–8.
Van der Pers, J. N. C. 1978: Bioerosion by *Polydora* (Polychaeta, Sedentaria, Vermes) off Heligoland, Germany. *Geol. Mijnbouw* **57**, pp. 465–78.
Van Diggelen, J. 1976: Is the Earth expanding? *Nature* **262**, pp. 675–6.
Van Donk, J. 1976: O^{18} record of the Atlantic Ocean for the entire Pleistocene period. In *Investigation of Late Quaternary Paleoceanography and Paleoclimatology*, ed. R. M. Cline and J. D. Hays, Geol. Soc. Am. Mem. **145**, pp. 147–63.
Van Flandern, T. C. 1979: Gravitation and the expansion of the Earth. *Nature* **278**, p. 821.
Van Woerkom, A. J. J. 1953: The astronomical theory of climatic changes. In *Climatic Change*, ed. H. Shapley, (Harvard University Press, Cambridge, Mass.), pp. 147–57.

Varnes, D. J. 1950: Relation of landslides to sedimentary features. In *Applied Sedimentation*, ed. P. D. Trask (Wiley, New York), pp. 229–46.

——. 1958: Landslide types and processes. In *Landslides and Engineering Practice* (Highway Res. Board, Washington, DC) spec. publ. **29**, pp. 20–47.

——. 1975: Slope movements in the western United States. In *Mass Wasting*, ed. E. Yatsu, A. J. Ward and F. Adams, 4th Guelph Symp. Geomorph., Geo Abstr. (University of East Anglia, Norwich), pp. 1–17.

Vedder, J. G. and Norris, R. M. 1963: Geology of San Nicolas Island California. *US Geol. Surv. Prof. Paper* **369**, pp. 1–65.

Veeh, H. H. 1966: Th^{230}/U^{238} and U^{234}/U^{238} ages of Pleistocene high sea level stand. *J. Geophys. Res.* **71**, pp. 3379–86.

—— and Chappell, J. 1970: Astronomical theory of climatic change: support from New Guinea. *Science* **167**, pp. 862–5.

—— and Valentine, J. W. 1967: Radiometric ages of Pleistocene fossils from Cayucos, California. *Geol. Soc. Am. Bull.* **78**, pp. 547–50.

Verbeck, G. J. and Klieger, P. 1957: Studies of salt scaling of concrete. *Highway Research Board Bull.* **150**, pp. 1–13.

—— and Landgren, R. 1960: Influence of physical characteristics of aggregates on frost resistance of concrete. *Am. Soc. Test. Mat. Proc.* **60**, pp. 1063–79.

Verma, V. K. 1973: Map parameters for correlation of marine terraces in California. *Geol. Soc. Am. Bull.* **84**, pp. 2737–42.

Vernekar, A. D. 1972: Long-period global variations of incoming solar radiation. *Am. Meteor. Soc. Monogr.* **12** (34), pp. 1–21.

Verstappen, H. Th. 1960: On the geomorphology of raised coral reefs and its tectonic significance. *Z. Geomorph.* NF **4**, pp. 1–28.

Vigdorchik, M. E. 1981: Insolation of the Arctic from the global ocean during glaciations. In *Sea Level, Ice, and Climatic Change*, Proc. Canberra Symp. Dec. 1979. Int. Assoc. Hydrol. Sci. Publ. **131**, pp. 303–22.

Viles, H. A. 1984: Biokarst. *Progr. Phys. Geogr.* **8**, pp. 523–42.

Viner-Brady, N. E. V. 1955: Folkestone Warren landslips: remedial measures 1948–1954. *Proc. Inst. Civil Eng.* **4**, pp. 429–41.

Visher, S. S. 1945: Climatic maps of geologic interest. *Geol. Soc. Am. Bull.* **56**, pp. 713–36.

Vita-Finzi, C. and Cornelius, P. F. S. 1973: Cliff sapping by molluscs. *J. Sed. Petrol.* **43**, pp. 31–2.

Vogt, J. H. L. 1907: Uber die schrage senkung und die spatere schrage hebung des landes im nordlichen Norwegen. *Norsk Geol. Tidsskr.* **1**, pp. 3–47.

Vogt, T. 1918: Om recente og gamle strandlinjer i fast fjeld. *Norsk Geol. Tidsskr.* **4**, pp. 107–27.

Voisey, A. H. 1934: The physiography of the Middle North Coast District of New South Wales. *Proc. J. Roy. Soc. NSW* **68**, pp. 83–103.

Von Engeln, O. D. 1942: *Geomorphology* (Macmillan, New York).

Von Richthofen, F. 1882: *China* (Reimer, Berlin) **2**, pp. 768–73.

——. 1886: Führer für Forschungsreisende (Jänecke, Hannover).

Vos, M. A. and Moddle, D. A. 1976: Repetitive wetting and drying as a test of weathering resistance. *Can. Inst. Mining Metallurgical Bull.* **69**, pp. 103–8.

Wada, S. and Kokubu, N. 1973: Chemical composition of maritime aerosols. *Geochem. J.* **6**, pp. 131–9.

Walcott, R. I. 1970: Isostatic response to loading of the crust in Canada. *Can. J. Earth Sci.* **7**, pp. 716–27.
——. 1972a: Late Quaternary vertical movements in eastern North America: quantitative evidence of glacio-eustatic rebound. *Rev. Geophys. Space Phys.* **10**, pp. 849–84.
——. 1972b: Past sea levels, eustasy, and deformation of the Earth. *Quat. Res.* **2**, pp. 1–14.
——. 1975: Recent and late Quaternary changes in water level. *Eos.* **56**, pp. 62–72.
——. 1980: Rheological models and observational data of glacio-isostatic rebound. In *Earth Rheology, Isostasy, and Eustasy*, ed. N. -A. Morner (Wiley, Chichester), pp. 3–10.
Walker, P. 1960: A soil survey of the county of Cumberland, Sydney Region, NSW *Soil Surv. Unit NSW Bull.* **2**.
Wallet, A. and Ruellan, F. 1950: Trajectories of particles within a partial clapotis. *Houille Blanche* **5**, pp. 483–9.
Walton, K. 1963: Geomorphology. In *The North-East of Scotland*, ed. A. C. O'Dell and J. Mackintosh (Nelson, Edinburgh), pp. 16–31.
Ward, W. H. 1945: The stability of natural slopes. *Geogr. J.* **105**, pp. 170–97.
——. 1948: A coastal landslip. *Proc. 2nd Int. Conf. Soil Mech. Foundation Eng. (Rotterdam)* **2**, pp. 33–8.
——. 1962: Coastal cliffs: report of a symposium. *Geogr. J.* **128**, pp. 303–20.
Ward, W. T. 1965: Eustatic and climatic history of the Adelaide area, South Australia. *J. Geol.* **73**, pp. 592–602.
——. 1971: Postglacial changes in level of land and sea. *Geol. Mijnbouw* **50**, pp. 703–18.
——. 1973: Correlations of Pleistocene shorelines in Gippsland, Australia, and Oahu, Hawaii. *Geol. Soc. Am. Bull.* **84**, pp. 3087–92.
——. Ross, P. J. and Colquhoun, D. J. 1971: Interglacial high sea levels and absolute chronology derived from shoreline elevations. *Palaeogeogr., Palaeoclim., Palaeoecol.* **9**, pp. 77–99.
Warme, J. E. 1970: Traces and significance of marine rock borers. In *Trace Fossils*, ed. T. P. Crimes and J. C. Harper (Seel House Press, Liverpool), pp. 515–25.
——. 1975: Borings as trace fossils and the processes of marine bioerosion. In *The Study of Trace Fossils*, ed. R. W. Frey, (Springer-Verlag, Berlin), pp. 181–227.
—— and Marshall, N. F. 1969: Marine borers in calcareous terrigenous rocks of the Pacific coast. *Am. Zool.* **9**, pp. 765–74.
Washburn, A. L. 1973: *Periglacial Processes and Environments* (Edward Arnold, London).
—— and Stuiver, M. 1962: Radiocarbon-dated postglacial delevelling in northeast Greenland and its implications. *Arctic* **15**, pp. 66–73.
Waters, R. S. 1964: The Pleistocene legacy to the geomorphology of Dartmoor. In *Dartmoor Essays*, ed. I. G. Simmons, Devonshire Assoc. Adv. Sci. Liter. Art, Exeter, pp. 73–96.
Watson, J. N. P. 1976: Coast of the mighty landslip: the Axmouth–Lyme Regis undercliffs. *Country Life* **4119**, p. 1550–2.
Watson, J. P. 1964: A soil catena on granite in southern Rhodesia. *J. Soil Sci.* **15**, pp. 238–57.

Watts, S. H. 1979: Some observations on rock weathering, Cumberland Peninsula, Baffin Island. *Can. J. Earth. Sci.* **16**, pp. 977–83.

Wayman, C. H. 1967: Adsorption on clay mineral surfaces. In *Principles and Applications of Water Chemistry*, ed. S. D. Faust and N. C. Hunter (Wiley, New York), pp. 127–67.

Wayne, C. J. 1974: Effect of artificial sea grass on wave energy and near-shore sand transport. *Trans. Gulf Coast Assoc. Geol. Soc.* **24**, pp. 279–82.

Webb, S. D. and Tessman, N. 1967: Vertebrate evidence of a low sea level in the middle Pliocene. *Science* **156**, p. 379.

Weggel, J. R. and Maxwell, W. H. C. 1970: Numerical model for wave pressure distributions. *Am. Soc. Civil Eng. Proc. Waterways, Harbors and Coastal Div.* **96**, pp. 623–42.

Wehmiller, J. F. *et al.* 1977: Correlation and chronology of Pacific coast marine terrace deposits of continental United States by fossil amino acid stereochemistry technique evaluation, relative ages, kinetic model ages, and geologic implications. *US Geol. Surv. Open-File Rept.* 77–680, pp. 1–191.

Welch, J. 1929: *Littorina* perforations in indurated chalk. *Ir. Naturalist J.* **2**, pp. 131–9.

Wellman, H. W. and Wilson, A. T. 1965: Salt weathering, a neglected geological erosive agent in coastal and arid environments. *Nature* **205**, pp. 1097–8.

——. 1968: Salt weathering or fretting. In *The Encyclopedia of Geomorphology*, ed. R. W. Fairbridge, (Dowden, Hutchinson and Ross, Stroudsburg, Penn.), pp. 968–70.

Wentworth, C. K. 1938: Marine bench-forming processes–water layer weathering. *J. Geomorph.* **1**, pp. 5–32.

——. 1939: Marine bench-forming processes: ii. Solution benching. *J. Geomorph.* **2**, pp. 3–25.

——. 1944: Potholes, pits and pans: subaerial and marine. *J. Geol.* **52**, pp. 117–30.

—— and Palmer, H. S. 1925: Eustatic bench of islands of the north Pacific. *Geol. Soc. Am. Bull.* **36**, pp. 521–44.

Werenskiold, W. 1953: The strandflat of Spitsbergen. *Geogr. Tidsskr.* **52**, pp. 302–9.

Wesson, P. S. 1978: *Cosmology and Geophysics* (Adam Higler, Bristol).

West, R. G. 1968: *Pleistocene Geology and Biology* (Longmans, Green and Co., London).

——. 1972: Relative land-sea level changes in southeastern England during the Pleistocene. *Phil. Trans. Roy. Soc. London* **272A**, pp. 87–98.

Wey, J. 1920: Die energie der meereswellen als grundlage zur berechnung von molen. *Jahrbuch Hafenbautechn.* **3**, pp. 201–36.

Weyer, E. M. 1978: Pole movement and sea levels. *Nature* **273**, pp. 18–21.

Weyl, P. K. 1959: Pressure solution and the force of crystallization: a phenomenological theory. *J. Geophys. Res.* **64**, pp. 2001–25.

——. 1967: The solution behaviour of carbonate materials in sea water. *Proc. Int. Conf. Trop. Oceanogr. (Miami Beach, Florida)*, pp. 178–228.

Whalley, W. B., Douglas, G. R. and McGreevy, J. P. 1982: Crack propagation and associated weathering in igneous rocks. *Z. Geomorph.* **26**, pp. 33–54.

Whitaker, W. 1911: In *Royal Commission on Coast Erosion and Afforestation* (HMSO, London) iii, p. 6.
White, S. E. 1976: Is frost action really only hydration shattering? A review. *Arctic Alpine Rev*. **8**, pp. 1–6.
Whiting, J. D. 1974: The frost resistance of concrete subjected to a de-icing agent. Univ. Windsor Ontario, M.Sc. thesis.
Whittow, J. B. 1965: The interglacial and post-glacial strandlines of North Wales. In *Essays in Geography for Austin Miller*, ed. J. B. Whittow and P. D. Wood (University of Reading, Reading), pp. 94–117.
Wilhelmy, H. 1958: *Klimamorphologie der Massengesteine* (G. Westermann, Braunschweig, W. Germany).
——. 1964: Cavernous rock surfaces (tafoni) in semi-arid and arid climates. *Pakistan Geogr. Rev*. **19**, pp. 9–13.
Williams, A. T. and Davies, P. 1980: Man as a geological agent: the sea cliffs of Llantwit Major, Wales, UK. *Z. Geomorph.* Suppl. Band 34, pp. 129–41.
Williams, L. 1964: Regionalization of freeze–thaw activity. *Ann. Assoc. Am. Geogr*. **54**, pp. 597–611.
Williams, P. J. le B. 1975: Biological and chemical aspects of dissolved organic material in sea water. In *Chemical Oceanography*, ed. J. P. Riley and G. Skirrow (Academic Press, London) **2**, pp. 301–63.
Williams, R. B. G. and Robinson, D. A. 1981: Weathering of sandstone by the combined action of frost and salt. *Earth Surface Processes and Landforms* **6**, pp. 1–9.
Williams, R. G. 1980: Sun–weather effects. *Nature* **285**, p. 71.
Williams, W. 1982: Speleothem dates, Quaternary terraces and uplift rates in New Zealand. *Nature* **298** (5871), pp. 257–60.
Williams, W. W. 1956: An east coast survey: some recent changes in the coast of East Anglia. *Geogr. J*. **122**, pp. 317–34.
Wilson, A. T. 1964: Origin of the ice ages: an ice shelf theory for Pleistocene glaciation. *Nature* **201**, pp. 147–9.
——. 1969: The climatic effects of large-scale surges of ice sheets. *Can. J. Earth Sci*. **6**, pp. 911–8.
Wilson, G. 1952: The influence of rock structures on coastline and cliff development. *Geol. Assoc. Proc*. **63**, pp. 20–48.
Wilson, L. 1968: Frost action. In *The Encyclopedia of Geomorphology*, ed. R. W. Fairbridge (Dowden, Hutchinson, and Ross, Stroudsberg, Penn.), pp. 369–81.
Wiman, S. 1963: A preliminary study of experimental frost weathering. *Geogr. Annlr*. **45**, pp. 113–21.
Winkler, E. M. 1973: *Stone: Properties, Durability in Man's Environment* (Springer-Verlag, New York).
——. 1979: Role of salts in development of granitic tafoni, South Australia: a discussion. *J. Geol*. **87**, pp. 119–20.
—— and Singer, P. C. 1972: Crystallization pressure of salts in stone and concrete. *Geol. Soc. Am. Bull*. **83**, pp. 3509–14.
—— and Wilhelm, E. J. 1970: Salt burst by hydration pressures in architectural

stone in an urban atmosphere. *Geol. Soc. Am. Bull.* **81**, pp. 567–72.

Woelke, C. E. 1957: The flatworm *Pseudostylochus ostreophagus* Hyman, a predator of oysters. *National Shellfish Assoc. Proc.* **47**, pp. 62–7.

Wollast, R., Garrels, R. M. and MacKenzie, F. T. 1980: Calcite–seawater reactions in ocean surface waters. *Am. J. Sci.* **280**, pp. 831–48.

Wollin, G., Ericson, D. B. and Ewing, M. 1971: Late Pleistocene climates recorded in Atlantic and Pacific deep-sea sediments. In *The Late Cenozoic Glacial Ages*, ed. K. K. Turekian (Yale University Press, New Haven, Conn.), pp. 200–14.

Wood, A. 1959: The erosional history of the cliffs around Aberystwyth. *Liverpool and Manchester Geol. J.* **2**, pp. 271–87.

——. 1962: Coastal cliffs: report of a symposium (discussion). *Geogr. J.* **128**, pp. 307–9.

——. 1968: Beach platforms in the chalk of Kent, England. *Z. Geomorph.* **12**, pp. 107–13.

——. 1974: Submerged platform of marine abrasion around the coasts of south-western Britain. *Nature* **252** (5484), p. 563.

Woodcock, A. H. 1953: Salt nuclei in marine air as a function of altitude and wind force. *J. Meteor.* **10**, pp. 362–71.

Woodring, W. P., Bramlette, M. N. and Kew, W. S. W. 1946: Geology and paleontology of Palos Verdes Hills, California. *US Geol. Surv. Prof. Paper* **207**, pp. 1–145.

Woodroffe, C. D., Stoddard, D. R., Harmon, R. S. and Spencer, T. 1983: Coastal morphology and Late Quaternary history, Cayman Islands, West Indies. *Quat. Res.* **19**, pp. 64–84.

Woodruff, F., Savin, S. M. and Douglas, R. G. 1981: Miocene stable isotope record: a detailed deep Pacific ocean study and its paleoclimatic implications. *Science* **212**, pp. 665–8.

Woods, A. J. 1980: Geomorphology, deformation, and chronology of marine terraces along the Pacific coast of central Baja California, Mexico. *Quat. Res.* **13**, pp. 346–64.

Wooldridge, S. W. and Morgan, R. S. 1937: *The Physical Basis of Geography* (Longmans, London).

Worsley, T. R., Nance, D. and Moody, J. B. 1984: Global tectonics and eustasy for the past 2 billion years. *Marine Geol.* **58**, pp. 373–400.

Wray, J. L. 1977: *Calcareous Algae* (Elsevier, Amsterdam).

Wright, L. W. 1967: Some characteristics of the shore platforms of the English Channel coast and the northern part of the North Island New Zealand. *Z. Geomorph.* **11**, pp. 36–46.

——. 1969: Shore platforms and mass movement: a note. *Earth Sci. J.* **3**, pp. 44–50.

——. 1970: Variation in the level of the cliff/shore platform junction along the south coast of Great Britain. *Marine Geol.* **9**, pp. 347–53.

Wu, J. 1981: Evidence of sea spray produced by bursting bubbles. *Science* **212**, pp. 324–6.

Wyrwoll, K. H. 1977: Late Quaternary events in Western Australia. *Search* **8**, pp. 32–4.

Yasuda, T., Gota, S. and Tsuchiya, Y. 1982: On the relation between changes of integral quantities of shoaling waves and breaking inception. *Proc. 18th Conf. Coastal Eng*., pp. 22–37.

Yates, R. A. 1965: Physiographical evolution. In *Wales*, ed. E. G. Bowen (Metheun, London), pp. 19–52.

Yonge, C. M. 1936: Mode of life, feeding, digestion and symbiosis with zooxanthellae in the Tridacnidae. *Sci. Rept. Great Barrier Reef Expedition 1928—29*, Brit. Museum (Nat. Hist.) London, **1** (11), pp. 283–321.

——. 1951: Marine boring organisms. *Research* **4**, pp. 162–7.

——. 1955: Adaption to rock boring in *Botula* and *lithophaga* (Lamellibranchia, Mytilidae), with a discussion on the evolution of this habit. *Quart. J. Microsc. Sci*. **96**, pp. 383–410.

——. 1958: Observations on *Petricola carditoides* (Conrad). *Proc. Malacol. Soc. London* **33**, pp. 25–31.

——. 1963: Rock-boring organisms. In *Mechanics of Hard Tissue Destruction*, ed. R. F. Sognnaes, Am. Assoc. Adv. Sci. Publ. **75**, pp. 1–24.

Yoshikawa, T., Ota, Y., Yonekura, N., Okada, A. and Iso, N. 1980: Marine terraces and their tectonic deformation on the northeast coast of the North Island, New Zealand. *Geogr. Rev. Japan* **53**, pp. 238–62.

Young, A. 1972: *Slopes* (Oliver and Boyd, Edinburgh).

Young, R. W. 1972: Shore platforms cut in latite: Cathedral Rocks, Kiama, NSW *Austr. Geogr*. **12**, pp. 49–50.

Zagwijn, W. H. 1974: The palaeogeographic evolution of the Netherlands during the Quaternary. *Geol. Mijnbouw* **53**, pp. 369–85.

——. 1975: Variations in climate as shown by pollen analysis, especially in the Lower Pleistocene of Europe. In *Ice Ages: Ancient and Modern*, ed. A. E. Wright, and F. Moseley (Seel House Press, Liverpool), pp. 137–52.

Zaneveld, J. S. and Verstappen, H. Th. 1952: A recent investigation about the geomorphology and the flora of some coral islands in the Bay of Djakarta. *Indon. J. Sci. Res*. **1**, pp. 58–68.

Záruba, Q. and Mencl, V. 1969: *Landslides and Their Control* (Elsevier, Amsterdam).

Zenkovitch, V. P. 1949: Some factors in the formation of marine terraces. *Dokl. Acad. Nauk. SSSR*. **65**, pp. 53–5. (French translation in *Centre d'et de Doc. Paleont*. no. 229).

——.1967: *Processes of Coastal Development* (Oliver and Boyd, Edinburgh).

Zeuner, F. E. 1952a: Pleistocene shore lines. *Geol. Rundschau* **40**, pp. 39–50.

——. 1952b: *Dating the Past: An Introduction to Geochronology*. (Methuen, London), 3rd edn.

——. 1958: Criteria for the determination of mean sea level for Pleistocene shorelines. *Quaternaria* **5**, pp. 143–7.

——. 1959: *The Pleistocene Period* (Hutchinson, London).

Zezza, F. 1981: Morfogenesi litorale e fenomeni d'instabilità della costa a falesia del Gargano tra Vieste e Manfredonia. *Geol. Applic. Idrogedogia (Bari Univ. Inst. Geol. Applicata)* **16**, pp. 193–226.

Index

This index lists subject and geographical references. Significant discussions of topics are boldfaced.